Nonconvex Optimization in Mechanics

Nonconvex Optimization and Its Applications

Volume 21

The titles published in this series are listed at the end of this volume.

Nonconvex Optimization in Mechanics

Algorithms, Heuristics and Engineering Applications by the F.E.M.

by

E.S. Mistakidis
Institute of Steel Structures,
Department of Civil Engineering,
Aristotle University,
Thessaloniki, Greece

and

G.E. Stavroulakis
Institute of Applied Mechanics,
Department of Civil Engineering,
Carolo Wilhelmina Technical University,
Braunschweig, Germany

KLUWER ACADEMIC PUBLISHERS
DORDRECHT / BOSTON / LONDON

A C.I.P. Catalogue record for this book is available from the Library of Congress.

ISBN 0-7923-4812-5

Published by Kluwer Academic Publishers,
P.O. Box 17, 3300 AA Dordrecht, The Netherlands.

Sold and distributed in the U.S.A. and Canada
by Kluwer Academic Publishers,
101 Philip Drive, Norwell, MA 02061, U.S.A.

In all other countries, sold and distributed
by Kluwer Academic Publishers,
P.O. Box 322, 3300 AH Dordrecht, The Netherlands.

Printed on acid-free paper

Printed in the Netherlands

This book is dedicated
to our teacher,
Professor P.D. Panagiotopoulos

Contents

List of Algorithms

Preface

Nonconvexity and nonsmoothness arise in a large class of engineering applications. In many cases of practical importance the possibilities offered by optimization with its algorithms and heuristics can substantially improve the performance and the range of applicability of classical computational mechanics algorithms. For a class of problems this approach is the only one that really works.

The present book presents in a comprehensive way the application of optimization algorithms and heuristics in smooth and nonsmooth mechanics. The necessity of this approach is presented to the reader through simple, representative examples. As things become more complex, the necessary material from convex and nonconvex optimization and from mechanics are introduced in a self–contained way.

Unilateral contact and friction problems, adhesive contact and delamination problems, nonconvex elastoplasticity, fractal friction laws, frames with semi-rigid connections, are among the applications which are treated in details here. Working algorithms are given for each application and are demonstrated by means of representative examples.

The interested reader will find helpful references to up-to-date scientific and technical literature so that to be able to work on research or engineering topics which are not directly covered here.

This book is primarily addressed to engineers who would like to take advantage of modern computational analysis techniques by getting a deeper theoretical understanding of established and new modelling methods and solution algorithms. The authors hope that mathematicians will also appreciate the wealth of engineering applications, which are formulated here in the language of optimization, and that they will be motivated to contribute their mathematical knowledge for the further development in this area.

The book is divided into four parts.

The first part presents a number of engineering problems involving nonmonotone stress-strain or reaction-displacement laws, leading to nonconvex energy functions. The difficulties introduced by the nonconvexity and nonsmoothness is demonstrated by very simple examples involving one or two variables.

The second part gives the mathematical background of applied convex and nonconvex optimization. After a short review of the literature, new algorithms inspired from engineering applications are proposed for the solution of the mathematical problem.

The third part deals with the mathematical formulation of convex and nonconvex, smooth and nonsmooth problems in mechanics and engineering. Chapter 3 deals with convex problems, while in Chapter 4 nonconvex problems are presented. The presentation of the problems follows the same sequence in both Chapters in order to facilitate the comparison of nonconvex problems with respect to the corresponding convex problems. Starting from the basic relations of mechanics, a detailed formulation is presented for a large class of engineering problems. First, interface problems are treated and the corresponding variational relations are obtained. Then the treatment is extended in order to include material nonlinearities.

In the last Chapter of this part (Chapter 5), some problems of optimal design of structures are briefly treated. After the theoretical formulation of the problem, certain algorithms are presented for its numerical treatment.

The fourth part presents algorithms specialized for the treatment of a large class of nonconvex, nonsmooth optimization problems appearing in mechanics and engineering. The algorithms are presented in Chapter 6 and described in all possible detail; certain hints on their computer implementation are given. In Chapter 7 a large number of engineering applications is presented. After the detailed examination of the behaviour of the algorithms on very simple examples, realistic engineering applications are presented with large number of unknowns, exhibiting the potential of the optimization approach.

The authors are deeply greatful to Professor P.D. Panagiotopoulos (Aristotle University of Thessaloniki, Greece and RWTH Aachen, Germany) for his encouragement to publish this book, for his helpful suggestions and his continuous support. We would also like to thank Professor K.T. Thomopoulos and Professor C.C. Baniotopoulos for the helpful discussions and their advices concerning the engineering oriented topics and Dr. O.K. Panagouli for her help on the topics concerning the application of fractal geometry to mechanics.

The second author greatfully acknowledges the support of Professor Dr. H. Antes (Carolo Wilhelmina Technical University of Braunschweig, Germany) and the financial support of the European Union (TMR Grant ERBFMBICT 960987), which made the completion of this project possible.

Finally, many thanks are due to our editors at Kluwer and especially to the managing editor of this Series, Professor P. Pardalos, for their cooperation during the preparation of this book.

We must apologize to those whose work was inadvertently neglected in compiling the literature of this book. We would be glad to receive feedback from our readers, either in the form of criticism and corrections or in the form of proposals for improvements.

EURIPIDIS S. MISTAKIDIS, GEORGIOS E. STAVROULAKIS
THESSALONIKI - BRAUNSCHWEIG, JUNE 1997

Guidelines for the Readers

The present book discusses the links between numerical optimization and computational mechanics with an emphasis on nonconvex and nonsmooth problems and on the corresponding structural analysis applications. The physical causes of nonconvexity and nonsmoothness are presented by means of several examples of engineering importance. The modelling follows the mathematical optimization theories and, as far as possible, the terminology and notation of mechanics. The algorithms take advantage of the numerical optimization techniques, although, at this stage, recourse to engineering motivated heuristics is unavoidable.

Engineers will use the material of this book to evaluate the difficulty of the problems they meet, to understand when and why their methods work and to optimally identify the appropriate methods to be used for each specific case. Applied mathematicians working on optimization and mathematical modelling research workers will find a large number of problems where they can apply their techniques or test their new developments.

The first Part introduces the readers to the nonconvex engineering applications which are discussed in this book and will be of interest for everybody.

A short introduction to applied optimization, mainly addressed to engineers, is given in Part two.

The formulation of convex and nonconvex problems in mechanics, with and without smoothness, is given in Part three. The material is arranged in order of increasing complexity. Each kind of difficulty (e.g., lack of differentiability, nonconvexity, etc.) is introduced by means of simple models and it is integrated to the previously discussed topics. For all applications which are not treated in details in this book the reader can find a sufficient number of references to continue his or her own study. The material of this Part is recommended for all readers.

The numerical algorithms which are proposed for the solution of the problems are described and tested on a number of engineering applications in Part four. The computer implementation details clarify the methods and algorithms of the previous Parts and in some cases provide motivation for new approaches. Moreover, the reader will find several modern engineering applications whose treatment is, partially, a topic of current research activity, but whose significance has already been acknowledged by modern design specifications and engineering codes. Engineers will appreciate this technical information on solution methods for modern problems.

All the numerical applications have been solved carefully and thus their results may also serve as model applications for the development and testing of new techniques by the interested readers.

I Nonconvexity in Engineering Applications

1 NONCONVEXITY IN ENGINEERING APPLICATIONS

1.1 NONCONVEXITY IN ENGINEERING

Elements involving nonconvex and/or nonsmooth energy potentials appear in several mechanical problems. The nonconvexity of the energy potential appears as a result of the introduction of a nonmonotone, possibly multivalued stress-strain or reaction-displacement law. Consider for example the nonmonotone reaction (S_i) - displacement (u_i) diagram of Fig. 1.1a which leads to the nonconvex energy potential of Fig. 1.1b. Similar is the situation with the sawtooth stress-strain law of Fig. 1.1c which appears in reinforced concrete under tension (Scanlon's diagram) and leads to the potential energy function of Fig. 1.1d. The same effects may appear also in other problems of structural mechanics.

A common case is that of a frame structure with connections obeying to a nonlinear moment-rotation law. This type of behaviour appears in almost every kind of civil engineering structure (concrete, steel, composite or timber) and is a result of pure or incomplete "cooperation" between the various structural elements (which is sometimes intentional) and incorporates various local instability effects such as buckling, crashing and cracking (Colson, 1992, Wald, 1994). Fig. 1.1e gives the typical moment-rotation diagram of a steel beam-to-column connection. In most cases the diagram has a significant softening branch leading to the nonconvex energy potential of Fig. 1.1f.

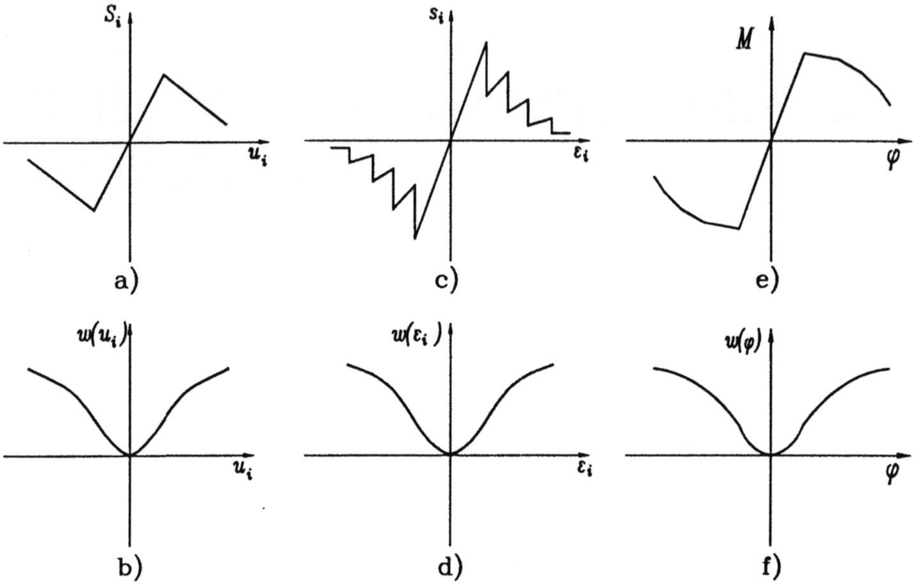

Figure 1.1. Nonmonotone laws and the corresponding nonconvex superpotentials

The same phenomena appear also in a large class of interface problems. We can mention here the reaction-displacement (or relative displacement) diagram of Fig. 1.2a which results in the tangential sense to the interface, if two bodies are in adhesive contact, i.e. if they are glued by an adhesive material. This material can sustain a small tension or compression and then it has either brittle fracture (dotted line) or semibrittle fracture. In the case of brittle fracture the law has complete vertical branches, i.e. it is a multivalued law. The same effects may appear at the interface of sandwich beams and plates. Also the nonmonotone variants of the well-known friction law of Coulomb (Fig. 1.2b) or the friction law between reinforcement and concrete of Fig. 1.2c can be mentioned. Notice also in this respect, that experimental results confirm the fractal nature of the friction laws in a number of mechanical problems involving a stick-slip process (Fig. 1.2d, Feder and Feder, 1991), increasing thus the difficulty of the theoretical and numerical treatment of such problems.

Another large category of problems involving nonmonotone, multivalued laws is related to composite materials. In particular, complete stress-strain or force-displacement diagrams in composite laminates exhibiting such softening behaviour have been recently obtained as consequence of the use of advanced

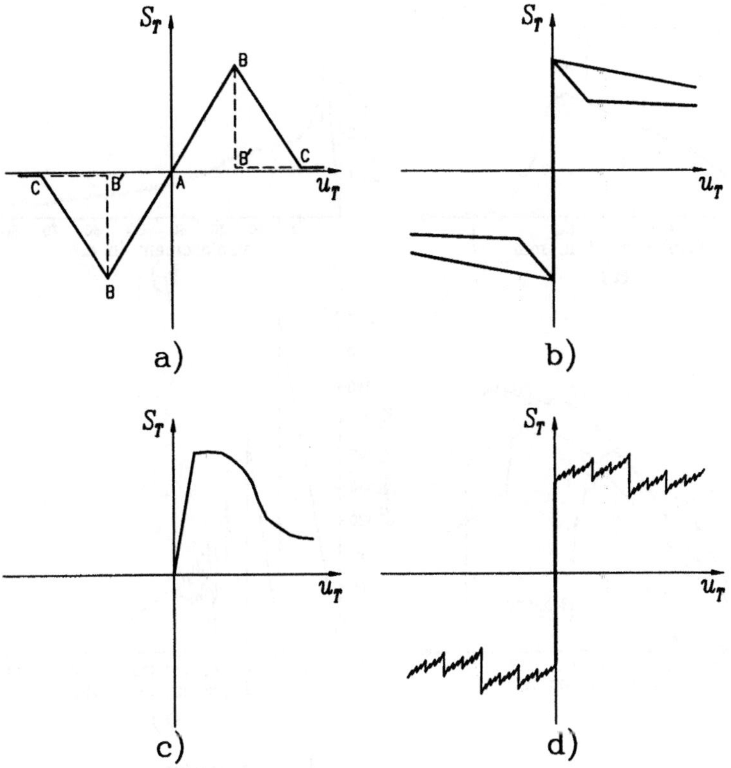

Figure 1.2. Nonmonotone interface laws

technology testing machines that minimize the range of instability effects during experimental tests.

The following typical experimental complete laws, which have been obtained during testing of laminated specimens, can be mentioned: the force-displacement diagram of Fig. 1.3a has been obtained from four point bending tests in short-fibre-reinforced glass composites (Roeder and Mayer, 1993) whereas the diagram of Fig. 1.3b has been obtained during pull-out tests on SiC fibre reinforced Aluminum-Silicon-Glass composites (Rausch et al., 1993).

Nonmonotone laws with complete vertical branches have been obtained during three point bending tests in composites with glass or ceramic matrix, reinforced with carbon-glass fibres (Pannhorst, 1993, Fig. 1.3c) and in glass-fibre reinforced polycarbonate composites (Hampe, 1989, Fig. 1.3d). All these nonmonotone complete laws provide the brittle crack initiation loading and information on whether failure is progressive or abrupt. Nonmonotone laws have

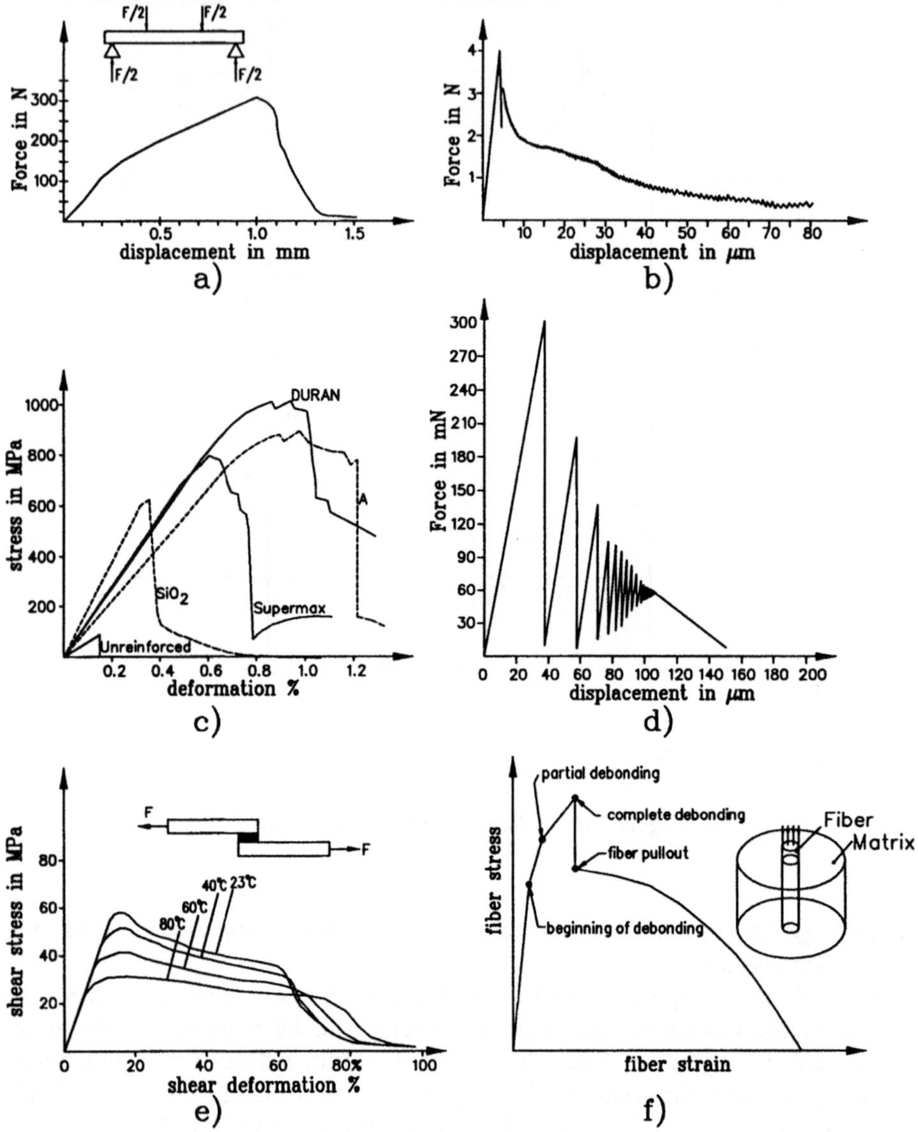

Figure 1.3. Nonmonotone stress-strain or reaction-displacement diagrams in composites

also been obtained during shear deformation tests (Fig. 1.3e, Schlimmer, 1993) and stretch deformation tests (Faupel, 1989) on various composite materials. Similar diagrams (Fig. 1.3f) have been obtained from pull-out tests in carbon fibre composites (CFC) (Peters, 1989). Nonmonotone stress-strain laws describe also the interlaminar fracture of composite laminates (Allix et al., 1995) as well as the debonding phenomena in composite materials (Corigliano and Bolzon, 1995). See also on this topic Green and Bowyer, 1981, Roman et al., 1981.

It has been observed, in the analysis of structural laminates, that a continuous stress redistribution occurs due to the successive failure of plies until total failure. The result of a study concerning stress and strain in the lamina for the case of a brittle system with poor interlaminar properties, lead to nonmonotone laws similar to the ones of Fig. 1.3c with complete vertical branches which represent the ply unloading completely until it carries no more load (Shidharan, 1982). Extensive testing programs, performed in order to evaluate the effect of the matrix resin on the impact damage tolerance of graphite-epoxy composite laminates, led also to a family of nonmonotone load-deflection curves with complete vertical branches (Williams and Rhodes, 1982). Similar complete load-deflection diagrams including vertical jumps, representing independent failure of each layer, obtained during testing of clad metals, have also been reported (Schwartz, 1984). Finally, hybrid laminated elements which have as constituents unidirectional composites exhibit a nonmonotone overall behaviour under tensile and bending loading (Wiedemann, 1996).

Diagrams exhibiting softening branches appear also in other engineering problems. In composite beams for example, experimental results show that the shear force – shear displacement diagram of the shear connectors exhibits a softening branch, representing the progressive cracking of the concrete around the shear connector and the plastification of the connector's material. (Aribert and Aziz, 1985, Gattesco, 1990, Oehlers and Johnson, 1987, Kuhn and Bucuner, 1986, Mistakidis et al., 1994).

The same type of diagrams relates also the moment with the rotational capacity of rectangular thin-walled steel beams, where the softening branch has its nature in the combination of material nonlinearity with severe local buckling of the cross-section's walls (Reck et al., 1976, Yener and Pekoz, 1980, Unger, 1973, Wang and Yeh, 1975, Grundy, 1990, Rasmussen and Hancock, 1993, Kecman, 1983, Thomopoulos et al., 1996). In this case the structural engineer should be very careful in the procedure of the redistribution of the beam's moments. The reader may also consult the recent review article Kyriakides, 1994 for other structural instability phenomena which can be modelled by means of nonmonotone laws.

Moreover, energies with a lot of local minima are also used for the modelling of microstructural changes, hysteresis effects and material phase transformation problems, for instance for the modelling of piezoelectric or shape memory alloys (see, among others, Krasnosel'skii and Pokrovskii, 1983, Visintin, 1994, Stillinger, 1995, James, 1996). Finally, oscillatory and stick-slip fracture instabilities in polymers and composites are among the effects which can be modelled by means of nonconvex potentials (see, for instance, the recent contributions of Webb and Aifantis, 1995, Webb and Aifantis, 1997a, Webb and Aifantis, 1997b).

1.2 SUPERPOTENTIAL PROBLEMS

The usual problem in structural analysis consists in finding an equilibrium point in the space of the displacements field under a given value of the field of forces and boundary conditions. Along the process of imposing a field of displacements, the mechanical system undergoes mechanical distortions (strains), resists by developing local counteractive forces at an infinitesimal level (stresses) and absorbs the work supplied by the loading field: this process can be described by means of a potential functional. Finding the equilibrium of a structural system is equivalent to finding the minimum of the potential function expressing the total work of the system, i.e. that absorbed by the structure minus that supplied by the loading field.

When the mechanical system remains in the normal operational framework, i.e. the loads are not severe enough to cause structural damage and no other complex physical mechanisms (material behaviour, unilateral contact, geometric nonlinearity, etc.) contribute to the energy balance of the system, the potential functional is quadratic and the respective unconstrained quadratic minimization problem constitutes a simple, everyday practice problem for the engineering community. In all other cases nonquadratic potentials arise.

Because of the non-quadratic form of the epigraph of the potential function, one is compelled to use a Newton-Raphson type non-linear solution technique. However, there is often the case that the classical methods of linear or linearized analysis encounter forbidding difficulties both in the formulation and in the numerical approximation of problems involving nonmonotone, possibly multivalued stress - strain or reaction - displacement laws such as the ones introduced in the previous Section (Moreau et al., 1988, Panagiotopoulos, 1983, Panagiotopoulos, 1985, Panagiotopoulos, 1988b, Panagiotopoulos, 1991, Baniotopoulos and Panagiotopoulos, 1987, Gilbert and Warner, 1978). This can be due to the fact that either the stress-strain laws in the interior of the elastic body or the respective boundary conditions are multivalued, i.e. complete vertical branches may be present in the one-dimensional case. Then, the respective energy functionals (superpotentials) involved, are nonconvex and nonsmooth.

The variational forms for such problems are termed hemivariational inequalities (Panagiotopoulos, 1983, Panagiotopoulos, 1985, Panagiotopoulos, 1988b, Panagiotopoulos, 1991, Panagiotopoulos, 1988a). The respective nonconvex energy functions, are called nonconvex superpotentials in accordance to the case of monotone mechanical behaviour described by convex superpotentials, where variational inequalities are obtained (Duvaut and Lions, 1972, Glowinski et al., 1981, Panagiotopoulos, 1985).

The theory of variational inequalities is closely related to the notion of convex, nondifferentiable superpotentials introduced into mechanics by J.J. Moreau (Moreau, 1963, Moreau, 1968) for monotone possibly multivalued boundary conditions and constitutive laws. On the other hand, the hemivariational inequalities were introduced as a generalization of the variational inequalities, by P.D. Panagiotopoulos (Panagiotopoulos, 1983, Panagiotopoulos, 1985, Panagiotopoulos, 1993). They are directly related to nonconvex superpotentials and describe nonmonotone boundary or constitutive laws.

From the theoretical point of view this transition to nonconvex problems requires new tools for the study of the arising problems. For instance, due to the lack of convexity of the energy functions, the mathematical study regarding the existence of solution for the hemivariational inequalities, must be exclusively based on weak compactness arguments (Panagiotopoulos, 1985, Panagiotopoulos, 1991). This makes things a bit more difficult than the case of variational inequalities where monotonicity and respective arguments can be applied (Duvaut and Lions, 1972, Panagiotopoulos, 1985). Under certain assumptions for the nonsmooth superpotential involved (Panagiotopoulos, 1985, Panagiotopoulos, 1993, Naniewicz and Panagiotopoulos, 1995), the aforementioned hemivariational inequalities are equivalent to nonsmooth, nonconvex potential or complementary energy minimization problems.

The theory of hemivariational inequalities leads to the results that local minima of the potential or the complementary energy of the structure represent equilibrium positions of the problem. It is possible, however, that certain solutions of the problem may not be local minima but other more general type of points which make the potential or the complementary energy "substationary". They are solutions of the differential inclusion $0 \in \bar{\partial}\Pi(u)$, where Π is the potential energy, u is the displacement vector and $\bar{\partial}$ denotes the generalized gradient of Clarke-Rockafellar (Rockafellar, 1979, Clarke, 1983). Analogous substationarity problem can be formulated in terms of the complementary energy of the structure Π^c and the stresses s, i.e. $0 \in \bar{\partial}\Pi^c(s)$.

Therefore, the substationarity points obtained (all the local minima, certain local maxima or saddle points) constitute all the possible solutions of a hemivariational inequality. This is a generalization of the minimum potential or complementary energy theorems which hold in the case of variational equalities and inequalities (Duvaut and Lions, 1972, Panagiotopoulos, 1985).

The determination of the full set of solutions of a substationarity problem, even when only smooth functionals are involved, remains as yet, an open problem and constitutes an area of active research in the field of computational mechanics. This holds indeed for a global optimization problem as well, the latter being only a particular case of the general substationarity problem (Fletcher, 1990, Gill et al., 1981, Pardalos and Rosen, 1987, Strodiot and Nguyen, 1988).

Moreover, one must keep in mind that the questions posed in mechanics are sometimes different from the ones of global optimization. A global minimum of the potential energy of a structure may be of relatively less importance if a given loading history is not sufficient to drive the structure near this equilibrium point. Stable or unstable equilibria, i.e., local minima or substationary points attainable by the given loading path may be of importance in this latter case.

In addition, the problems encountered in mechanics usually have a very large dimension, of the order of several thousands of degrees of freedom. Moreover, many types of mechanical problems are history or path dependent. Therefore, only the stable and unstable solutions on the specific loading path are of importance. Consequently, the existing nonconvex optimization algorithms can provide only a partial remedy for engineering problems.

Although the progress in the theoretical study of the existence and approximation questions for hemivariational inequalities is considerable (cf. Panagiotopoulos, 1991, Panagiotopoulos, 1993, Panagiotopoulos and Stavroulakis, 1990, Miettinen and Haslinger, 1992, Motreanu and Panagiotopoulos, 1996 Haslinger and Panagiotopoulos, 1995, Fundo, 1997, Naniewicz and Panagiotopoulos, 1995, Naniewicz, 1989, Naniewicz, 1993 and the references given therein), relatively few methods exist for the numerical treatment. The following approaches have appeared in the literature:

- The dissipation method (cf. e.g. Frémond, 1987), which has advantages from the standpoint of the physical interpretation of the nonmonotonicity.
- Bundle - type methods of Nonsmooth Optimization (Lemarechal and Mifflin, 1978, Kiwiel, 1985, Strodiot and Nguyen, 1988, Schramm and Zowe, 1992, Mäkelä and Neittaanmäki, 1992, Miettinen et al., 1995).
- Regularization techniques resulting in a sequence of variational equalities.
- Nonlinear solvers with step-length control to remain stable around limit points based on a direct enforcement of the nonmonotone stress-strain diagrams (Crisfield, 1986, Crisfield and Wills, 1988, Crisfield, 1991, Riks, 1972, Schellekens and DeBorst, 1991).

Some of these methods can find only very limited use in practical applications, due to the fact that their effectiveness fails rapidly with increasing problem size (order of one hundred of unknowns), due mainly to numerical stability problems. Moreover, they have inherent difficulties regarding the treatment of the complete vertical branches of the stress-strain or reaction-displacement diagrams.

1.3 TYPICAL, REPRESENTATIVE EXAMPLES

In this Section, the difficulties we face with the introduction of nonmonotone laws in engineering problems, is demonstrated by simple examples. For this reason, we consider the simple structure of Fig. 1.4a. The structure consists of the rigid body G which comes in contact with a rigid support. Between the body and the support frictional or adhesive contact conditions arise for which we assume that the nonmonotone law of Fig. 1.4b holds. Moreover, the body is connected through an elastic element with a rigid support. The modulus of elasticity for this element is $E = 4$ and has unit length L and cross-section area A, all in compatible units. The structure is loaded with the forces P_x and P_y at G. We want to determine the possible equilibrium positions of the system.

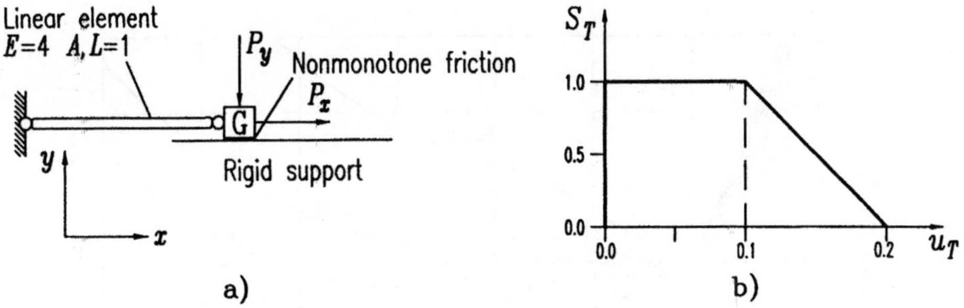

Figure 1.4. A model unilateral contact problem with nonmonotone friction

According to the elementary theory of mechanics, the nonconvex energy function Π_{nc} for this problem, is the sum of the convex superpotential of the linear element and of the nonconvex superpotential j of the nonmonotone law of Fig. 1.4b, minus the energy supplied by the loading, i.e.

$$\Pi_{nc} = \Pi_q(u_x)+j(u_x) = \frac{1}{2}u_x\frac{EA}{L}u_x - P_x u_x + j(u_x) = 2u_x^2 - P_x u_x + j(u_x). \quad (1.1)$$

It is easily verified that the nonmonotone law $g(u_x)$ of Fig. 1.4b has the form:

$$g(u_x) = \begin{cases} [0,+1] & \text{if } u_x = 0 \\ +1.0 & \text{if } 0 < u_x \leq 0.1 \\ 1 - 10(u_x - 0.1) & \text{if } 0.1 < u_x \leq 0.2 \\ 0 & \text{if } 0.2 < u_x \end{cases} \quad (1.2)$$

and the respective nonconvex superpotential $j(u_x)$ the form:

$$j(u_x) = \begin{cases} 0 & \text{if } u_x = 0 \\ u_x & \text{if } 0 < u_x \leq 0.1 \\ u_x - 5(u_x - 0.1)^2 & \text{if } 0.1 < u_x \leq 0.2. \\ 0.15 & \text{if } 0.2 < u_x \end{cases} \tag{1.3}$$

Fig. 1.5 depicts the function Π_{nc} for several values of the external loading. In all the cases, the potential energy function in non-convex and non-smooth. Indeed, the non-smoothness is caused by the vertical branch of the diagram of Fig. 1.4b at the point $u_x = 0$. As a result we get the corresponding non-differentiability point at the diagram of Fig. 1.5. Moreover, the nonconvexity of the superpotentials, is caused by the softening form of the diagram of Fig. 1.4b.

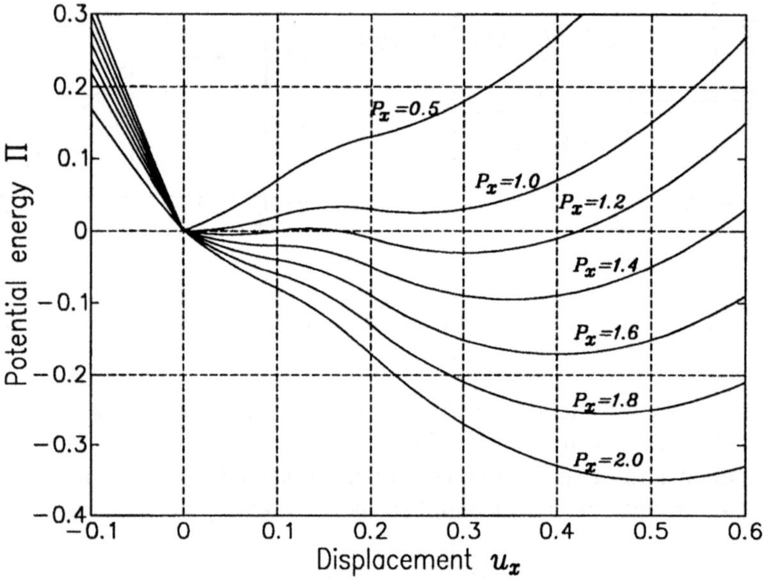

Figure 1.5. Nonconvex-nonsmooth energy potentials

The substationarity points of the nonconvex function Π_{nc} for a given load level, can be found, in the present simple problem, by taking the derivatives with respect to u_x in the respective intervals. We notice that although the function is always nonconvex, it does not have multiple solutions for all the loadings. For example, for $P = 1.0$ the problem has a unique solution $u_x = 0$. The same happens for $P = 1.05$ where the unique solution is $u_x = 0.05$. Multiple solutions exist for $1.1 \leq P_x \leq 1.5$.

Of course, in this simple example, it is easy to find all the minima of the nonconvex function. But, if the problem involves even a few unknowns, the determination of all the minima of the function is a very difficult task. In this case, additional assumptions have to be introduced in order to find the solution (or the solutions) that will be finally realized, because all the solutions are mathematically acceptable.

As a second example we will consider the simple assembly of Fig. 1.6a. The structure consists of the linear spring elements 1 and 3 and of the nonlinear spring elements 2 and 4. For the elements 2 and 4 we assume that the force-displacement diagram of Fig. 1.6b holds. We notice that for $u \leq 0.1$ these elements behave linearly with k=10. The structure is loaded with the forces $P_x = 1$ and $P_y = 1$ at G. For the given loading we want to determine the equilibrium positions of the system.

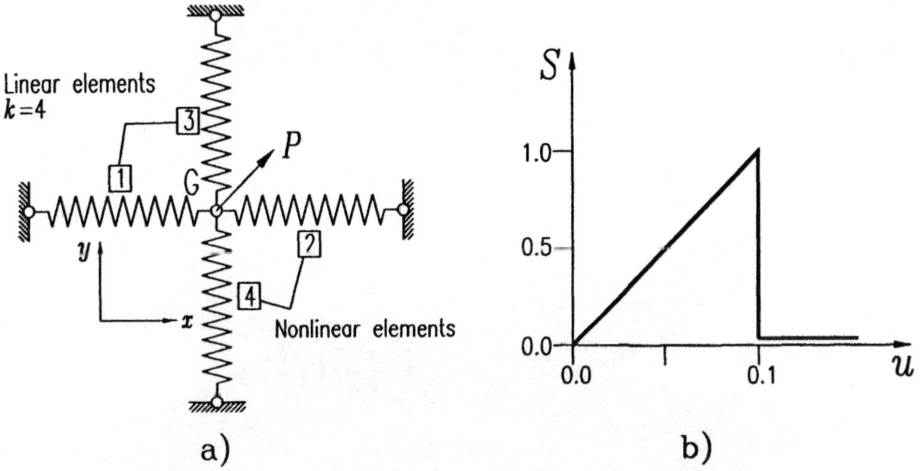

Figure 1.6. Simple structure with 2 degrees of freedom

With the given data we formulate the potential energy of the structure as a function of the displacements u_x, u_y of point G. As in the previous example, the total potential energy Π_{nc} is written as the sum of the convex superpotential of the linear elements and of the superpotentials of the nonlinear elements, minus the energy supplied by the loading, i.e.

$$\Pi_{nc} = \frac{1}{2}\mathbf{u}^T\mathbf{K}\mathbf{u} + j(u_x) + j(u_y) - \mathbf{u}^T\mathbf{P} \tag{1.4}$$

where $\mathbf{u} = [u_x \quad u_y]^T$, $\mathbf{P} = [P_x \quad P_y]^T$ and $\mathbf{K}$ is the stiffness matrix of the structure which has the very simple form $\mathbf{K} = \begin{bmatrix} 4 & 0 \\ 0 & 4 \end{bmatrix}$.

The nonconvex, nonsmooth function Π_{nc}, for $u_x, u_y \in [0, 0.5]$ is depicted in Fig. 1.7. Notice here the nondifferentiability lines caused by the vertical branch of the nonmonotone diagram of Fig. 1.6b. As it is more clear from the contours of the nonconvex function given in the lower part of the figure, the function has nine critical points, one of which is a global minimum, three are local minima and one is local maximum.

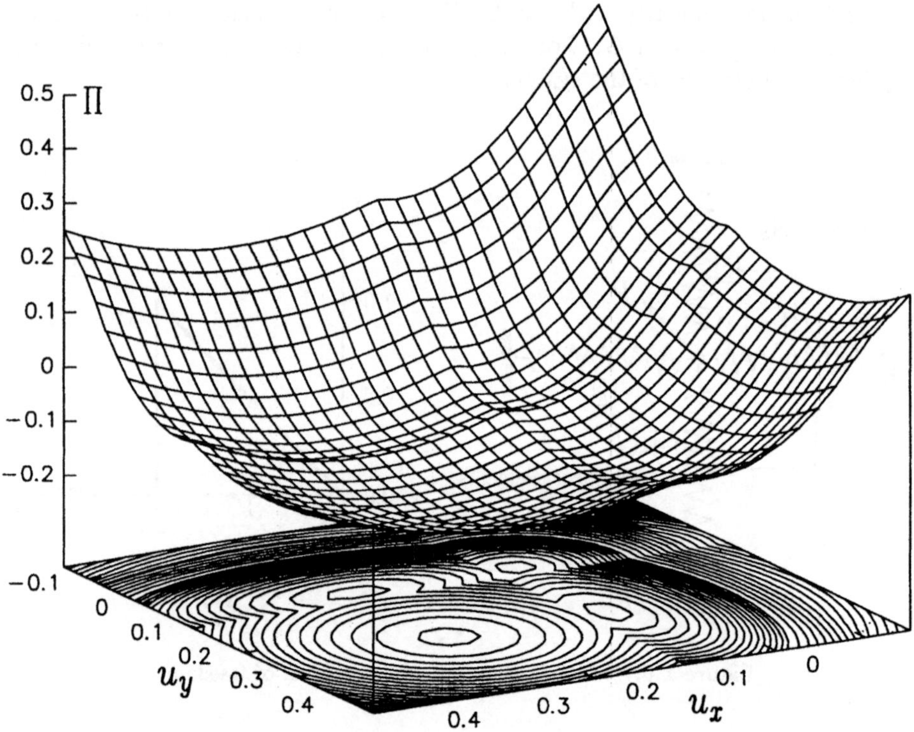

Figure 1.7. Nonconvex, nonsmooth energy surface

1.4 NONCONVEX OPTIMIZATION BY SOLVING CONVEX SUBPROBLEMS

As it became clear in the previous Section, the task of locating all the substationarity points of the nonconvex energy function, i.e. finding all the equilibrium configurations of the structure can be a demanding task even in cases with two unknowns. Furthermore, the difficulty of the problem can somehow be reduced by considering that only solutions attainable along a loading path are of importance. Moreover, real engineering problems often involve thousands of unknowns making the complete mathematical solution of the problem almost impossible. For this reason, in engineering applications, the sole mathematical examination of the problem has to be abandoned and more heuristic methods have to be investigated, which take into account the engineering characteristics of the problems at hand. The present book has exactly this task, to fill in the gap between the mathematical theory of nonconvex optimization and the need to calculate accurate solutions in complex, real life, engineering problems. In this framework, new methods have been developed, which, on one hand take into account the exact mechanical characteristics of the considered problem and on the other hand take advantage of their engineering nature.

The aim of the present book is to demonstrate new methods for the solution of the nonconvex optimization problems in mechanics and engineering. The characteristic of the developed methods is that they reduce the nonconvex optimization problem into a number of convex minimization subproblems. This is equivalent to the approximation of the hemivariational inequality describing the problem involving the nonmonotone laws by a number of variational inequalities that involve monotone laws only. Two general categories of methods can be distinguished in which:

- the nonconvex problem is proved to be equivalent to an appropriate combination of convex problems by means of quasidifferentiability (Stavroulakis, 1993, Dem'yanov et al., 1996) and
- the nonconvex energy problem is approximated locally by an appropriately chosen sequence of convex subproblems (Mistakidis and Panagiotopoulos, 1993, Mistakidis and Panagiotopoulos, 1994, Mistakidis and Panagiotopoulos, 1997).

Then the problem is formulated as a finite set of variational inequalities or convex optimization problems on different domains of the displacements space. Subsequently, each variational inequality can be treated effectively by specialized convex minimization algorithms (Fletcher, 1990, Tzaferopoulos, 1993, Womersley and Fletcher, 1986). The verified reliability and high convergence rates of the latter make a precious advantage. An additional advantage is that their combination with the methods developed in nonlinear finite element analysis, with efficient iterative minimization procedures and specialized preconditioning schemes may lead to the construction of algorithms that are able to operate on large scale systems (order of thousands of unknowns) for the nonmonotonicities considered.

References

Allix, O., Ladevéze, P., and Corigliano, A. (1995). Damage analysis of interlaminar fracture specimens. *Composite Structures*, 31:61–74.

Aribert, J. M. and Aziz, U. A. (1985). Calcul des poutres mixtes jusq' à l' état ultime avec un effect de soulèvement à l' interface acier-béton. *Construction métallique*, 4(4):3–36.

Baniotopoulos, C. C. and Panagiotopoulos, P. D. (1987). A hemivariational approach to the analysis of composite material structures. In Paipetis, S. A. and Papanicolaou, G. C., editors, *Engineering Applications of New Composites*. Omega Publications, London.

Clarke, F. H. (1983). *Optimization and nonsmooth analysis*. J. Wiley, New York.

Colson, A., editor (1992). *COST-C1: Proceedings of the first state of the art workshop*. Ecole Nationale Supérieure des Arts et Industries de Strasbourg, Strasbourg.

Corigliano, A. and Bolzon, G. (1995). Numerical simulation of debonding phenomena in composite materials. In Owen, D. R. J. and Oñate, E., editors, *Computational plasticity. Fundamentals and applications*, pages 1179–1190, Swansea. Pineridge Press.

Crisfield, M. A. (1986). Snap-through and snap-back response in concrete structures and the dangers of under-integration. *Computer Method in Applied Mechanics and Engineering*, 22:751–767.

Crisfield, M. A. (1991). *Non-linear finite element analysis of solids and structures*. J. Wiley, Chichester.

Crisfield, M. A. and Wills, J. (1988). Solution strategies and softening materials. *Computer Method in Applied Mechanics and Engineering*, 66:267–289.

Dem'yanov, V. F., Stavroulakis, G. E., Polyakova, L. N., and Panagiotopoulos, P. D. (1996). *Quasidifferentiability and nonsmooth modelling in mechanics, engineering and economics*. Kluwer Academic, Dordrecht.

Duvaut, G. and Lions, J. L. (1972). *Les inéquations en méchanique et en physique*. Dunod, Paris.

Faupel, F. (1989). Metall-Polymer-Schichtsysteme: mechanische Spannungen, Haftung und Diffusion an der Grenzfläche. In Brockmann, W., editor, *Haftung bei Verbundwerkstoffen und Werkstoffverbunden*. DGM Informationsgesellschaft, Oberursel.

Feder, H. J. S. and Feder, J. (1991). Self-organized criticality in a stick-slip process. *Phys. Rev. Lett.*, 66:2669–2672.

Fletcher, R. (1990). *Practical methods of optimization*. J. Wiley, Chichester.

Frémond, M. (1987). Contact unilatéral avec adhérence : une théorie du premier gradient. In DelPierro, G. and Maceri, F., editors, *Unilateral problems in structural analysis - 2*, volume 304 of *CISM Courses and Lectures*. Springer, Wien - New York.

Fundo, M. (1997). Hemivariational inequalities in subspaces of $L^p(\Omega)(p \geq 3)$. *Nonlin. Anal. Theo. Meth. Appl.*, (to appear).

Gattesco, N. (1990). Long-span steel and concrete beams with partial shear connection. *Studi e Ricerche*, 12:243–266.

Gilbert, R. I. and Warner, R. F. (1978). Tension stiffening in reinforced concrete slabs. *J. Struct. Div. ASCE*, 104(12):1885 - 1900.

Gill, P. E., Murray, W., and Wright, M. H. (1981). *Practical optimization.* Academic Press, New York.

Glowinski, R., Lions, J. L., and Trémolières, R. (1981). *Numerical analysis of variational inequalities.* Studies in Mathematics and its Applications, Vol. 8. Elsevier, Amsterdam-New York.

Green, A. K. and Bowyer, W. H. (1981). The testing analysis of novel top-hat stiffener fabrication methods for use in GRP ships. In Marshall, I. H., editor, *Proc. 1st International Conf. on Composite Structures*, Barking-Essex. Applied Science.

Grundy, P. (1990). Effect of pre and post bucking behaviour on load capacity of continuous beams. *Thin-walled Structures*, 9(1-4):407–415.

Hampe, A. (1989). Messung der Scherfestigkeit in der Grenzfläche zwischen Polymer und Glas durch Einzelfaserexperimente. In Brockmann, W., editor, *Haftung bei Verbundwerkstoffen und Werkstoffverbunden.* DGM Informationsgesellschaft, Oberursel.

Haslinger, J. and Panagiotopoulos, P. D. (1995). Optimal control of systems governed by hemivariational inequalities. Existence and approximation results. *Nonlin. Anal. Theo. Meth. Appl.*, 24:105–119.

James, R. D. (1996). Hysteresis in phase transformations. In Kirchgässner, K., Mahrenholtz, O., and Memmicken, R., editors, *ICIAM 95*, pages 135–154, Berlin. Akademie Verlag.

Kecman, D. (1983). Bending collapse of rectangular and square tubes. *Int. J. of Mechanical Sciences*, 13(9-10):623–636.

Kiwiel, K. C. (1985). *Methods of descent for nondifferentiable optimization.* Springer, Berlin. Lecture notes in mathematics No. 1133.

Krasnosel'skii, M. A. and Pokrovskii, A. V. (1983). *Systems with hysteresis.* Nauka, Moscow. English translation Springer Verlag 1995.

Kuhn, J. M. and Bucuner, C. D. (1986). Effect of concrete placement on shear strength of headed studs. *Journal of Structural Engineering (ASCE)*, 112:1965–1970.

Kyriakides, S. (1994). Propagating instabilities in structures. *Advances in applied mechanics*, 30:68–191.

Lemarechal, C. and Mifflin, R., editors (1978). *Bundle methods in nonsmooth optimization.* Pergamon Press, Oxford.

Mäkelä, M. M. and Neittaanmäki, P. (1992). *Nonsmooth optimization: analysis and algorithms with applications to optimal control.* Word Scientific Publ. Co.

Miettinen, M. and Haslinger, J. (1992). Approximation of optimal control problems of hemivariational inequalities. *Numer. Funct. Anal. and Optim.*, 13:43–68.

Miettinen, M., Mkel, M. M., and Haslinger, J. (1995). On numerical solution of hemivariational inequalities by nonsmooth optimization methods. *Journal of Global Optimization*, 8(4):401–425.

Mistakidis, E. S. and Panagiotopoulos, P. D. (1993). Numerical treatment of nonmonotone (zig–zag) friction and adhesive contact problems with debonding. Approximation by monotone subproblems. *Computers and Structures*, 47:33–46.

Mistakidis, E. S. and Panagiotopoulos, P. D. (1994). On the approximation of nonmonotone multivalued problems by monotone subproblems. *Computer Methods in Applied Mechanics and Engineering*, 114:55–76.

Mistakidis, E. S. and Panagiotopoulos, P. D. (1997). On the search for substationarity points in the unilateral contact problems with nonmonotone friction. *J. of Math. and Comp. Modelling*, (to appear).

Mistakidis, E. S., Thomopoulos, K., Avdelas, A., and Panagiotopoulos, P. D. (1994). Shear connectors in composite beams: A new accurate algorithm. *Thin-Walled Structures*, 18:191–207.

Moreau, J. J. (1963). Fonctionnelles sous - différentiables. *C.R. Acad. Sc. Paris*, 257A:4117 – 4119.

Moreau, J. J. (1968). La notion de sur-potentiel et les liaisons unilatérales en élastostatique. *C.R. Acad. Sc. Paris*, 267A:954 – 957.

Moreau, J. J., Panagiotopoulos, P. D., and Strang, G., editors (1988). *Topics in nonsmooth mechanics.* Birkhäuser, Basel-Boston.

Motreanu, D. and Panagiotopoulos, P. D. (1996). On the eigenvalue problem for hemivariational inequalities: existence and multiplicity of solutions. *Math. Anal. Appl.*, 197:75–89.

Naniewicz, Z. (1989). On some nonconvex variational problems related to hemivariational inequalities. *Nonlin. Anal.*, 13:87–100.

Naniewicz, Z. (1993). On the existence of solutions to the continuum model of delamination. *Nonl. Anal. Theory, Meth. Appl.*, 20:481–507.

Naniewicz, Z. and Panagiotopoulos, P. D. (1995). *Mathematical theory of hemivariational inequalities and applications.* Marcel Dekker.

Oehlers, D. J. and Johnson, R. D. (1987). The strength of stud shear connections in composite beams. *The structural Engineer*, 65B(2):44–48.

Panagiotopoulos, P. D. (1983). Nonconvex energy functions. Hemivariational inequalities and substationary principles. *Acta Mechanica*, 42:160–183.

Panagiotopoulos, P. D. (1985). *Inequality problems in mechanics and applications. Convex and nonconvex energy functions.* Birkhäuser, Basel - Boston - Stuttgart. Russian translation, MIR Publ., Moscow 1988.

Panagiotopoulos, P. D. (1988a). Hemivariational inequalities and their applications. In Moreau, J. J., Panagiotopoulos, P. D., and Strang, G., editors, *Topics in Nonsmooth Mechanics.* Birkhäuser, Basel.

Panagiotopoulos, P. D. (1988b). Nonconvex superpotentials and hemivariational inequalities. Quasidiferentiability in mechanics. In Moreau, J. J. and Panagiotopoulos, P. D., editors, *Nonsmooth Mechanics and Applications*, volume 302 of *CISM Lect.*, pages 83–176. Springer, Wien - New York.

Panagiotopoulos, P. D. (1991). Coercive and semicoercive hemivariational inequalities. *Nonlin. Anal. Theo. Meth. Appl.*, 16:209 – 231.

Panagiotopoulos, P. D. (1993). *Hemivariational inequalities. Applications in mechanics and engineering.* Springer, Berlin - Heidelberg - New York.

Panagiotopoulos, P. D. and Stavroulakis, G. E. (1990). The delamination effect in laminated von Karman plates under unilateral boundary conditions. A variational - hemivariational inequality approach. *J. Elast.*, 23:69 – 96.

Pannhorst, W. (1993). Langfaserverstärkte Verbundwerstoffe mit glasiger oder glaskeramischer Matrix. In Leonhardt, G. and Ondraček, G., editors, *Verbundwerkstoffe und Werkstoffverbunde.* DGM Informationsgesellschaft, Oberursel.

Pardalos, P. M. and Rosen, J. B. (1987). *Constrained global optimization. Algorithms and applications*, volume 268 of *Lecture Notes in Computer Science.* Springer, Berlin.

Peters, P. W. M. (1989). Methoden zur Bestimmung der Faser / Matrix-Haftfestigkeit in FVW (CFK). In Brockmann, W., editor, *Haftung bei Verbundwerkstoffen und Werkstoffverbunden.* DGM Informationsgesellschaft, Oberursel.

Rasmussen, K. J. R. and Hancock, G. J. (1993). Design of cold-formed stainless steel tubular members. II: Beams. *J. Str. Div. ASCE*, 119 (ST8):2368–2386.

Rausch, G., Meier, B., and Grathwohl, G. (1993). Bestimmung von Grenzflächeneigenschaften faserverstärkter Verbundwerkstoffe mittels Eindruckmethoden. In Leonhardt, G. and Ondraček, G., editors, *Verbundwerkstoffe und Werkstoffverbunde.* DGM Informationsgesellschaft, Oberursel.

Reck, H., Pekoz, T., and Winter, G. (1976). Inelastic strength of cold-formed steel beams. *J. Str. Div. ASCE*, 101(ST11):2193–2204.

Riks, E. (1972). The application of Newton's method to the problem of elastic stability. *J. Appl. Mech.*, 39:1060 – 1066.

Rockafellar, R. T. (1979). *La théorie des sous-gradients et ses applications à l' optimization. Fonctions convexes et non-convexes.* Les Presses de l' Université de Montréal, Montréal.

Roeder, E. and Mayer, H. J. (1993). Unidirektionale Kurzfaserverstärkung von Gläsern mit Hilfe des Strangpreßverfahrens. In Leonhardt, G. and Ondraček, G., editors, *Verbundwerkstoffe und Werkstoffverbunde.* DGM Informationsgesellschaft, Oberursel.

Roman, I., Harlet, H., and Marom, G. (1981). Stress intensity factor measurements in composite sandwich structures. In Marshal, I. H., editor, *Proc. 1st Conf. on Composite Structures*, pages 633–645, London. Applied Science Publishers.

Schellekens, J. C. J. and DeBorst, R. (1991). Applications of linear and nonlinear fracture mechanics options to free edge delamination in laminated composites. *Heron*, 36:37 - 48.

Schlimmer, M. (1993). Zur Berechnung der Festigkeit von Werkstoffverbunden mit polymerer Zwischenschicht. In Leonhardt, G. and Ondraček, G., editors, *Verbundwerkstoffe und Werkstoffverbunde.* DGM Informationsgesellschaft, Oberursel.

Schramm, H. and Zowe, J. (1992). A version of the bundle idea for minimizing a nonsmooth function: conceptual idea, convergence analysis, numerical results. *SIAM J. Optimization*, 2:121–152.

Schwartz, M. M. (1984). *Composite materials handbook.* McGraw-Hill, New York.

Shidharan, N. S. (1982). Elastic and strength properties of continuous chopped glass fiber hybrid sheet molding compounds. In *Short fiber reinforced composite materials*, volume 787, pages 167–179. ASTM STP, Philadelphia.

Stavroulakis, G. E. (1993). Convex decomposition for nonconvex energy problems in elastostatics and applications. *European Journal of Mechanics A / Solids*, 12(1):1–20.

Stillinger, F. H. (1995). A topographic view of supercooled liquids and glass formation. *Science*, 267.

Strodiot, J. J. and Nguyen, V. H. (1988). On the numerical treatment of the inclusion $0 \in \partial f(x)$. In Moreau, J. J., Panagiotopoulos, P. D., and Strang, G., editors, *Topics in Nonsmooth Mechanics.* Birkhäuser, Basel.

Thomopoulos, K. T., Mistakidis, E. S., Koltsakis, E. K., and Panagiotopoulos, P. D. (1996). Softening behaviour of continuous thin-walled beams. Two numerical methods. *J. Construct. Steel Res.*, 36:1–13.

Tzaferopoulos, M. A. (1993). On an efficient method for the frictional contact problem of structures with convex energy density. *Computers and Structures*, 48(1):87–106.

Unger, B. (1973). Ein Beitrag zur Ermittlung der Tranglast von querbelasteten Durchlauftragern mit dünnwändigen Quershnitt, insbensondere von durchlaufenden Trapezblechen für Dack und Geschossdecken. *Der Stahlbau*, 42:20–24.

Visintin, A. (1994). *Differential models of hysteresis*. Springer Verlag, Berlin Heidelberg.

Wald, F., editor (1994). *COST-C1: Proceedings of the second state of the art workshop*. ECSP-EEC-EAEC, Brussels-Luxembourg.

Wang, S. T. and Yeh, S. S. (1975). Post local-buckling behaviour of continuous beams. *J. Str. Div. ASCE*, 100(ST6):1169–1188.

Webb, T. W. and Aifantis, E. C. (1995). Oscillatory fracture in polymeric materials. *International Journal of Solids and Structures*, 32(17-18):2725–2744.

Webb, T. W. and Aifantis, E. C. (1997a). Crack growth resistance curves and stick-slip fracture instabilities. *Mechanics Research Communications*, 24(2):123–129.

Webb, T. W. and Aifantis, E. C. (1997b). Loading rate dependence of stick-slip fracture in polymers. *Mechanics Research Communications*, 24(2):115–121.

Wiedemann, J. (1996). *Leichtbau. Band 2: Konstruktion. 2. Auflage*. Springer.

Williams, J. G. and Rhodes, M. D. (1982). Effect of resin on impact damage tolerance of graphite/epoxy laminates. In Daniel, I., editor, *6th International Conf. on Composite Materials, Testing and Design*. ASTM STP, Philadelphia.

Womersley, R. S. and Fletcher, R. (1986). An algorithm for composite nonsmooth optimization problems. *Journal of Optimization Theory and Applications*, 48:493–523.

Yener, M. and Pekoz, T. (1980). Inelastic load-carrying capacity of cold-formed steel beams. In *Proceedings of the 5th International Speciality Conference on Cold-Formed Steel Structures*. University of Missouri-Rolla.

II Applied Nonconvex Optimization Background

2 APPLIED NONCONVEX OPTIMIZATION BACKGROUND.

2.1 OPTIMIZATION PROBLEMS

The problem of finding an extremum of a given function over the space where the function is defined or over a subset of it, is called an optimization problem. In mechanics several "principles" which govern physical phenomena in general and the response of mechanical systems in particular are written in the form of an optimization problem. The principles of minimum potential energy in statics, the maximum dissipation principle in dissipative media and the least action principle in dynamics are some examples (see, among others, Hamel, 1949, Lippmann, 1972, Cohn and Maier, 1979, de Freitas, 1984, de Freitas and Smith, 1985, Panagiotopoulos, 1985, Hartmann, 1985, Sewell, 1987, Bažant and Cedolin, 1991). More general optimization problems which are not always based on physical considerations arise in several engineering problems, for instance problems of optimal design of structures, control, identification and reliability analysis of structures and mechanical systems.

On the other hand, a wealth of theoretical and numerical results exists in the area of mathematical optimization or mathematical programming. This field has been developed primarily to cope with complicated management and operational research problems and has nowadays found applications to the majority of the applied sciences. In particular, the computational mechanics applications which are studied in this book profit from the rigour of the algorithmic computational optimization techniques. The same is true for the theoretical results provided by the related fields of matrix analysis, convexity, nonsmooth analysis, approximation theory and computer science. All these results are indispensable for the development of effective algorithms and the numerical treatment of modern engineering mechanics problems. A short summary of results which will be referred in the sequel are collected in this Chapter. The presentation is more–or–less task–driven here; all mathematical proofs and details can be found, among others, in the references.

For a real–valued function f defined on an open set $\Omega \subset \mathbb{R}^n$ a local minimum (resp. global minimum) point $x^* \in \Omega$ is defined by the following relation

$$f(x) \geq f(x^*), \ \forall x \in O(x^*) \ (\text{resp. } \forall x \in \Omega), \tag{2.1}$$

where $O(x^*)$ denotes a neighbourhood of x^*.

By using the notion of the first derivative $f'(x^*)$ of f at x^* the function f has an extremum at point $x^* \in \Omega$ and in addition it is differentiable at this point, then the following relation holds true there:

$$f'(x^*) = 0. \tag{2.2}$$

For the relative extremum of a function f with respect to a subset U of $\Omega \subset \mathbb{R}^n$, the notion of Lagrange multipliers is used. Let the set U be defined by means of m functions $\phi_i : \Omega \to \mathbb{R}$, which are continuous with continuous first derivatives (i.e. $\phi_i \in C^1$, $i = 1, \ldots, m$), as:

$$U = \{x \in \Omega : \phi_i(x) = 0, \ i = 1, \ldots, m\} \subset \Omega, \tag{2.3}$$

and let the derivatives $\phi_i'(x^*), i = 1, \ldots, m$ be linearly independent. If f has a relative extremum at x^* with respect to the set U, then there exist m Lagrange multipliers $\lambda_i(x^*), i = 1, \ldots, m$ such that:

$$f'(x^*) + \sum_{i=1}^{m} \lambda_i(x^*) \phi_i'(x^*) = 0. \tag{2.4}$$

By assuming that f is twice differentiable at point $x^* \in \Omega$, then the necessary condition for a relative minimum of f at x^* reads

$$f''(x^*)(\overline{x}, \overline{x}) \geq 0, \ \forall \overline{x} \in \mathbb{R}^n. \tag{2.5}$$

In order to facilitate the understanding of the above conditions and to fix notations concerning the first and second derivatives f', f'', the gradient and the Hessian, ∇f and $\nabla^2 f$, the Taylor expansion formulae for a twice differentiable function $f : \mathbb{R}^n \to \mathbb{R}$ at point x^* are recalled here:

$$f(x^* + \overline{x}) = f(x^*) + f'(x^*)\overline{x} + \frac{1}{2}f''(x^*)(\overline{x}, \overline{x}) + \|\overline{x}\|_2^2 \epsilon(\overline{x}), \tag{2.6}$$

$$f(x^* + \overline{x}) = f(x^*) + (\nabla f(x^*), \overline{x}) + \frac{1}{2}((\nabla^2 f(x^*)\overline{x}, \overline{x}) + (\overline{x}, \overline{x})\epsilon(\overline{x}), \tag{2.7}$$

$$f(x^* + \overline{x}) = f(x^*) + (\nabla f(x^*))^T \overline{x} + \frac{1}{2}\overline{x}^T \nabla^2 f(x^*)\overline{x} + \overline{x}^T \overline{x}\epsilon(\overline{x}). \tag{2.8}$$

Here $(\cdot, \cdot)$ denotes the $\mathbb{R}^n$ Euclidean scalar product, $x^*, \overline{x} \in \mathbb{R}^n$, $\|x\|_2^2$ is the L_2 norm and $\epsilon(\overline{x})$ is a small scalar with $\lim_{\overline{x} \to 0} \epsilon(\overline{x}) = 0$.

As an example, for a quadratic functional defined by means of a symmetric $n \times n$ matrix K and a vector $p \in \mathbb{R}^n$, as:

$$f(x) = \frac{1}{2}(Kx, x) - (p, x), \tag{2.9}$$

one has

$$f'(x)^T = \nabla f(x) = Kx - p, \ \forall x \in \mathbb{R}^n. \tag{2.10}$$

If one identifies K with the stiffness matrix of a structural system, p with the loading vector and x with the (displacement) degrees of freedom for, say, a finite element model, then the solution of the system of linear equilibrium equations (2.10) is an extremum, through (2.2), of the potential energy function of the system (2.9). If, moreover, an extremum of (2.9) with respect to a set defined by a given $m \times m$ matrix A and a vector $b \in \mathbb{R}^m$ is sought, i.e., over the set:

$$X_{ad} = \{x \in \mathbb{R}^n \ : \ Ax = b\}, \tag{2.11}$$

then the necessary condition (2.4) leads to the following system

$$\begin{aligned} Kx + A^T\lambda &= p \\ Ax &= b. \end{aligned} \tag{2.12}$$

In the previous structural analysis interpretation, the constraint set (2.11) may define, for instance, displacement boundary conditions (discrete supports) for the structure. Accordingly, the Lagrange multipliers vector $\lambda \in \mathbb{R}^m$ can be identified with the reactions of the structure at the support. Obviously, the reactions appear at the support points, where the constraints are applied on the structure.

An effective way to distinguish between local and global minima (cf. (2.1)) is provided by the notion of convexity. A subset of a vector space is called convex if for any two points x_1, x_2 belonging to it, the closed segment $x = \lambda x_1 + (1-\lambda)x_2$, $0 \leq \lambda \leq 1$ belongs to the subset. A necessary condition for a differentiable function f to attain a relative minimum at x^* with respect to a convex subset U is given by Euler's inequalities (Ciarlet, 1989 p. 242):

$$f'(x^*)(\overline{x} - x^*) \geq 0, \ \forall \overline{x} \in U. \tag{2.13}$$

If moreover the function f is convex over U, i.e., if

$$f(\lambda x_1 + (1-\lambda)x_2) \leq \lambda f(x_1) + (1-\lambda) f(x_2), \ \forall x_1, x_2 \in U, \ \lambda \in [0,1], \tag{2.14}$$

then the inequality (2.13) is also a sufficient condition for minimality. For a differentiable (resp. twice differentiable) function f, convexity is equivalent to the following relations:

$$f(\overline{x}) \geq f(x^*) + f'(x^*)(\overline{x} - x^*), \ \forall \overline{x}, x^* \in U, \tag{2.15}$$

respectively

$$f''(x^*)(\overline{x} - x^*, \overline{x} - x^*) \geq 0, \ \forall \overline{x}, x^* \in U. \tag{2.16}$$

If moreover the inequalities (2.15), (2.16) are strict, then the function is strictly convex and uniqueness of the corresponding optimization problem is assured.

Recall that the quadratic function (2.9) is convex (resp. strictly convex) if matrix K is non–negative definite (resp. positive definite).

Theoretical and technical details pertaining to the previous discussion can be found in Stoer and Witzgall, 1970, Elster et al., 1977, Ciarlet, 1989, Aubin, 1993, Bertsekas, 1982, Fletcher, 1990, Murty, 1988, Mangasarian, 1994, among others.

2.2 NONSMOOTH CONVEX PROBLEMS

2.2.1 *Smooth, inequality constrained, convex problems*

Before dealing with the general form of optimality conditions for nondifferentiable, convex problems restricted by convex equality or inequality relations, let us discuss constrained optimization problems for smooth functions. Let us consider an inequality constrained optimization problem, defined over the set (cf. (2.3))

$$U = \{x \in \Omega : \phi_i(x) \leq 0, \ i = 1, \ldots, m\}. \tag{2.17}$$

The set U is assumed to be convex, which holds true if all functions ϕ_i, $i \in \{1,\dots,m\}$ are convex. Let the active index set at point $x^* \in \Omega$ be denoted as

$$\mathcal{I}(x^*) = \{i \in \{1,\dots,m\} \ : \ \phi_i(x^*) = 0\} . \tag{2.18}$$

Suppose that the functions ϕ_i, $i \in \mathcal{I}(x^*)$ are continuously differentiable at x^*, the function $f : \Omega \to \mathbb{R}$ whose minimum is sought is also continuously differentiable at x^* and that the remaining, inactive constraints $i \notin \mathcal{I}(x^*)$ are continuous at x^*.

The Lagrange multiplier conditions (2.4), given previously for the case of equality constrained problems (2.3), are generalized by the following Karush–Kuhn–Tucker condition, which covers inequality constraints (2.17): If f has a relative minimum at x^* with respect to the set U defined by (2.17) and if the previously given hypotheses (constraint qualifications) hold true, then there exist Lagrange multipliers $\lambda_i(x^*)$, $i \in \mathcal{I}(x^*)$, such that (cf. (2.4))

$$\begin{gathered} f'(x^*) + \sum_{i \in \mathcal{I}(x^*)} \lambda_i(x^*)\phi_i'(x^*) = 0, \\ \lambda_i(x^*) \geq 0, \ \forall i \in \mathcal{I}(x^*). \end{gathered} \tag{2.19}$$

Karush–Kuhn–Tucker condition (2.19) is a necessary optimality condition if U defined by (2.17) is a convex set and becomes a sufficient condition for a minimum if in addition function f is convex (see e.g. Ciarlet, 1989, p. 345).

In view of (2.17), condition (2.19) can be written as a set of equalities, inequalities and a complementarity condition as follows:

$$\begin{gathered} f'(x^*) + \sum_{i \in \mathcal{I}(x^*)} \lambda_i(x^*)\phi_i'(x^*) = 0, \\ \lambda_i(x^*) \geq 0, \ \phi_i(x^*) \leq 0, \ i \in \{1,\dots,m\}, \\ \sum_{i \in \mathcal{I}(x^*)} \lambda_i(x^*)\phi_i(x^*) = 0. \end{gathered} \tag{2.20}$$

Relations (2.20) constitute a nonlinear complementarity problem (NL.C.P.). If function f is quadratic and the set U is defined by linear or affine functions ϕ_i, $i = 1,\dots,m$, then relations (2.20) constitute a nonstandard linear complementarity problem (L.C.P., see Glowinski and LeTallec, 1989, Murty, 1988, Cottle et al., 1992, among others). Thus, for the minimum of the quadratic function (2.9) over the set (cf. (2.11))

$$X_{ad} = \{x \in \mathbb{R}^n \ : \ Ax - b \leq 0\} , \tag{2.21}$$

the following nonstandard linear complementarity problem must be solved (cf. (2.12)):

$$Kx + A^T\lambda \ = \ p, \tag{2.22}$$

$$Ax - b \leq 0, \ \lambda \geq 0, \ \lambda^T(Ax - b) = 0.$$

Here the constraint qualification mentioned earlier simply requires that the set (2.21) be nonempty.

Note that after the introduction of a set of nonnegative slack variables y, the following standard linear complementarity problem can be formulated from (2.22) (on the assumption that K is invertible, i.e., that f is strictly convex):

$$y = b - Ax = -AK^{-1}A^T\lambda + b - AK^{-1}p, \qquad (2.23)$$
$$y \geq 0, \ \lambda \geq 0, \ y^T\lambda = 0.$$

Note also that the special case of linear complementarity problems (2.23) has certain advantages for the theoretical and numerical study (cf. the box type (2.59) constrained problems that will be discussed later). For example, problem (2.23) can be solved by variants of the pivoting techniques developed in the area of linear programming (see e.g. Murty, 1988).

2.2.1.1 Lagrangians, saddle points, duality. At the end of this Section a short introduction to Lagrangians, saddle points and the theory of duality is presented. Restricting ourselves to the fundamental points of this much wider area, we are going to consider Lagrangians and saddle points as a consistent way of introducing constraints, of equality or inequality type, into a given optimization problem. Duality serves in this Section, intuitively speaking, as a means of simplifying inequality and equality constraints, which, in turn, permits the use of simple projection (relaxation) type solution algorithms in a primal–dual setting. From another point of view, which will be discussed in other parts of this book, Lagrangians permit a unified writing of engineering analysis problems (including mixed formulations). Duality, in turn, provides the link between the various variational (minimum, or in general stationary) principles in mechanics which, usually, have been developed separately from different scientists in the past and for these historical reasons, they are called "principles" (see e.g. Hamel, 1949, Washizu, 1968, Lippmann, 1972, Matthies et al., 1979, Oden and Reddy, 1982, Hartmann, 1985, Sewell, 1987).

Let us consider again the minimization of a convex, differentiable function $f(x)$ over the convex subset of $\mathbb{R}^n$ defined by the differentiable inequality constraints of (2.17). The Lagrangian function associated with this problem, $\mathcal{L}(x, \mu) : \mathbb{R}^n \times \mathbb{R}^m_+ \to \mathbb{R}$, reads:

$$\mathcal{L}(x, \mu) = f(x) + \sum_{i=1}^{m} \mu_i \phi_i(x). \qquad (2.24)$$

A point $(x^*, \lambda) \in \mathbb{R}^n \times \mathbb{R}^m_+$ is called a saddle point of the Lagrangian $\mathcal{L}(x, \mu)$ if point x^* is a minimum of the function $\mathcal{L}(\cdot, \mu) : \mathbb{R}^n \to \mathbb{R}$ and point λ is a maximum of the function $\mathcal{L}(x, \cdot) : \mathbb{R}^m_+ \to \mathbb{R}$, i.e., if

$$\sup_{\mu \in \mathbb{R}^m_+} \mathcal{L}(x^*, \mu) = \mathcal{L}(x^*, \lambda) = \inf_{x \in \mathbb{R}^n} \mathcal{L}(x, \lambda). \tag{2.25}$$

Note that vector λ at the saddle point contains the Lagrangian multipliers of the Karush–Kuhn–Tucker conditions (2.19). Furthermore, from the second equality in (2.25) we see that the original (primal) constrained optimization problem can be replaced by an unconstrained one, provided that the values of the Lagrange multipliers $\lambda \in \mathbb{R}^m_+$ are known. These multipliers can be found independently if one considers the first equality in (2.25) in the form of the dual problem:

$$\text{Find } \lambda \in \mathbb{R}^m_+ \ : \ G(\lambda) = \sup_{\mu \in \mathbb{R}^m_+} G(\mu), \tag{2.26}$$

with the function $G : \mathbb{R}^m_+ \to \mathbb{R}$ defined by:

$$G(\mu) = \inf_{x \in \mathbb{R}^n} \mathcal{L}(x, \lambda). \tag{2.27}$$

2.2.2 *Concise form of optimality conditions*

Let us consider again the inequality constrained set (2.17), defined by means of m continuously differentiable functions $\phi_i \in C^1(\mathbb{R}^n)$, $i \in \{1, \ldots, m\}$:

$$U = \{x \in \Omega \subset \mathbb{R}^n \mid \phi_i(x) \le 0, \ i = 1, \ldots, m\}. \tag{2.28}$$

A concise expression of the above outlined optimality conditions can be obtained if one uses conical approximations of the boundary of the constraint set (2.28) (see, e.g., Dem'yanov et al., 1996). Under appropriate regularity conditions the cone of admissible directions at a point $x^* \in \Omega \cap U$ coincides with the Bouligand cone (see Fig. 2.1) and the tangent cone, and can be expressed by:

$$\Gamma_U(x^*) = \{g \in \mathbb{R}^n \mid \nabla\phi_i(x^*)^T g \le 0, \ j \in \mathcal{J}_0(x^*)\}, \tag{2.29}$$

where, in turn, the active index set $\mathcal{J}_0(x^*)$ is defined by:

$$\mathcal{J}_0(x^*) = \{j \in \{1, \ldots, m\} \mid \phi_i(x^*) = 0\}. \tag{2.30}$$

By using the notion of the conjugate cone $\Gamma_U^+(x^*)$ to the cone $\Gamma_U(x^*)$:

$$\Gamma_U^+ = \{x^* \in \mathbb{R}^n \mid (x^*, y) \le 0, \ \forall y \in \Gamma_U\}, \tag{2.31}$$

we define the normal cone $\mathcal{N}_U(x^*)$ to the convex set U at the point $x^* \in U$ by the relation (see Fig. 2.1)

$$\mathcal{N}_U(x^*) = -\Gamma_U^+(x^*). \tag{2.32}$$

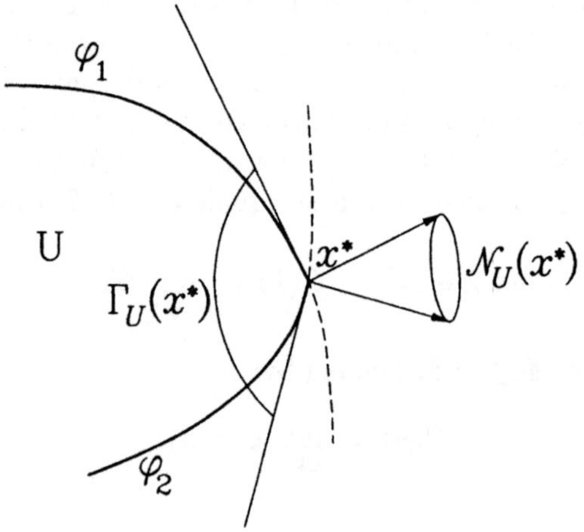

Figure 2.1. The normal and Bouligand cone

Using the previous notation, the optimality conditions for the minimization of a smooth, convex function $f(x)$ over a convex set U defined by means of smooth and convex inequality constraints as in (2.28), read (cf. Hiriart-Urruty and Lemaréchal, 1993):

Find $x \in U$ such that:

$$f'(x, x^* - x) \geq 0, \ \forall x^* \in U, \tag{2.33}$$

or equivalently

$$f'(x, y) \geq 0, \ \forall y \in \Gamma_U(x), \tag{2.34}$$

or equivalently

$$f'(x) \cap \Gamma_U^+(x) \neq \emptyset \Longleftrightarrow 0 \in f'(x) + \mathcal{N}_U(x), \tag{2.35}$$

where $f'(x, y) = f'(x)y = (\nabla f(x), y)$ is the directional derivative of f at point x in the direction of y.

2.2.3 Convex, nonsmooth optimization

2.2.3.1 Unconstrained optimization. Let us consider now first the case of unconstrained minimization of a convex, nonsmooth function. For a convex, nondifferentiable function f the notion of the first derivative must be generalized such as to take into account the fact that there exist points where the function may have cusps. Convex analysis provides us with a set–valued enhancement of the derivative which is based upon the inequality (2.15). A subgradient $w(x^*) \in \mathbb{R}^n$ is defined as a vector which satisfies the relation:

$$f(\overline{x}) \geq f(x^*) + (w(x^*), \overline{x} - x^*), \ \forall \overline{x} \in U. \tag{2.36}$$

The set of all such vectors forms the convex analysis subdifferential of f at x^*, i.e.,

$$\partial f(x^*) = \{w(x^*) \in \mathbb{R}^n \ : \ (w(x^*), \overline{x} - x^*) \leq f(\overline{x}) - f(x^*), \ \forall \overline{x} \in U\}\,. \tag{2.37}$$

For further reference we give here also the expression of the directional derivative of a nondifferentiable, subdifferentiable convex function, at point x in the direction y, denoted by $f'(x,y)$:

$$f'(x,y) = \max_{w \in \partial f(x)} (w,y), \ \forall y \in \mathbb{R}^n\,. \tag{2.38}$$

Accordingly, the first order optimality condition (2.2) reads:

$$0 \in \partial f(x^*). \tag{2.39}$$

Observe that due to (2.37), the previous relation can also be written as

$$(w(x^*), \overline{x} - x^*) \geq 0, \forall w(x^*) \in \partial f(x^*), \ \forall \overline{x} \in U, \tag{2.40}$$

which reveals the certain connection with the Euler's inequalities (2.13).

2.2.3.2 Constrained optimization. We may now generalize the optimality conditions (2.33)–(2.35) for the case of a convex, nondifferentiable function f. They read:
Find $x \in U$ such that:

$$\partial f(x) \cap \Gamma_U^+(x) \neq \emptyset \Longleftrightarrow 0 \in \partial f(x) + \mathcal{N}_U(x). \tag{2.41}$$

Besides the pioneering works of Moreau, 1963, Rockafellar, 1970, the interested reader may find more material on convex analysis and optimization in the following references: Aubin and Frankowska, 1991, Aubin, 1993, Clarke, 1983, Cottle et al., 1992, Dem'yanov and Vasiliev, 1985, Dem'yanov and Rubinov, 1995, Hiriart-Urruty and Lemaréchal, 1993, Murty, 1988, Shor, 1985, Rockafellar, 1982. Applications in mechanics and engineering can be found, among others, in Ekeland and Temam, 1976, Panagiotopoulos, 1985, Ciarlet, 1989, Friedman, 1982, Rodrigues, 1987, Hlavaček et al., 1988.

2.3 NONCONVEX PROBLEMS

A more challenging class of optimization problems arise if the convexity requirement is abandoned. In this case, even for differentiable functions, multiple minima may exist. The necessary optimality conditions described in Section 2.1 still hold true for differentiable functions; nevertheless they characterize all critical points of the function, i.e., local and global minima and maxima, saddle points etc. Second order conditions for twice differentiable functions must be used in this case to distinguish between the various cases. The theory of nonconvex or global optimization is currently an active field of research (see e.g. Horst and Tuy, 1990, Pardalos and Rosen, 1987, Horst et al., 1995 and many previously published books in the present Series).

Things are even more complicated for nonconvex and nondifferentiable functions. In this case the convex analysis tools outlined in the previous Section are no more applicable. Various extensions to cope with this situation have been developed. In the sequel we will review several notions which are going to be used in the models and the algorithms proposed in this book. They concern certain classes of functions which have sufficient structure to support set–valued extensions of the gradient operators, much like the convex analysis subdifferential. For this reason we call these problems structured optimization ones.

These structured optimization problems, which are considered in more details here, cover the majority of the problems arising in applications and the case of global optimization problems defined by smooth (differentiable) functions (or constraints). Moreover, they allow for the development of effective solution algorithms. On the other hand the classes of smooth, global optimization problems which will be considered, will be discussed in the next Section with the numerical algorithms and later on in this book.

2.3.1 Difference convex optimization

Let us consider a nonconvex, possibly nondifferentiable function $f : \mathbb{R}^n \to \mathbb{R}$ which can be decomposed into a difference of convex functions $f_1, f_2 : \mathbb{R}^n \to \mathbb{R}$, i.e.,

$$f(x) = f_1(x) - f_2(x). \tag{2.42}$$

Components f_1, f_2 may themselves be nondifferentiable functions. Moreover, the difference convex (d.c.) decomposition (2.42) may be local, i.e., valid in a neighbourhood of $x \in \mathbb{R}^n$, or global. In the first case, in an iterative algorithm, the decomposition must be updated during the iterations. For finite dimensional problems, as the ones treated in this book, there are several methods for the construction of the d.c. decomposition (2.42) for an arbitrary function f (see, e.g., Horst et al., 1995, Chapt. 4).

Note that these functions consist a specific case of quasidifferentiable functions, so the quasidifferentiable optimization theory which is outlined in the next Section is also applicable on d.c. optimization problems (see also Dem'yanov and Rubinov, 1985, Dem'yanov and Rubinov, 1995, Dem'yanov et al., 1996).

Note here that a d.c. decomposition is in general nonunique. Moreover, the rules for defining a d.c. decomposition of a composite function are, in general, not very well developed (see Hiriart-Urruty and Lemaréchal, 1993).

As expected, gradient or subgradient algorithms can be considered for the optimization of functions like (2.42). They try to use the special d.c. structure, and the underlying convexity and concavity information, as it will be outlined later in this Chapter. In accordance with the optimality conditions and the Lagrangian approach to duality which has been presented so far, we will give here elements of a d.c. critical point and duality theory, based on Toland, 1979, Stuart and Toland, 1980 and Auchmuty, 1989. Following Auchmuty, 1989, let all possible tangential directions of the function f at point x be defined by the (nonconvex) set:

$$\Theta f(x) = \{w = w_1 - w_2 \mid w_1 \in \partial f_1(x), w_2 \in \partial f_2(x)\}. \tag{2.43}$$

Let us consider the d.c. optimization problem:

$$\text{Find } x^* \in \mathbb{R}^n \ : \ f(x^*) = \min_{x \in \mathbb{R}^n} \{f_1(x) - f_2(x)\}. \tag{2.44}$$

Necessary optimality condition for problem (2.44) is that (Auchmuty, 1989)

$$0 \in \Theta f(x^*) \text{ or } \partial f_1(x^*) \cap \partial f_2(x^*) \neq \emptyset. \tag{2.45}$$

Note that condition (2.45) characterizes all critical points of the d.c. optimization problem (2.44). A necessary and sufficient optimality condition for a d.c. optimization problem reads (Hiriart-Urruty, 1985):

$$\partial_\varepsilon f_2(x^*) \subset \partial_\varepsilon f_1(x^*) \quad \forall \varepsilon \geq 0, \tag{2.46}$$

where $\partial_\varepsilon f_i(x^*)$ denote the ε-subdifferentials of the convex functions f_i, $i = 1, 2$ at the point x^*. Nevertheless, the latter condition cannot be easily used in practical algorithms, since for the most cases of convex functions we are not able to calculate all elements of the ε-subdifferential (see Dem'yanov et al., 1996, Stavroulakis and Polyakova, 1996a).

A duality theory can also be constructed for d.c. problems. Following Auchmuty, 1989 we define the Lagrangian function of type II, $\mathcal{L}_{II}(x,w) : \mathbb{R}^n \times \mathbb{R}^n \to \overline{\mathbb{R}}$ by:

$$\mathcal{L}_{II}(x,w) = f_1(x) + f_2^c(w) - (x,w), \tag{2.47}$$

where $f_2^c(w)$ denotes the convex conjugate function of $f_2(x)$:

$$f_2^c(w) = \max_{x \in \mathbb{R}^n} \left\{ (x,w) - f_2 \right\}. \tag{2.48}$$

The initial optimization problem is now transformed into the Lagrangian, critical point problem:

Find $x^* \in \mathbb{R}^n$ and $w^* \in \mathbb{R}^n$ such that

$$\mathcal{L}_{II}(x^*,w^*) \leq \mathcal{L}_{II}(x,w^*), \ \forall x \in \mathbb{R}^n \tag{2.49}$$

and

$$\mathcal{L}_{II}(x^*,w^*) \leq \mathcal{L}_{II}(x^*,w), \ \forall w \in \mathbb{R}^n. \tag{2.50}$$

Indeed from relations (2.47)–(2.50) we have

$$\begin{aligned} &\min_{x \in \mathbb{R}^n} \{f_1(x) - f_2(x)\} = \\ &= \min_{x \in \mathbb{R}^n} \left\{ f_1(x) - \max_{w \in \mathbb{R}^n} \{(w,x) - f_2(x)\} \right\} = \\ &= \min_{x \in \mathbb{R}^n} \left\{ f_1(x) + \min_{w \in \mathbb{R}^n} \{-(w,x) + f_2(x)\} \right\} = \\ &= \min_{x \in \mathbb{R}^n,\ w \in \mathbb{R}^n} \{f_1(x) + f_2^c(x) - (w,x)\} = \\ &= \min_{x \in \mathbb{R}^n,\ w \in \mathbb{R}^n} \mathcal{L}_{II}(x,w). \end{aligned}$$

Note here that in the paper of Auchmuty, 1989 and in the concrete applications given in the sequel, the more general case of a d.c. function with $f_2(\Lambda x)$ is considered, where Λ is a given transformation matrix. For simplicity of the presentation we do not use this form in this introductory Section.

In contrast to the classical Lagrangian function (2.24), defined earlier for convex optimization problems, the Lagrangian of type II is in general a nonconvex function, but it is convex in each one of its arguments when the other is kept constant (i.e. the functions $\mathcal{L}_{II}(\cdot, w)$ and $\mathcal{L}_{II}(x, \cdot)$ are convex). Moreover, the Lagrangian formulation for d.c. problems does not lead to saddle point problems, as it is obvious from the previous discussion.

Note that optimization problems with inequality constraints defined by d.c. functions can be formulated as well. In general, it is possible to transform a given problem into several typical forms: either the d.c. structure appears into the function to be minimized, or, into the function(s) which define the subsidiary constraints. In every case the difficulty introduced by the nonconvex structure of a d.c. function (2.42) remains. More details can be found in Hiriart-Urruty, 1985, Polyakova, 1986, Thach, 1987, Horst and Tuy, 1990, Tuy et al., 1994, Horst et al., 1995, Thach and Konno, 1997.

2.3.2 Quasidifferentiable optimization

A more general class of nonconvex, nondifferentiable functions which nevertheless permit a unified treatment of their differentiability properties and, thus, they can be used in optimization problems, constitutes the class of quasidifferentiable functions, in the sense of V.F. Dem'yanov (see, among others, Dem'yanov and Rubinov, 1980, Dem'yanov and Vasiliev, 1985, Dem'yanov and Rubinov, 1985, Jahn, 1994, Dem'yanov and Rubinov, 1995, Dem'yanov et al., 1996).

A function f defined on an open set $X \subset \mathbb{R}^n$ and directionally differentiable at a point $x \in X$ is called quasidifferentiable if there exists an ordered pair of convex compact sets $[U, V]$ in $\mathbb{R}^n \times \mathbb{R}^n$ such as they produce the directional derivative of the function by means of the following formula

$$f'(x, y) = \max_{w \in U}(w, y) + \min_{w \in V}(w, y), \ \forall y \in \mathbb{R}^n . \tag{2.51}$$

The class of equivalent pairs of convex compact sets $[U, V]$ of (2.51) is called the quasidifferential of f at x and it is denoted by $\mathcal{D}f(x)$. The set U is called the subdifferential of f and is denoted by $\underline{\partial} f(x)$ and the set V is called the *superdifferential* of f and is denoted by $\overline{\partial} f(x)$, i.e. $\mathcal{D}f(x) = [\underline{\partial} f(x), \overline{\partial} f(x)]$.

Observe here that quasidifferentiable functions, although nonconvex in general, permit us to express their differentiability properties by using convex analysis techniques (cf. (2.38)). Observe also, as in the case of d.c. functions, that the quasidifferentials are not uniquely defined. Here and in the rest of this book, we are not going to deal with this point. We assume that the decompositions used are defined at the level of the formulation of the engineering problem and they do not change thereafter. Sometimes the decomposition is imposed by the nature of the engineering problem.

Let us also mention that a d.c. function (2.42) is quasidifferentiable with e.g., $\mathcal{D}f(x) = [\partial f_1(x), -\partial f_2(x)]$, where the convex subdifferentials of the constituent functions $f_1(x)$ and $f_2(x)$ are used.

In contrast to what happens with general nonconvex functions, there exist calculus rules for the construction of the quasidifferentials (thus, also, of a d.c. decomposition) of composite functions, sums, differences, products etc. of a finite number of functions (see Dem'yanov and Rubinov, 1985).

Let us give here the optimality conditions for an unconstrained optimization problem of a quasidifferentiable function:

Find x^* such that

$$-\overline{\partial} f(x^*) \subset \underline{\partial} f(x^*). \tag{2.52}$$

Analogous rules for constrained optimization problems can be found in Dem'yanov and Rubinov, 1985, Dem'yanov and Rubinov, 1995, Dem'yanov et al., 1996.

2.4 A REVIEW OF OPTIMIZATION ALGORITHMS AND HEURISTICS

A short review of optimization algorithms and heuristics is given in the rest of this Chapter in order to facilitate the discussion on the specific engineering algorithms given later in this book. The various algorithms are schematically described, starting from the relaxation schemes of convex, smooth, unconstrained optimization and lead, through Lagrangian multiplier methods and nonsmooth problems, to stochastic and neural network based optimization algorithms for the general nonconvex optimization problem.

2.4.1 Convex optimization algorithms

2.4.1.1 Smooth unconstrained problems. First and second order methods for smooth unconstrained problems are reviewed. The notion of strict convexity together with the one of ellipticity assure the existence of one unique minimum of the following optimization problem defined, in general, over a nonempty, convex closed subset U_{ad} of a Hilbert space U:

$$\text{Find } x \in U \text{ s.t.} f(x) = \inf_{x^* \in U_{ad} \subset U} f(x^*). \tag{2.53}$$

Recall that a real valued functional defined on a Hilbert space U is elliptic if it is continuously differentiable in U and if there exists a real constant $c > 0$ such that

$$(\nabla f(x^*) - \nabla f(x), x^* - x) \geq c\|x^* - x\|^2, \ \forall x^*, x \in U. \tag{2.54}$$

Moreover, an elliptic functional $f : U \rightarrow \mathbf{R}$ is strictly convex and coercive if it satisfies the inequality:

$$f(x^*) - f(x) \geq (\nabla f(x), x^* - x) + \frac{c}{2}\|x^* - x\|^2, \ \forall x^*, x \in U. \tag{2.55}$$

Recall that the notion of elliptic functionals on a Hilbert space U is an extension of the notion of quadratic functionals with positive definite matrix over $\mathbf{R}^n$. In this case U is replaced by $\mathbf{R}^n$ (cf. (2.9) with K positive definite; then c in (2.54) is the smallest eigenvalue of K, see e.g. Ciarlet, 1989, p. 291). In engineering applications the appropriate Hilbert space depends on the problem and the discretization applied.

A generic iterative solution algorithm which converges for a coercive, convex, differentiable functional $f(u)$, reads:

Algorithm 2.1 *Iterative First Order Minimization.*

1. For iteration $k = 0, 1, \ldots$, for starting vector $x^{(0)}$.

2. Define a descent direction vector $d^{(k)}(x^{(k)})$.

3. Solve the one–dimensional minimization problem: Find $\rho(x^{(k)}, d^{(k)}) \in \mathbf{R}$ such that

$$f(x^{(k)} + \rho(x^{(k)}, d^{(k)})d^{(k)}) = \inf_{\rho \in \mathbf{R}} f(x^{(k)} + \rho d^{(k)}) \tag{2.56}$$

   in order to find the step size $\rho(x^{(k)}, d^{(k)})$.

4. Update current point

$$x^{(k+1)} = x^{(k)} + \rho(x^{(k)}, d^{(k)})d^{(k)}, \tag{2.57}$$

   and, if convergence is not obtained, continue with step 2.

A relaxation–type technique for $U = \mathbf{R}^n$ simply uses all coordinate axis directions $e^{(1)}, \ldots, e^{(n)}$ as "descent" directions in step 2 of Algorithm (2.1). For the quadratic functional (2.9) it simply reduces to the Gauss–Seidel iterative solution technique for the solution of the linear system of equations $Kx = p$ which corresponds to the optimality condition of the problem (2.10).

A gradient solution technique uses the direction $d^{(k)}(x^{(k)}) = -\nabla f(x^{(k)})$ instead.

If one uses higher order gradient information it is possible to enhance the effectiveness of the previously outlined first order schemes. For instance, for a twice continuously differentiable function f, one may replace steps 2–4 of the previous algorithm by defining the update formula:

$$x^{(k+1)} = x^{(k)} - \left\{H(x^{(k)})\right\}^{-1} \nabla f(u^{(k)}), \tag{2.58}$$

where $H(x^{(k)})$ is the Hessian $\nabla^2 f(x^{(k)})$ of the function f, for the Newton iterative algorithm, or a positive definite approximation of it, for the various quasi–Newton schemes.

If second order gradient information is not available, or if their calculation is expensive, or even if vector–valued arithmetic operations are performed effectively in a given computing environment, then a number of directions $\ldots, d^{(k-3)}, d^{(k-2)}, d^{(k-1)}$, calculated in the previous cycles of the algorithm can be used for the estimation of H^{-1} in (2.58) and, thus, for the determination of $d^{(k)}$, $\rho(x^{(k)}, d^{(k)})$ in (2.56), (2.57); this is the general class of conjugate gradient algorithms (see, e.g., Gill et al., 1981, Fletcher, 1990).

2.4.1.2 Constrained problems. Unfortunately, the simple scheme of Algorithm (2.1) does not converge for general constrained optimization problems, with the exception of box–type inequality constraints of the form:

$$U_{ad} = \{x \in \mathbf{R}^n : \alpha_i \leq x_i \leq \beta_i,\ i = 1, \ldots, n\}, \tag{2.59}$$

where $\alpha_i, \beta_i \in \mathbf{R}^1$ and one may have $\alpha_i = -\infty$ and/or $\beta_i = +\infty$. In this case the relaxation method of (2.56), (2.57) is accompanied by a projection over the set (2.59), i.e., for instance (2.57) reads:

$$x^{(k+1)} = P_{U_{ad}} \left(x^{(k)} + \rho(x^{(k)}, d^{(k)}) d^{(k)}\right). \tag{2.60}$$

For instance, the quadratic minimization problem with $f(x)$ defined in (2.9) and $X_{ad} = \{x \in \mathbf{R}^n : x \geq 0\} = \mathbf{R}^n_+$, may be solved by the simple projection gradient scheme

$$x_i^{(k+1)} = \max\left\{x_i^{(k)} - \rho_k (K x^{(k)} - p)_i, 0\right\},\ i = 1, \ldots, n. \tag{2.61}$$

In mechanics, early attempts to solve unilateral contact problems have been based on (2.61) (see, e.g., Panagiotopoulos, 1985, Glowinski and LeTallec, 1989).

The importance of the simple projection schemes like (2.60), (2.61), lies in the fact that more complicated constrained optimization problems may be transformed, through duality and the introduction of Lagrange multipliers, so that they have inequality constraints of the simple form (2.59).

One should observe that in the dual problem (2.26) the involved constraints are simple, box-type ones; thus simple projected iterative solution techniques can be used for the solution of this problem. Thus, a primal-dual solution strategy where the primal problem is transformed, for given values of the Lagrangian multipliers, into an unconstrained optimization problem, whereas the Lagrangian multipliers are estimated by the dual optimization problem (2.26), with simple inequality constraints, seems to be meaningful. In fact, Uzawa's method is based on this idea:

Algorithm 2.2 *Uzawa's Algorithm.*

1. For initial values $x^{(0)}, \lambda^{(0)}$ and
   for iterations $k = 1, 2, \ldots$

2. Keep $\lambda^{(k-1)}$ constant and calculate $x^{(k)}$ as solution of the unconstrained minimization problem:

$$f(x^{(k)}) + \sum_{i=1}^{m} \lambda_i^{(k-1)} \phi_i(x^{(k)}) = \inf_{x \in \mathbf{R}^n} \left\{ f(x) + \sum_{i=1}^{m} \lambda_i^{(k-1)} \phi_i(x) \right\}. \quad (2.62)$$

3. Keep $x^{(k)}$ constant and calculate $\lambda^{(k)}$ from the optimization problem (cf. (2.26)):

$$\lambda_i^{(k)} = \max\left\{ \lambda_i^{(k)} + \rho \phi_i(x^{(k)}), 0 \right\}, \; i = 1, \ldots, m. \quad (2.63)$$

4. If convergence is not achieved, continue iterations from step 2.

Here, for the writing of the simple update rule (2.63), we have used the fact that from the minimality conditions of (2.27) (cf. (2.19), (2.24)), $G(\mu)$ in (2.26) is a linear function of μ with gradient composed of the components $(\nabla G(\mu))_i = \phi_i(x^{(k)})$, $i = 1, \ldots, m$. Thus (2.63) is the simple projection gradient scheme (2.61) for the solution of the minimization problem for the function $-G(\mu)$ (cf. (2.26)).

For the quadratic optimization problem (Q.P.P.) (2.9), (2.21), Uzawa's algorithm reads:

Algorithm 2.3 *Uzawa's Algorithm for Q.P.P.*

1. For initial values $x^{(0)}, \lambda^{(0)}$ and
   for iterations $k = 1, 2 \ldots$

2. For $\lambda^{(k-1)}$ given, solve:

$$Kx^{(k)} + A^T \lambda^{(k-1)} \phi_i = p. \tag{2.64}$$

3. For $x^{(k)}$ given, update λ by means of the scheme:

$$\lambda_i^{(k)} = \max \left\{ \lambda_i^{(k)} + \rho(Ax^{(k)} - b)_i, 0 \right\}. \tag{2.65}$$

4. If convergence is not achieved, continue iterations from step 2.

One may observe that Uzawa's scheme is actually a gradient algorithm for the dual optimization problem.

All the previous results can easily be modified to take into account the presence of equality constraints. For more details, the reader is referred, among others, to Bertsekas, 1982, Ciarlet, 1989. For details on more complicated relaxation schemes the reader may consult, e.g., Glowinski et al., 1981, Hlavaček et al., 1988.

Linear and nonlinear complementarity problems have been extensively studied and applied for the solution of several engineering and economic applications, see among others Harker and Pang, 1990, Cottle et al., 1992, Kojima et al., 1992, Nesterov and Nemirovskii, 1993, Ferris and Pang, 1995, Todd, 1996.

Finally, recent developments on the application of adaptive multigrid methods for the selective, controlled refinement of the discretization and the numerical solution of large scale problems can be found in Kornhuber, 1997 and the references given therein.

2.4.1.3 Nonsmooth problems. Nonsmooth convex problems may arise either by a nondifferentiable, convex function $f(\cdot)$ or, if one introduces equality and inequality constraints defined by the set U_{ad}, by means of exact penalty techniques. The latter case corresponds to the writing of the problem (2.53) in the following form:

$$\text{Find } x \in U \text{ s.t.} f(x) = \inf_{x^* \in U} f(x^*) + \mathbf{I}_{U_{ad}}(x^*), \tag{2.66}$$

where the indicator function of the convex set U_{ad} has been used, which is defined as:

$$\mathbf{I}_{U_{ad}}(\mathbf{x}) = \begin{cases} 0, & \text{if } \mathbf{x} \in U_{ad} \\ +\infty, & \text{otherwise.} \end{cases} \tag{2.67}$$

Recall that optimality conditions for (2.66) are written in turn as:

Find vector $x \in \mathbb{R}^n$ such that:

$$0 \in \partial(f(x) + \mathbf{I}_{U_{ad}}(x)) = \partial f(x) + \mathcal{N}_{U_{ad}}(x). \tag{2.68}$$

A gradient is not uniquely defined for a nondifferentiable function at the points of nondifferentiability (kinks). Thus, gradient–type optimization schemes, as the ones outlined previously, are not directly applicable here. Moreover, the optimality condition is written in the form of an inclusion and not of an equation and this should be taken into account in the algorithm for both the evaluation of a descent direction and in the stopping criterion. Accordingly, second order information (Hessian) cannot be defined in the classical way.

On the other hand, it should be admitted that the sets where the potential function is nondifferentiable consist of isolated points in $\mathbb{R}$, or of segments in $\mathbb{R}^2$, etc. Therefore it was reasonable that the first attempts to solve nondifferentiable problems were based on smooth optimization techniques.

The use of a single, arbitrary element $g^{(k)}$ of the subdifferential in step 2 of Algorithm (2.1), i.e., of $d^{(k)} = g^{(k)} \in \partial f(X^{(k)})$, leads to subgradient optimization methods (Shor, 1985). A more refined strategy would be based on the use of all the first order information which is included in the subgradient. A search direction which is a steepest descent direction for function $f(x)$, is provided by the solution of the convex minimization (sub)–problem:

$$-d^{(k)} = s^{(k)} = \arg \min_{g^{(l)} \in \partial f(x^{(k)})} \|g^{(k)}\|. \tag{2.69}$$

Here $\|.\|$ denotes an appropriate norm in $\mathbb{R}^n$, for instance the Euclidean norm. A complete characterization of the set $\partial f(x^{(k)})$ is needed for the solution of (2.69). This characterization exists indeed for known, structured convex and nonsmooth functions of mechanics (i.e. for composite functions, sums of given max–type functions etc., cf. e.g. Fletcher, 1990, Womersley and Fletcher, 1986). In turn, nondifferentiability must be considered for the solution of the one–dimensional line search subproblem in the step 3 of the Algorithm (2.1) as well.

Nevertheless, the lack of continuity of the subdifferential operator causes convergence problems in the above outlined scheme (see, e.g., Hiriart-Urruty and Lemaréchal, 1993, vol. I, p.363, Dem'yanov and Vasiliev, 1985). Roughly speaking the subdifferential operator is unable to describe what is happening in the vicinity of a given point. Moreover, it changes discontinuously, it is set–valued for certain values of its domain and, numerically, one has practically no chance to detect these points exactly due to numerical inaccuracies in a computer implementation. Theoretically, nonlocal extensions of the subdifferential operator, like the ∂_ϵ (ϵ–subdifferential), may be used in connection with (2.69).

Nevertheless, in general it is difficult to calculate this set–valued operator explicitly even for simple problems (cf. Dem'yanov and Vasiliev, 1985). Thus, practically, two methods remain for the treatment of the problem: the bundle optimization concept, where an approximation of ∂_ϵ is iteratively constructed along the steps of an iterative scheme and the hypodifferentiable optimization algorithm, where, for a given class of functions, a Hausdorff continuous operator is constructed, which can be used for numerical purposes.

The bundle optimization algorithm introduces a polyhedral approximation of $\partial f(x^{(k)})$ in (2.69), which, in turn, is defined by information accumulated along the iteration steps $i = 0, \ldots, k$ of the algorithm. In this case the minimal assumption is done that for each point $x^{(k)}$ the value of $f(x^{(k)})$ and of one subgradient $g^{(k)} \in \partial f(x^{(k)})$ is available. From the "bundle" of information:

$$x^{(i)},\ f(x^{(i)}),\ g^{(i)} \in \partial f(x^{(i)}),\ i = 0, 1, \ldots, k,$$

an appropriate local approximation of the function is constructed, which in various implementations, takes into account both informations from the vicinity of the current point $x^{(i)}$, as well as second order informations of the function. An exposition of this idea is given, e.g., in Hiriart-Urruty and Lemaréchal, 1993, Schramm and Zowe, 1992, Mäkelä and Neittaanmäki, 1992.

2.4.2 Nonconvex optimization algorithms

Nonconvexity causes problems to the classical methods outlined in the previous Section, even for smooth (differentiable) problems. Quasi-Newton methods which only consider descent directions of the function may have problems to find minimizers. The consideration of both descent directions and directions of negative curvature has been proposed in Forsgren and Ringertz, 1993 for the numerical treatment of this problem. Moreover, the multiplicity of the solution plays in this case a more important role. In fact, multiplicity is expected to occur and the question of which minimum should be considered arises (some local ones or the global minimum). Therefore, the importance of optimization methods based on local information about the function and/or the constraints is fading.

One should note here that, although the subject of global optimization, i.e., the search for the global minimum or of some local minima of a nonconvex problem has gained importance in the recent years, the complexity it implies is not always necessary for the engineering applications considered here. For the most of the applications in structural analysis, global minima of some potential energy may of course be of importance for the ultimate strength capacity (or behaviour) of a structure, but, usually, loading is imposed along a predetermined loading path. Thus, solutions (i.e. minima) caused by external actions belonging to this path, or points in their neighbourhood, if one considers uncertainties and imperfections, are most probably of importance for a structural analysis application. The problem cannot be simplified in other kind of engineering problems, for instance in problems of structural optimization.

Having the previous observation in mind we will restrict our attention here in relatively easily implementable methods of nonconvex optimization, which does not promise to solve all global optimization problems but, nevertheless, have a potential to tackle large–scale nonconvex problems arising in practical engineering applications. The most general of them are based on difference convex approximation and optimization techniques, which are outlined in the sequel. They cover both smooth and nonsmooth problems. The more general case of quasidifferentiable optimization has been treated in a recent monograph in the same Series (Dem'yanov et al., 1996) and therefore will not be discussed in more details here. Algorithms which are not directly based on classical optimization concepts and heuristics will be treated in the next Section.

2.4.2.1 Difference convex (d.c.) problems. Several methods have been proposed for the treatment of unconstrained optimization problems with d.c. functions, which try to use the structure of (2.42). For recent developments in this area, which are mainly focussed on solving the arising global optimization problem, i.e., in the determination of the global optimum of a d.c. optimization problem (2.44), the reader is referred to Horst and Tuy, 1990, Horst et al., 1995 among others.

Here we will present two methods which try to solve the corresponding critical point problems (cf. (2.45)). In most cases, this goal is sufficient for the engineering applications treated in this book. These two methods are based on the decomposition of the problem into a series of convex subproblems, by using the structure of the d.c. function (2.42).

The first method has been proposed in Auchmuty, 1989 and is based on the Lagrangian critical point problem (2.49)–(2.50).

Algorithm 2.4 *D.C. Critical Point Algorithm 1 (due to Auchmuty)*

1. For initial values $x^{(0)}, w^{(0)}$ and
   for iterations $k = 1, 2, \ldots$

2. Keep $w^{(k-1)}$ constant and calculate $x^{(k)}$ as solution of the minimization problem:

$$\mathcal{L}_{II}(x^{(k)}, w^{(k-1)}) = \min_{x \in \mathbf{R}^n} \left\{ \mathcal{L}_{II}(x, w^{(k-1)}) \right\}. \tag{2.70}$$

3. Keep $x^{(k)}$ constant and calculate $w^{(k)}$ from the optimization problem:

$$\mathcal{L}_{II}(x^{(k)}, w^{(k)}) = \min_{w \in \mathbf{R}^n} \left\{ \mathcal{L}_{II}(x^{(k)}, w) \right\}. \tag{2.71}$$

4. If convergence is not achieved, continue iterations from step 2.

Recall here that in the definition of the Lagrangian function of type II, $\mathcal{L}_{II}(x, w)$ (see (2.47)) the convex conjugate function of f_2 was involved. This fact may induce additional constraints in the space of the w's, which should be taken into account in step 3 of Algorithm (2.4). For the sake of simplicity of the schematic presentation of the algorithm this point has not been included here. On the other hand, in general, the calculation of the convex conjugate functions f_2^c (cf. (2.48)), which are needed in the previous algorithm, is not always an easy task. Within a finite element environment, if one uses collocation methods (or lower degree elements) for the discretization of variables w, the methods discussed in Glowinski and LeTallec, 1989, LeTallec, 1990 for mixed finite element formulations in elastoplasticity and viscoelasticity can be applied. One should mention here the interesting discussion of the degree of nonconvexity and of sparsity in difference convex optimization problems and in the d.c. duality theory, in the sense of Toland, given in Thach and Konno, 1997.

A second critical point algorithm, proposed by Polyakova (see Dem'yanov et al., 1996, Stavroulakis and Polyakova, 1996a, Stavroulakis and Polyakova, 1996b), does not involve conjugate functions and in this respect is more advantageous. For the description of this algorithm let us first define the function

$$\varphi(x, z) = f_1(x) - f_2(z) - < w(z), x - z > .$$

If the constituent function $f_1(\cdot)$ is strongly convex, then function $\varphi(\cdot, z)$ is strongly convex at x for every fixed $z \in \mathbf{R}^n$ and $w(z) \in \partial f_2(z)$ and thus we can write

$$\varphi(x) \leq \varphi(x, z) = f_1(x) - f_2(z) - < w(z), x - z > . \tag{2.72}$$

Recall that, due to the arbitrariness of the d.c. decomposition (2.42), component functions $f_1(\cdot)$, $f_2(\cdot)$ can be considered to be strongly convex (if not, just add in each one of them an appropriate strongly convex function).

The idea of the relaxation algorithm is that, with respect to an arbitrary point x_0 and to every element $w(x_0) \in \partial f_2(x_0)$, one considers the following convex optimization problem:

$$\inf_{x \in \mathbf{R}^n} \varphi(x, x_0) = \varphi(x_1, x_0).$$

Put

$$\varphi_0(x) = f_1(x) - f_2(x_0) - < w(x_0), x - x_0 >$$

and find

$$\min_{x \in \mathbf{R}^n} \varphi_0(x) = \varphi_0(x_1). \tag{2.73}$$

The minimum in (2.73) is attained at a point x_1, and this point is unique. In this case

$$0 \in \partial\varphi_0(x_1) = \partial f_1(x_1) - w(x_0). \tag{2.74}$$

Thus

$$w(x_0) \in \partial f_1(x_1) \cap \partial f_2(x_0).$$

This process is a relaxation–type one and for every step we get either the point x_k for which the condition

$$\partial f_1(x^*) \cap \partial f_2(x^*) \neq \emptyset \tag{2.75}$$

is satisfied, or

$$\varphi(x_k) < \varphi(x_{k+1}).$$

Algorithm 2.5 *D.C. Critical Point Algorithm 2 (due to Polyakova)*

1. For initial values $x^{(0)}, w^{(0)} \in \partial f_2(x^{(0)})$ and for iterations $k = 1, 2, \ldots$, if condition (2.75) holds at point x_0 then stop.

2. Choose an element $w_k \in \partial f_2(x_k)$ and calculate $x^{(k)}$ as solution of the minimization problem:

$$\min_{x \in \mathbf{R}^n} \{\varphi_k(x)\} = \varphi_k(x_{k+1}), \tag{2.76}$$

   where

$$\varphi_k(x) = f_1(x) - f_2(x_k) - < w_k, x - x_k > .$$

3. If convergence is not achieved, continue iterations from step 2.

For more details on the last algorithm the reader is referred to (Dem'yanov et al., 1996, Stavroulakis and Polyakova, 1996a, Stavroulakis and Polyakova, 1996b).

For the more general case of quasidifferentiable functions and the corresponding quasidifferentiable optimization problems the reader is referred to Dem'yanov and Vasiliev, 1985, Dem'yanov and Rubinov, 1985, Dem'yanov and Rubinov, 1995, Dem'yanov et al., 1996 and the references given therein.

2.4.3 Engineering motivated heuristics for nonconvex optimization problems

In this Section a heuristic algorithm is presented for the solution of the substationarity problem of a nonconvex function $\Pi(x)$ which is composed of a convex part $f_c(x)$ and a nonconvex part $w(x)$, i.e.

$$\Pi(x) = f_c(x) + w(x). \tag{2.77}$$

This case is very common in mechanics where we deal with problems involving a number of elements with convex energy density and a number of elements with nonconvex energy density.

The substationarity problem of $\Pi(x)$ can be written equivalently in the form of the differential inclusion

$$0 \in \bar{\partial}\Pi(x) \tag{2.78}$$

where $\bar{\partial}$ denotes the generalized gradient of Clarke-Rockafellar (Rockafellar, 1979, Clarke, 1983), which is an extension of the subdifferential of convex analysis for nonconvex problems.

For convenience of the reader we recall here a definition of the Clarke's generalized gradient for a function $f(\mathbf{x})$ of the $\mathcal{PC}^r$ class, with $r > 1$. Moreover, we assume that function $f(\mathbf{x})$ is a continuous selection (CS) of k C^r functions, i.e., that f is continuous and $\forall \mathbf{x} \in \mathbf{R}^n$ there exists $i \in \{1, \ldots, k\}$ such that $f(\mathbf{x}) = f_i(\mathbf{x})$. In this case we denote $f \in CS\ \{f_1, \ldots, f_k\}$.

One may then show (cf. Bartels et al., 1995, Kuntz and Scholtes, 1995) that f is locally Lipschitz continuous and Bouligand differentiable, with the $B-$ derivative at a point $\mathbf{x}_0 \in \mathbf{R}^n$ in the direction $\mathbf{d} \in \mathbf{R}^n$ being a continuous selection of the functions $\nabla f_i(\mathbf{x}_0)^T\mathbf{d}$, $i \in \widehat{\mathcal{I}}(\mathbf{u}_0)$. Here $\widehat{\mathcal{I}}(\mathbf{x}_0)$ denotes the essentially active index set $\widehat{\mathcal{I}}(\mathbf{x}_0) = \{i \in \mathcal{I}(\mathbf{x}_0) \mid \mathbf{x}_0 \in \text{cl}(\text{int}(\{\mathbf{x} \in U \mid f(\mathbf{x}) = f_i(\mathbf{x})\}))\}$. Moreover, "cl" (resp. "int") abbreviates the closure (resp. the interior) of a set. Clarke's generalized subdifferential is now given by

$$\partial_{Cl} f(\mathbf{x}_0) = \text{conv}\{\nabla f_i(\mathbf{x}_0) \mid i \in \widehat{\mathcal{I}}(\mathbf{x}_0)\} \tag{2.79}$$

where "conv" stands for the convex hull.

Note, furthermore, that here and in Panagiotopoulos, 1985, Panagiotopoulos, 1993 and in all previous publications of the authors, the Clarke's generalized differential is denoted by $\overline{\partial}$, i.e. $\overline{\partial} = \partial_{Cl}$.

For further definitions of the Clarke's notion the reader is referred, among others, to Clarke, 1983, Panagiotopoulos, 1985, p. 143, Dem'yanov et al., 1996, pp. 45, 75. A first attempt to use continuous selection potential functions in mechanics can be found in Stavroulakis and Rohde, 1996.

The aim of the proposed algorithm is to avoid the nonconvex minimization problem by minimizing a sequence of appropriately defined convex functions in which the nonconvex part $w(x)$ has been replaced by the convex function $p^{(i)}(x)$. Consider first the following minimization problems

$$\min\left\{\Pi_c^{(i)} = f_c(x) + p^{(i)}(x)\right\} \tag{2.80}$$

where in each step the convex function $p^{(i)}(x)$ is selected such that the following relation is fulfilled

$$\partial p^{(i)}(x) = \bar{\partial} w(x) \tag{2.81}$$

at the point $x^{(i-1)}$. In this case the function $\Pi_c^{(i)}$ is a sum of convex functions and thus a convex function itself.

Then, the nonconvex minimization problem is written in the form

$$\begin{aligned}\min\{\Pi(x)\} &= \min\left\{f_c(x) + p^{(i)}(x) + [w(x) - p^{(i)}(x)]\right\} = \\ &= \min\left\{\Pi_c^{(i)} + [w(x) - p^{(i)}(x)]\right\}\end{aligned} \tag{2.82}$$

or equivalently, the differential inclusion (2.78) in the form

$$\begin{aligned}0 \in \bar{\partial}\Pi(x) = \bar{\partial}(f_c(x) + w(x)) = \bar{\partial}(f_c(x) + p^{(i)}(x) + w(x) - p^{(i)}(x)) \subset \\ \subset \bar{\partial}(f_c(x) + p^{(i)}(x)) + \bar{\partial}(w(x) - p^{(i)}(x)) = \\ = \partial(f_c(x) + p^{(i)}(x)) + \bar{\partial}(w(x) - p^{(i)}(x)) = \\ = \partial\Pi_c^{(i)} + \bar{\partial}(w(x) - p^{(i)}(x)).\end{aligned} \tag{2.83}$$

Based on the previous decomposition of the convex and nonconvex parts, the following approximation scheme can be formulated:

$$\begin{aligned}\min\left\{\Pi^{(1)}(x)\right\} &= \min\left\{\overbrace{\Pi_c^{(1)}(x)}^{1-step} + \overbrace{[w(x^{(0)}) - p^{(0)}(x^{(0)})]}^{0-step}\right\} \\ &= \min\left\{\Pi_c^{(1)}(x)\right\} + C^{(1)} \\ &\ \vdots \\ \min\left\{\Pi^{(i)}(x)\right\} &= \min\left\{\overbrace{\Pi_c^{(i)}(x)}^{i-step} + \overbrace{[w(x^{(i-1)}) - p^{(i-1)}(x^{(i-1)})]}^{(i-1)-step}\right\} \\ &= \min\left\{\Pi_c^{(i)}(x)\right\} + C^{(i)}\end{aligned}$$

$$\vdots$$

$$\begin{aligned}\min\left\{\Pi^{(n)}(x)\right\} &= \min\left\{\overbrace{\Pi_c^{(n)}(x)}^{n-step}+\overbrace{[w(x^{(n-1)})-p^{(n-1)}(x^{(n-1)})]}^{(n-1)-step}\right\}\\ &= \min\left\{\Pi_c^{(n)}(x)\right\}+C^{(n)}. \end{aligned} \tag{2.84}$$

In the above iterative procedure, the first part is referred to the current step, whereas the second part is a scalar number obtained from the solution of the previous step. Using the above decomposition, in each step i, only the convex part $\Pi_c^{(i)}(x) = f_c(x) + p^{(i)}(x)$ is minimized and the obtained solution $x^{(i)}$ is used for the selection of a new convex approximation $p^{(i+1)}(x)$ of the nonconvex function $w(x)$ through (2.81). This convex function will be used in the next step of the iterative procedure. Also, the constant term $C^{(i+1)}$ is calculated, at the equilibrium point $x^{(i)}$.

If we suppose that in the last case we have the convergence of the iterative scheme, then we have that $|x^{(n)} - x^{(n-1)}| \leq \varepsilon_1$, where ε_1 is an appropriately small number. On the assumption that $x^{(n)}$ is not a nondifferentiability point of $w(x)$, then $|C^{(n)} - C^{(n-1)}| \leq \varepsilon_2$ where ε_2 is also an appropriately small number. In the case that $x^{(n)}$ is a nondifferentiability point of $w(x)$ then the condition $|x^{(n)} - x^{(n-1)}| \leq \varepsilon_1$ is not sufficient to have convergence but the condition $|C^{(n)} - C^{(n-1)}| \leq \varepsilon_2$ should also be satisfied.

Finally, we can write for n large enough that

$$\text{argmin}\,\{\Pi(x)\} = \text{argmin}\left\{\Pi_c^{(n)}(x)\right\} \tag{2.85}$$

where in the left hand side we mean the local minimum sought.

By means of the previous relations it is easily verified that a solution of the initial minimization problem of $\Pi(x)$ can be obtained using the proposed iterative scheme but the full proof of convergence remains still an open problem. However, in the various numerical experiments we have performed for special forms of nonconvex superpotentials and always for scalar nonmonotone multivalued laws, convergence was always achieved in a small number of steps.

Here we shall show the graphical meaning of the iterative scheme defined by (2.80), (2.81). Let us assume the one-dimensional nonconvex function $\Pi(x) = f_c(x) + w(x)$ which results as a sum of the convex function $f(x)$ and of the nonconvex function $w(x)$ (see Fig. 2.2a). The nonconvex function $w(x)$ results from the integration of the nonmonotone law $g(x)$ (see Fig. 2.2b). In an engineering problem this law can be understood, for example, as a nonlinear stress-strain relation or as a nonlinear boundary condition. In this case, the nonconvex function $w(x)$ is the superpotential of the law $g(x)$.

In the first step we approximate the nonmonotone law g with the fictitious monotone law $h^{(1)}$ of Fig. 2.3. This monotone law gives rise to the convex function $p^{(1)}$. The minimization of $\Pi_c^{(1)}(x) = f_c(x) + p^{(1)}(x)$ (which is a sum of convex functions) gives as a unique solution the value $x^{(1)}$. This is not a solution of the minimization of $\Pi(x)$ because the solution point does not lie on the nonmonotone law $g(x)$. We select now a new convex function $p^{(2)}$ such that relation (2.81) is fulfilled. For the one-dimensional case treated here this relation is equivalent to $h^{(2)}(x^{(1)}) = g(x^{(1)})$. One possible such monotone law is the one depicted in Fig. 2.3. The minimization of $\Pi_c^{(2)}(x) = f_c(x) + p^{(2)}(x)$ yields the solution $x^{(2)}$ and this procedure is continued until the difference between the solutions of two consecutive iterations is small enough. In this case, as it is obvious from (2.84), the minimizer of $\Pi_c^{(n)}(x)$ will also minimize $\Pi(x)$.

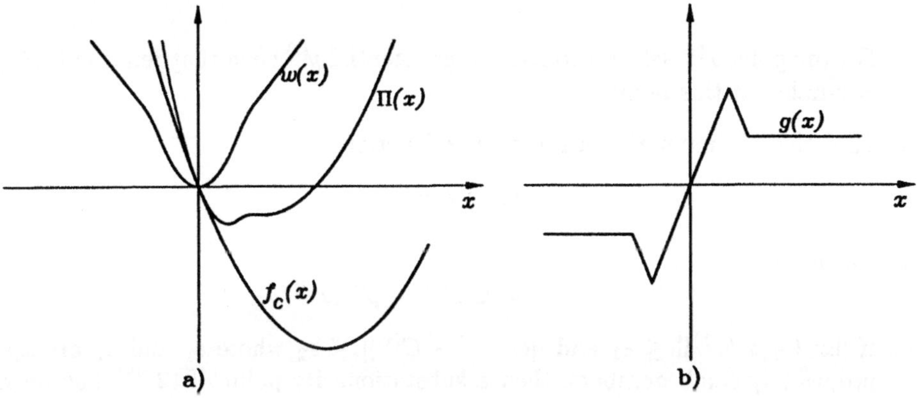

Figure 2.2. The nonconvex function $\Pi(x) = f_c(x) + w(x)$

We have to notice here that the possibility presented in Fig. 2.3 for the approximation of the nonmonotone law is only one of the different monotone laws that would approximate the nonmonotone one. In general, the convex superpotentials that approximate the nonconvex superpotential are selected in such a way that the computational effort for the solution of the arising convex problem is minimized. This task depends on the particular nonconvex functions to be approximated.

Thus, the following heuristic algorithm can be formulated:

Algorithm 2.6 *Heuristic nonconvex optimization algorithm (due to Mistakidis)*

1. Select a starting point $x^{(0)}$ and initialize i to 1.

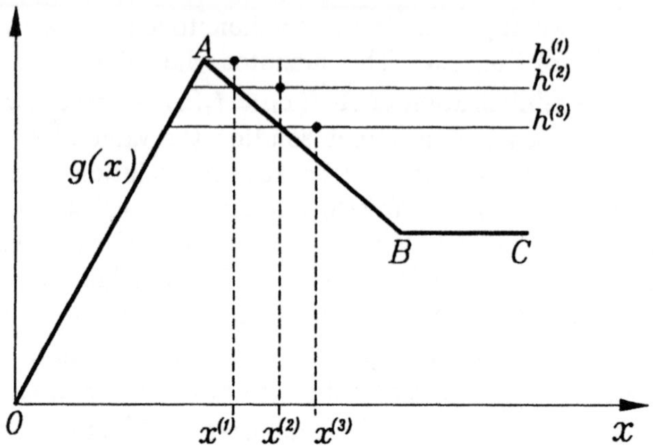

Figure 2.3. Graphical explanation of the heuristic nonconvex minimization algorithm

2. For the point $x^{(i)}$ select a convex superpotential $p^{(i)}$ such that relation (2.81) is fulfilled at this point.

3. Find the minimum $x^{(i)}$ of the convex function

$$f_c(x) + p^{(i)}(x).$$

4. Calculate

$$C^{(i+1)} = w(x^{(i)}) - p^{(i)}(x^{(i)}).$$

5. If $||x^{(i)} - x^{(i-1)}|| \leq \varepsilon_1$ and $||C^{(i+1)} - C^{(i)}|| \leq \varepsilon_2$ where ε_1 and ε_2 are appropriately small numbers, then a substationarity point of (2.77) has been determined and terminate the algorithm, else set $i = i+1$ and repeat step 2.

Given a starting point $x^{(0)}$, the above algorithm is able to detect a substationarity point of $\Pi(x)$ and not only a local minimum point, as it becomes obvious from (2.83). Different starting points of the iterative procedure, may lead to different substationarity points of $\Pi(x)$. Although from the mathematical point of view the task of determining the starting point for this algorithm seems to be guided only by experience, in engineering problems the determination of the starting points results from a procedure with certain mechanical explanation, as it will be demonstrated in the following Chapters.

2.4.4 Soft computing techniques

This Section deals with "soft" computing techniques for the solution of optimization problems and related problems in engineering. Some aspects from the dynamic systems approach to optimization and neural network techniques will be discussed. All these methods have two advantages when compared with the classical, computational optimization techniques, which have been considered in the previous Sections. First, they have the potential to deal with general nonconvex problems and to avoid, nevertheless in a stochastic, noncontrollable way, local minima and related effects. Moreover, these methods are appropriate for use in a distributed or parallel computing environment, a fact that makes them especially promising in view of recent hardware developments.

2.4.4.1 Dynamical systems and optimization. Several problems of linear algebra, systems theory and optimization can be solved by appropriate reformulation as dynamical systems, either in continuous time or in discrete time, see e.g., Helmke and Moore, 1994, Cichocki and Unbehauen, 1993. This point of view offers a unified framework for the numerical treatment of a variety of problems, including eigenvalue end eigenvector extraction algorithms, singular value decomposition (SVD), least square problems and certain optimization problems. For instance, the iterative, unconstrained, quasi–Newton optimization algorithm (cf. (2.56)–(2.58))

$$x^{(k+1)} = x^{(k)} - \mathbf{A}^{(k)} \nabla f(x^{(k)}), \tag{2.86}$$

where $\mathbf{A}^{(k)}$ is a symmetric positive definite matrix, can be formulated as a discrete approximation of a continuous–time trajectory from an initial point $x^{(0)}$ to the minimum of a dynamical system determined by the system of differential equations:

$$\frac{dx}{dt} = -M \nabla f(x),\ x(0) = x^{(0)}. \tag{2.87}$$

Here $x(t) \in \mathbb{R}^n$ and $M(x,t)$ is an $n \times n$ symmetric positive definite matrix with elements generally dependent on the value $x(t)$ and the time t. It is well known that as far as M is positive definite, system (2.87) is stable. Moreover, a relative minimum of function $f(u)$ is a local attractor of the system of differential equations (2.87), while a relative maximum is a local repeller (Wiggins, 1990).

For practical implementation of this dynamical system approach by means of an electronic circuit, which should directly model system (2.87), several assumptions must be done. For example, an appropriate choice of constant values for M would allow for the design of such a circuit with fixed characteristic values.

An inverse problem can also be posed: find appropriate values for M such that the resulting dynamical system (2.87) converges to a given point x and/or minimizes an appropriate error function $f(x)$. This problem is connected with the supervised and unsupervised learning algorithms for neural network models (Cichocki and Unbehauen, 1993).

A dynamical system approach to optimization allows for several extensions to cope with a certain extend of nonconvexity and nondifferentiability problems. A first extension, which is in accordance with the engineering intuition and practice, consists in the consideration of the system of second order differential equations (instead of (2.87)):

$$\delta(t)\frac{d^2x}{dt^2} = -\gamma(t)T\frac{dx}{dt} - M\nabla f(x),\ x(0) = x^{(0)},\ \frac{dx(0)}{dt} = \dot{x}^{(0)}. \tag{2.88}$$

Here $\delta(t)$, $\gamma(t)$ are real–valued positive functions and T, M are $n\times n$ symmetric, positive definite matrices. The engineering interpretation of system (2.88) is quite straightforward, see, e.g., Cichocki and Unbehauen, 1993, p.102: the solution process is simulated by the movement of a particle of mass $\delta(t)$ in the n–dimensional space under the effect of the potential force $-M\nabla f(x)$ and of the frictional dissipative force $-\gamma(t)T\frac{dx}{dt}$, with $\gamma(t)$ being the friction coefficient (cf., the method of heavy ball, Křížek and Neitaanmäki, 1990, p. 84).

The dynamical system (2.88) has several potential advantages to avoid local minima of $f(x)$ and to arrive at global solutions of the minimization problem with a nonconvex function $f(x)$. This is due to the inertial effects which are present in the relation (2.88) and which give the ability to the algorithm to overcome local minima and to find (possibly) a global minimum. Moreover, damping effects allow for more stable numerical integration algorithms to be used. These advantages are due to the introduction of the inertial terms in (2.88) and the additional flexibility of the iterative process given by the existence of a second set of initial values (the one concerning the initial velocities $\dot{x}^{(0)}$). In computational mechanics this technique is known under the name of dynamic relaxation (Zienkiewicz and Taylor, 1991).

Besides the inertial terms, which have been used in the second–order dynamical system (2.88), elements of stochastic optimization can be used to help the dynamical system avoid local minima of the goal function $f(x)$ and eventually solve a global, nonconvex optimization problem. For instance, a slightly and randomly perturbed energy function $\tilde{f}(x, N)$ can be used in the gradient system (2.87), where:

$$\tilde{f}(x, N) = f(x) + c(t)x^T N(t), \tag{2.89}$$

$N^T = [N_1(t), \dots, N_n]^T$ is a vector composed of independent noise sources and $c(t)$ is the parameter which controls the magnitude of the applied noise. By using this randomly perturbed function a stochastic gradient dynamical system arises, instead of (2.87), which has the form:

$$\frac{dx}{dt} = -M\left[\nabla f(x) + c(t)N(t)\right]. \tag{2.90}$$

A scheme which gradually reduces the applied noise, i.e., $c(t) \to 0$, as $t \to +\infty$, can eventually lead to a global solution of a nonconvex problem. A systematic way of introducing stochastic perturbations into the optimization problem and, accordingly, into the dynamical system which is used for its solution, is provided by the simulated annealing algorithm.

Closing this short introduction one should mention that the dynamical system approach to optimization can be extended to cope with constrained optimization problems and nondifferentiable optimization problems. The latter case follows a subgradient like approach in the sense that one element of the generally multivalued gradient of $f(x)$ is used to drive the dynamical system. Convergence is of course based on the damping property of the dynamical system. Moreover, the dynamical system interpretation of iterative solution algorithms in optimization can be extended to include the Lagrangian (primal–dual) schemes for both the convex and the difference convex optimization problems, as they have been reviewed in the previous Sections.

Beyond the specific models outlined in the sequel in connection with the neural network approach, the interested reader can find more material on the links between dynamical systems, optimization and matrix analysis with possible neural network implementations, among others, in Helmke and Moore, 1994, Pyne, 1956, Chua and Lin, 1984.

2.4.4.2 Neural network techniques. Following the dynamical system approach to optimization and data processing and given that electronic circuits can be designed to emulate the dynamical behaviour of a given, maybe nonlinear dynamical system, the potential of applying electronic devices for on–line solution of large scale problems seems straightforward. In fact, one of the aspects of neural network computing is exactly this application. An iterative algorithm is best suited for neural network implementation if it can be resolved into a number of relatively simple steps which can be executed by separate elements (distributed), which are interconnected by appropriate connection lines. The high parallelization of this scheme makes it also appropriate for a parallel computer implementation, if a hardware implementation is not available or at the development stage.

In this Section we follow the Hopfield neuron model which is today the most popular dynamic model of artificial neurons. The circuit for this neuron (Fig. 2.4) consists of a capacitor C_j, resistors R_{ji} and a nonlinear amplifier with sigmoid transfer functions Ψ. The output v_j is a function of these sigmoid transfer functions, i.e. $v_j = \Psi(u_j)$, where u_j is an internal signal called the internal potential. Fig. 2.5 gives three common types of Ψ: the hard limiter (Fig. 2.5a), the threshold logic nonlinearity (Fig. 2.5b) and the sigmoidal nonlinearity (Fig. 2.5c) (Lippmann, 1987).

It is assumed that the amplifier provides two symmetrical outputs (i.e. voltage signal v_j and its inverse $-v_j$) in order to ensure that all resistors simulating synaptic weights have positive values. This means that a positive synaptic weight is realized by connecting the resistor R_{ji} to $+v_i$ and a negative weight by connecting R_{ji} to the associated signal $-v_i$. The current I_j represents the bias (or the independent external input signal).

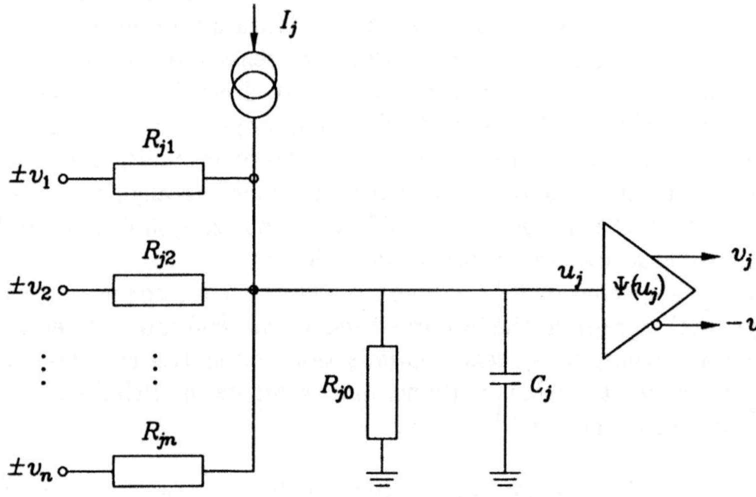

Figure 2.4. Hopfield model of a dynamic basic neuron cell

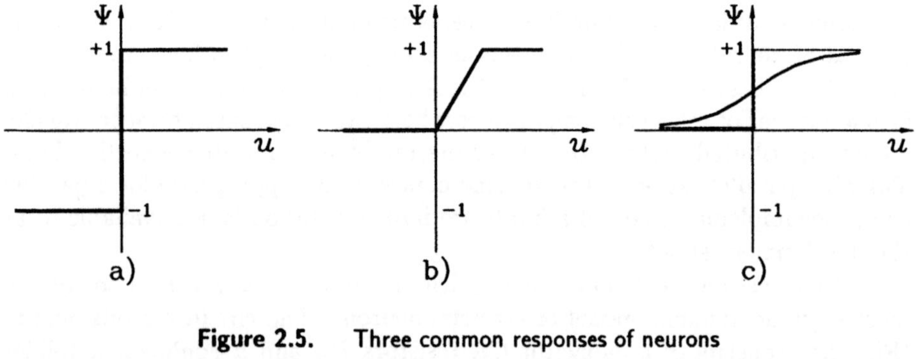

Figure 2.5. Three common responses of neurons

The basic Hopfield model (Fig. 2.6), appropriate for treatment of unconstrained optimization problems, is implemented by interconnecting an array of resistors, nonlinear amplifiers with symmetric outputs and external bias current sources. Thus, the neural network consists of n interconnected artificial Hopfield neurons as the one described above.

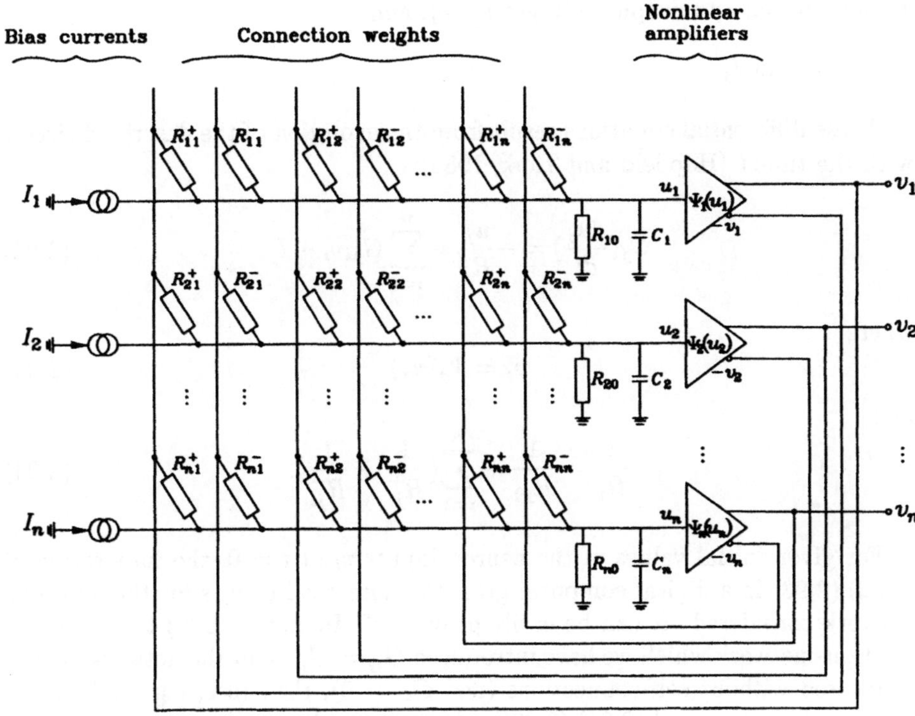

Figure 2.6. Hopfield neural network for unconstrained optimization

The output of each artificial neuron is transmitted to the other neurons which perform the same basic action. This procedure continues until a so-called "stable state" of the network is achieved, which corresponds to the minimum of a characteristic network energy function.

This stable state can be determined from the solution of a system of differential equations (Cichocki and Unbehauen, 1993) which have as parameters the values of

- the resistors R_{ji},
- the capacitors C_j,
- the conductances G_{ji} representing the synaptic weight from neuron i to neuron j defined as

$$G_{ji} = \frac{1}{R_{ji}^{+}} - \frac{1}{R_{ji}^{-}} \quad \text{with } R_{ji}^{\pm} > 0, \tag{2.91}$$

- the internal and output voltages u_j, v_j, and
- the current I_j.

These differential equations result from the evolution of the described circuit with the time t (Hopfield and Tank, 1985):

$$C_j(\frac{du_j}{dt}) = -\frac{u_j}{R_j} + \sum_{i=1}^{n} G_{ji}v_i + I_j \tag{2.92}$$

where

$$v_j = \Psi_j(u_j) \tag{2.93}$$

and

$$\frac{1}{R_j} = \frac{1}{R_{j0}} + \sum_{i=1}^{n}(\frac{1}{R_{ji}^{+}} + \frac{1}{R_{ji}^{-}}). \tag{2.94}$$

For given initial values of the neuron inputs v_i at $t = 0$, the integration of (2.92),(2.93) in a digital computer gives the numerical results for the fictitious network considered. It can be easily proved (cf. Hopfield, 1982) that if in the fictitious network which we have introduced $G_{ij} = G_{ji}$, and the network is fully connected without self-connections (i.e., $G_{ii} = 0$), then (see, e.g., Zell, 1994, p. 199), the solution of (2.92),(2.93) converges to solutions having constant the outputs v_j of all neurons.

These solutions (the stable states) make stationary the quantity

$$E = -\frac{1}{2}\sum_{j,i=1}^{n} G_{ji}v_jv_i + \sum_{j=1}^{n}(\frac{1}{R_j})\int_0^{v_j} \Psi_j^{-1}(v_j)dv_j - \sum_{j=1}^{n} I_jv_j \tag{2.95}$$

which is also called the Liapunov function of the system.

Thus, the neural network approach, by means of Hopfield networks, to an unconstrained optimization problem consists in the determination of the synapses' conductivities G_{ji}, the neuron input currents I_i, the total resistances R_i, and the functions Ψ_j in (2.95) such that a stable state will be the solution of the minimum problem concerning the function (2.95).

Analogous procedures can be followed for the solution of linear and nonlinear optimization problems and of inequality constrained problems. In these cases the neural network approach consists in transforming the optimization problem into a system of differential equations and then constructing an appropriate neural network (hardware device) for the solution of this system. For more details the interested reader is referred to Chiu et al., 1991, Maa and Shanblatt,

1992a, Maa and Shanblatt, 1992b, Ritter et al., 1990, Kosko, 1992, Rojas, 1992, Cichocki and Unbehauen, 1993, Zell, 1994 and the references given therein.

Engineering applications of solution algorithms based on neural networks are presented among others in: Vanluchene and Roufei, 1990, Berke and Hajela, 1992, Kortesis and Panagiotopoulos, 1993, Theocaris and Panagiotopoulos, 1993, Avdelas et al., 1995, Meade and Sonneborn, 1996, Papadrakakis et al., 1996.

It should be noted that the problems of defining an appropriate network which is able to solve a specific problem, or other types of networks which are appropriate for the solution of, for instance, inverse problems, are not discussed in this short review.

2.4.4.3 Stochastic approaches. Simulated annealing. The previously outlined idea of introducing stochastic elements in the optimization algorithm (see (2.90)) can be applied in the Hopfield neural network model as well. More precisely, the adoption of a stochastic activation function within a neural network of the same topology (nodes, connectivity etc.) with the one used previously, leads to the Boltzmann engine model. Moreover, the activation function of the neural network algorithm (or even other variables of the network) can change during the solution of the network (or the simulation of it).

This flexibility of the network, may also enhance the ability of the network to solve more complicated optimization problems (up to problems of combinatorial optimization, see Hopfield and Tank, 1986).

This technique, which is related to the simulated annealing method, imitates the slow cooling process of a melted metal. At the first stages of this process, with high temperature, the atoms of the material have a high energy level. Thus, they can move relatively free and statistically they can occupy all the available positions of the configuration space. Later on, during the slow cooling, the energy is gradually reduced so that the system attains an equilibrium configuration at a point with a minimum energy (for more details see, among others, Kirkpatrick et al., 1983, Hopfield and Tank, 1986, Zell, 1994, Chapt. 18, Papageorgiou, 1991).

Let us consider a simple ramp-style activation function (cf. (2.93))

$$v_j = \Psi_j(u_j) = \begin{cases} 0 & \text{if} \quad u_j \leq 0 \\ 1 & \text{if} \quad u_j > 0, \end{cases} \tag{2.96}$$

which is approximated by the smooth sigmoidal activation function:

$$\Psi_j(u_j) = \frac{1}{1 + e^{-u_j}}. \tag{2.97}$$

The previous possible extensions can be outlined as follows. First a stochastic activation function may be considered:

$$v_j = \Psi_j(u_j) = \begin{cases} 0 & \text{with possibility} \quad \eta = \frac{1}{1+e^{-\beta u_j}} \\ 1 & \text{with possibility} \quad 1-\eta \end{cases} . \tag{2.98}$$

For a large number of the real parameter β ($\beta \to \infty$) one gets the deterministic activation function (2.96). Furthermore, a parametrized activation function is considered

$$\Psi_j(u_j;\beta) = \frac{1}{1+e^{-\beta u_j}}. \tag{2.99}$$

Here the inverse of the parameter β, $T = 1/\beta$ can be interpreted to be the temperature in the previously given simulated annealing physical interpretation. Thus for $T \to 0$ the ramp activation function (2.96) is approximated.

Roughly speaking the idea behind deterministic (resp. stochastic or simulated) annealing approaches is that one solves the neural network by using the activation function (2.99) (resp. (2.98)) by starting from a large temperature value T and by letting this value go to zero. The reader is referred to the specialized literature for more details on these models which have a potential application for solving difficult nonconvex (even combinatorial or discrete variable) optimization problems (see, among others Davis, 1988, Aarts and Korst, 1989, vanLaarhoven and Aarts, 1989).

References

Aarts, H. L. E. and Korst, J. (1989). *Simulated annealing and Boltzmann machines : a stochastic, approach to combinatorial optimization and neural computing.* Wiley, Chichester.

Aubin, J. P. (1993). *Optima and equilibria.* Springer Verlag, Berlin.

Aubin, P. and Frankowska, H. (1991). *Set-valued analysis.* Birkhäuser, Berlin-Heidelberg.

Auchmuty, G. (1989). Duality algorithms for nonconvex variational principles. *Num. Functional Analysis and Optimization,* 10:211–264.

Avdelas, A. V., Panagiotopoulos, P. D., and Kortesis, S. (1995). Neural networks for computing in elastoplastic analysis of structures. *Meccanica,* 30:1–15.

Bartels, S. G., Kuntz, L., and Scholtes, S. (1995). Continuous selections of linear functions and nonsmooth critical point theory. *Nonlinear analysis,* 24(3):385–408.

Bažant, Z. P. and Cedolin, L. (1991). *Stability of structures. Elastic, inelastic, fracture and damage theories.* Oxford University Press, New York, Oxford.

Berke, L. and Hajela, P. (1992). Applications of artificial neural nets in structural mechanics. *Structural Optimization,* 4:90–98.

Bertsekas, D. P. (1982). *Constrained optimization and Lagrange multiplier methods.* Academic Press, New York.

Chiu, C., Maa, C. Y., and Shanblatt, M. A. (1991). Energy function analysis of dynamic programming neural networks. *IEEE Transactions on Neural Networks,* 2(4):418–426.

Chua, L. O. and Lin, G. N. (1984). Nonlinear programming without computation. *IEEE Trans. Circuits and Systems,* 31:182–188.

Ciarlet, P. G. (1989). *Introduction to numerical linear algebra and optimization.* Cambridge University Press, Cambridge.

Cichocki, A. and Unbehauen, R. (1993). *Neural networks for optimization and signal processing.* J. Wiley and Sons and Teubner, Chichester - Stuttgart.

Clarke, F. H. (1983). *Optimization and nonsmooth analysis.* J. Wiley, New York.

Cohn, M. Z. and Maier, G., editors (1979). *Engineering plasticity by mathematical programming.* Pergamon Press, Oxford.

Cottle, R. W., Pang, J. S., and Stone, R. E. (1992). *The linear complementarity problem.* Academic Press, Boston.

Davis, L. (1988). *Genetic algorithms and simulated annealing.* Pitman, London.

de Freitas, J. A. T. (1984). A general methodology for nonlinear structural analysis by mathematical programming. *Engineering Structures*, 6:52–60.

de Freitas, J. A. T. and Smith, D. L. (1985). Energy theorems for elastoplastic structures in a regime of large displacements. *J. de mecanique theorique et appliqee*, 4(6):769–784.

Dem'yanov, V. F. and Rubinov, A. M. (1980). On quasidifferentiable functionals. *Soviet Math. Dokl.*, 21:14–17.

Dem'yanov, V. F. and Rubinov, A. M. (1985). *Quasidifferentiable calculus.* Optimization Software, New York.

Dem'yanov, V. F. and Rubinov, A. M. (1995). *Introduction to constructive nonsmooth analysis.* Peter Lang Verlag, Frankfurt-Bern-New York.

Dem'yanov, V. F., Stavroulakis, G. E., Polyakova, L. N., and Panagiotopoulos, P. D. (1996). *Quasidifferentiability and nonsmooth modelling in mechanics, engineering and economics.* Kluwer Academic, Dordrecht.

Dem'yanov, V. F. and Vasiliev, L. N. (1985). *Nondifferentiable optimization.* Optimization Software, New York.

Ekeland, I. and Temam, R. (1976). *Convex analysis and variational problems.* North-Holland, Amsterdam.

Elster, K. H., Reinhardt, R., Schaeule, M., and Donath, G. (1977). *Einführung in die nichtlineare Optimierung.* BSB B.G. Teubner Verlagsgesellschaft, Leipzig.

Ferris, M. C. and Pang, J. S. (1995). Engineering and economic applications of complementarity problems. *SIAM Review*, (submitted). Math. Progr. Tech. Report 95-07 Comp. Sci. Dept. Univ. of Wisconsin Madison.

Fletcher, R. (1990). *Practical methods of optimization.* J. Wiley, Chichester.

Forsgren, A. and Ringertz, U. (1993). On the use of a modified Newton method for nonlinear finite element analysis. *Computer Methods in Applied Mechanics and Engineering*, 110:275–283.

Friedman, A. (1982). *Variational principles and free boundary problems.* J. Wiley, New York.

Gill, P. E., Murray, W., and Wright, M. H. (1981). *Practical optimization.* Academic Press, New York.

Glowinski, R. and LeTallec, P. (1989). *Augmented Lagrangian and operator–splitting methods for nonlinear mechanics.* SIAM, Philadelphia.

Glowinski, R., Lions, J. L., and Trémolières, R. (1981). *Numerical analysis of variational inequalities.* Studies in Mathematics and its Applications, Vol. 8. Elsevier, Amsterdam-New York.

Hamel, G. (1949). *Theoretische Mechanik.* Springer Verlag, Berlin.

Harker, P. T. and Pang, J. S. (1990). Finite dimensional variational inequality and nonlinear complementarity problems: a survey of theory, algorithms and applications. *Mathematical Programming*, 48:161–220.

Hartmann, F. (1985). *The mathematical foundation of structural mechanics.* Springer Verlag, Berlin.

Helmke, U. and Moore, J. B. (1994). *Optimization and dynamical systems.* Springer, London.

Hiriart-Urruty, J. B. (1985). *Generalized differentiability, duality and optimization for problems dealing with differences of convex functions*, volume 256 of *Lect. Notes in Economics and Mathematical Systems*, pages 37–50. Springer.

Hiriart-Urruty, J. B. and Lemaréchal, C. (1993). *Convex analysis and minimization algorithms I.* Springer, Berlin-Heidelberg.

Hlavaček, I., Haslinger, J., Nečas, J., and Lovisek, J. (1988). *Solution of variational inequalities in mechanics*, volume 66 of *Appl. Math. Sci.* Springer.

Hopfield, J. J. (1982). Neural networks and physical systems with emergent collective computational abilities. *Proc. of the Nat. Acad. of Sciences*, 79:2554–2558.

Hopfield, J. J. and Tank, D. W. (1985). "Neural" computation of decisions in optimization problems. *Biol. Cybern.*, 52:141–152.

Hopfield, J. J. and Tank, D. W. (1986). Computing with neural circuits. *Science*, 233:625–633.

Horst, R., Pardalos, P. M., and Thoai, N. Y. (1995). *Introduction to global optimization.* Kluwer Academic, Dordrecht Boston.

Horst, R. and Tuy, H. (1990). *Global optimization.* Springer, Berlin - Heidelberg.

Jahn, J. (1994). *Introduction to the theory of nonlinear optimization.* Springer Verlag, Berlin, Heidelberg.

Kirkpatrick, S., Gelatta, C. D., and Vecchi, M. P. (1983). Optimization by simulated annealing. *Science*, 220:671–680.

Kojima, M., Megiddo, N., and Ye, Y. (1992). An interior point potential reduction algorithm for the linear complementarity problem. *Mathematical Programming*, 54:267–279.

Kornhuber, R. (1997). *Adaptive monotone multigrid methods for nonlinear variational problems.* B.G. Teubner, Stuttgart.

Kortesis, S. and Panagiotopoulos, P. D. (1993). Neural networks for computing in structural analysis: methods and prospects of applications. *International Journal for Numerical Methods in Engineering*, 36:2305–2318.

Kosko, B. (1992). *Neural networks and fuzzy systems. A dynamical system approach to machine intelligence.* Prentice Hall, New York.

Kuntz, L. and Scholtes, S. (1995). Qualitative aspects of the local approximation of a piecewise differentiable function. *Nonlinear Analysis. Theory, Methods and Applications*, 25(2):197–215.

Křížek, M. and Neitaanmäki, P. (1990). *Finite element approximation of variational problems and applications.* Longman Scientific and Technical, Essex U.K.

LeTallec, P. (1990). *Numerical analysis of viscoelastic problems.* Masson, Springer, Paris, Berlin.

Lippmann, H. (1972). *Extremum and variational principles in mechanics.* Springer, CISM Courses and Lectures 54, Wien.

Lippmann, R. (1987). An introduction to computing with neural nets. *IEEE ASSP Magazine*, April:4–22.

Maa, C. Y. and Shanblatt, M. A. (1992a). Linear and quadratic programming neural network analysis. *IEEE Transactions on Neural Networks*, 3(4):580–594.

Maa, C. Y. and Shanblatt, M. A. (1992b). A two-phase optimization neural network. *IEEE Transactions on Neural Networks*, 3(6):1003–1009.

Mäkelä, M. M. and Neittaanmäki, P. (1992). *Nonsmooth optimization: analysis and algorithms with applications to optimal control.* Word Scientific Publ. Co.

Mangasarian, O. L. (1994). *Nonlinear programming.* SIAM, Philadelphia.

Matthies, H. G., Strang, G., and Christiansen, E. (1979). The saddle point of a differential problem. In Glowinski, R., Rodin, E., and Zienkiewicz, O., editors, *Energy methods in finite element analysis*, New York. J. Wiley and Sons.

Meade, A. J. J. and Sonneborn, H. C. (1996). Numerical solution of a calculus of variations problem using the feedforward neural network architecture. *Advances in Engineering Software*, 27(3):213–225.

Moreau, J. J. (1963). Fonctionnelles sous - différentiables. *C.R. Acad. Sc. Paris*, 257A:4117 – 4119.

Murty, K. G. (1988). *Linear complementarity, linear and nonlinear programming.* Heldermann, Berlin.

Nesterov, Y. E. and Nemirovskii, A. S. (1993). *Interior point polynomial methods in convex programming: theory and algorithms.* SIAM, Philadelphia PA.

Oden, J. T. and Reddy, J. N. (1982). *Variational methods in theoretical mechanics.* Springer Verlag, Berlin.

Panagiotopoulos, P. D. (1985). *Inequality problems in mechanics and applications. Convex and nonconvex energy functions.* Birkhäuser, Basel - Boston - Stuttgart. Russian translation, MIR Publ., Moscow 1988.

Panagiotopoulos, P. D. (1993). *Hemivariational inequalities. Applications in mechanics and engineering.* Springer, Berlin - Heidelberg - New York.

Papadrakakis, M., Papadopoulos, V., and Lagaros, N. D. (1996). Structural reliability analysis of elastic-plastic structures using neural networks and Monte Carlo simulation. *Computer Methods in Applied Mechanics and Engineering*, 136(1-2):145–164.

Papageorgiou, M. (1991). *Optimierung.* Oldenbourg, München.

Pardalos, P. M. and Rosen, J. B. (1987). *Constrained global optimization. Algorithms and applications*, volume 268 of *Lecture Notes in Computer Science.* Springer, Berlin.

Polyakova, L. N. (1986). On minimizing the sum of a convex function and a concave function. *Mathematical Programming Study*, 29:69–73.

Pyne, I. B. (1956). Linear programming on an analogue computer. *Trans. AIEE, Part I*, 75:139–143.

Ritter, H., Martinez, J., and Schulten, K. (1990). *Neuronale Netze.* Adison-Wesley, Bonn.

Rockafellar, R. T. (1970). *Convex analysis.* Princeton University Press, Princeton.

Rockafellar, R. T. (1979). *La théorie des sous-gradients et ses applications à l' optimization. Fonctions convexes et non-convexes.* Les Presses de l' Université de Montréal, Montréal.

Rockafellar, R. T. (1982). *Network flows and monotropic optimization.* J. Wiley, New York.

Rodrigues, J. F. (1987). *Obstacle problems in mathematical physics.* North Holland, Amsterdam.

Rojas, R. (1992). *Theorie der neuronalen Netze.* Springer Verlag, Berlin.

Schramm, H. and Zowe, J. (1992). A version of the bundle idea for minimizing a nonsmooth function: conceptual idea, convergence analysis, numerical results. *SIAM J. Optimization*, 2:121–152.

Sewell, N. J. (1987). *Maximum and minimum principles. A unified approach with applications.* Cambridge University Press, Cambridge.

Shor, N. Z. (1985). *Minimization methods for nondifferentiable functions.* Springer, Berlin.

Stavroulakis, G. E. and Polyakova, L. N. (1996a). Difference convex optimization techniques in nonsmooth computational mechanics. *Optimization. Methods and Software*, 7(1):55–87.

Stavroulakis, G. E. and Polyakova, L. N. (1996b). Nonsmooth and nonconvex structural analysis algorithms based on difference convex optimization techniques. *Structural Optimization*, 12:167–176.

Stavroulakis, G. E. and Rohde, A. (1996). Stability of structures with quasidifferentiable energy functions. In Sotiropoulos, D. and Beskos, D., editors, *2nd Greek Conf. on Computational Mechanics*, pages 406–413, Chania.

Stoer, J. and Witzgall, C. (1970). *Convexity and optimization in finite dimensions I.* Springer Verlag, Berlin Heidelberg.

Stuart, C. A. and Toland, J. F. (1980). A variational method for boundary value problems with discontinuous nonlinearities. *J. London Math. Soc. (2)*, 21:319–328.

Thach, P. T. (1987). D.c. sets, d.c. functions and nonlinear equations. *Mathematical Programming*, 58:415–428.

Thach, P. T. and Konno, H. (1997). On the degree and separability of nonconvexity and applications to optimization problems. *Mathematical Programming*, 77:23–47.

Theocaris, P. S. and Panagiotopoulos, P. D. (1993). Neural networks for computing in fracture mechanics. Methods and prospects of applications. *Computer Methods in Applied Mechanics and Engineering*, 106:213–228.

Todd, M. J. (1996). Potential-reduction methods in mathematical programming. *Mathematical Programming*, 76:3–45.

Toland, J. F. (1979). A duality principle for nonconvex optimization and the calculus of variations. *Arch. Rat. Mech. Analysis*, 71:41–61.

Tuy, H., Zam, B. T., and Dan, N. D. (1994). Minimizing the sum of a convex function and a specially structured nonconvex function. *Optimization*, 28(3-4):237–248.

vanLaarhoven, P. J. M. and Aarts, E. H. L. (1989). *Simulated annealing : theory and applications.* Kluwer, Dordrecht.

Vanluchene, R. D. and Roufei, S. (1990). Neural networks in structural engineering. *Microcomputers in Civil Engineering*, 5:207–215.

Washizu, K. (1968). *Variational methods in elasticity and plasticity.* Pergamon Press, Oxford.

Wiggins, S. (1990). *Introduction to applied nonlinear dynamical systems and chaos.* Springer, New York-Berlin.

Womersley, R. S. and Fletcher, R. (1986). An algorithm for composite nonsmooth optimization problems. *Journal of Optimization Theory and Applications*, 48:493–523.

Zell, A. (1994). *Simulation Neuronaler Netze.* Addison-Wesley, Bonn Paris Reading MA.

Zienkiewicz, O. C. and Taylor, R. L. (1991). *The finite element method. Vol. II: Solid and fluid mechanics, dynamics and non–linearity.* McGraw-Hill.

III Superpotential Modelling and Optimization in Mechanics with and without Convexity and Smoothness

3 CONVEX SUPERPOTENTIAL PROBLEMS. VARIATIONAL EQUALITIES AND INEQUALITIES.

3.1 VARIATIONAL PROBLEMS IN MECHANICS

A systematic way for the derivation of variational principles in mechanics goes through the consideration of a potential energy or a complementary energy function. The classical set of possibly nonlinear equations of mechanics from the one side, i.e., the compatibility equations, the equilibrium equations and the material laws, and, from the other side, the optimality conditions of the mathematical optimization theory are integrated in this approach. In fact, the governing relations of the problem either are taken into account in the derivation of the problem or they are produced from the optimality conditions of the associated energy optimization problem.

For classical systems without inequality constraints and for smooth (differentiable) potential functions, the necessary optimality conditions, i.e., the well-known relation that the gradient of the potential energy should be equal to zero at the optimum, lead to the governing relations of the mechanical problem. In the variational formulation we write that the directional derivative of the potential energy must be equal to zero for all directions emanating from the solution (i.e. the equilibrium) point. The latter statement means that the product of the gradient of the potential energy with the small (virtual) variations of the function's argument is zero; thus a *variational equality* problem arises. For historical reasons and since the most frequently used function is the potential energy function of a system written in terms of displacement variables and the gradient of this function plays the role of a stress or force vector, the above mentioned relation is called the *principle of virtual work.* In a dynamic analysis framework where the potential is expressed in terms of velocities, the term *principle of virtual power* is used instead.

Convex differentiable potentials are very favourable with respect to the previously outlined optimization approach to structural analysis. For instance, linear elastic (or linearized) problems can be produced from quadratic potential energy functions which, if instabilities are excluded, are convex functions.

In the presence of inequality constraints or of nondifferentiable potentials, the above method of study requires certain modifications. The optimality condition for a convex nondifferentiable function, possibly under inequality constraints is no more a simple equation; it is a set–valued equation or a convex differential inclusion, where the set–valued generalization of the classical gradient, the subdifferential of convex analysis, appears. Equivalently, the directional derivative of a nondifferentiable function cannot be written as a linear function of the virtual variations (the direction) or, in the presence of inequality restrictions in the space of configuration variables, not all variations are permitted. Thus, *variational inequality* problems are formulated. Unilateral contact problems are typical examples of structural analysis problems with kinematic inequality constraints which physically describe the no–penetration restriction of the unilateral contact mechanism. Nondifferentiable potentials, the so–called superpotentials, arise in multi–modulus elasticity problems, in holonomic (or step–wise holonomic) plasticity models, in static friction problems etc., as it will be discussed in the sequel with concrete examples.

3.1.1 Small displacement smooth elastostatics: a convex, differentiable optimization problem

A small displacement elastostatic problem is considered here. An elastic structure occupying a region $\Omega \subset \mathbb{R}^n$, $n = 2,3$, with boundary denoted by Γ is considered. The structure is referred to the orthogonal Cartesian coordinate system $Ox_1x_2(x_3)$. The purpose of this Section is to outline the variational and potential formulation of the elastostatic analysis problem under the hypothesis of small displacements (i.e. the kinematics are linearized) and with linear and nonlinear elastic material laws. The link with the convex (quadratic for linear material) optimization is pointed out.

The structure is discretized by means of m_1 finite elements and the stress and deformation vectors of the finite element assemblage are denoted by $\mathbf{s}, \mathbf{e}$ with elements denoted by s_i, e_i, $i = 1, \ldots, m$. Here m depends on the number of independent stresses (resp. strains) of each finite element of the structure (natural stresses and strains in the sense of Argyris, 1965). Moreover, let $\mathbf{u}$ be the n–dimensional vector of nodal displacements and $\mathbf{p}$ be the energy corresponding n–dimensional vector of nodal forces.

The static analysis problem is described by the following relations:

- *Stress equilibrium equations*:

$$\mathbf{G}\mathbf{s} = \mathbf{p} \tag{3.1}$$

where $\mathbf{G}$ is the equilibrium matrix of the discretized structure.

- *Strain–displacements compatibility equations*:

$$\mathbf{e} = \mathbf{G}^T\mathbf{u}. \tag{3.2}$$

- *Linear material constitutive law* for the structure:

$$\mathbf{e} = \mathbf{e}_0 + \mathbf{F}_0\mathbf{s}, \tag{3.3}$$

or

$$\mathbf{s} = \mathbf{K}_0(\mathbf{e} - \mathbf{e}_0). \tag{3.4}$$

Here $\mathbf{F}_0$ and $\mathbf{K}_0 = \mathbf{F}_0^{-1}$ are the natural and stiffness flexibility matrices of the unassembled structure and $\mathbf{e}_0$ is the initial deformation vector.

- Classical *equality boundary conditions* written in the general form:

$$\mathbf{E}\mathbf{u} = \mathbf{u}_0 \tag{3.5}$$

and

$$\mathbf{Z}\mathbf{s} = \mathbf{F} \tag{3.6}$$

where $\mathbf{E}$ and $\mathbf{Z}$ are appropriately defined transformation matrices and $\mathbf{u}_0$, $\mathbf{F}$ denote the known nodal boundary displacements (support) and boundary loading (traction). Note that in any point of the boundary, only one of (3.5), (3.6) can be active.

For the discretized structure one may write the *virtual work* and the *complementary virtual work* equation. The first one expresses the equality between the work produced by the internal stresses and the work of the applied loading and it takes the form:

$$\mathbf{s}^T(\mathbf{e}^* - \mathbf{e}) = \mathbf{p}^T(\mathbf{u}^* - \mathbf{u}), \ \forall \mathbf{e}^*, \mathbf{u}^*, \text{ s.t. } (3.2), (3.5) \text{ hold.} \tag{3.7}$$

The *complementary virtual work* involves variations of the stresses and reads:

$$\mathbf{e}^T(\mathbf{s}^* - \mathbf{s}) = \mathbf{u}^T(\mathbf{p}^* - \mathbf{p}) \ , \ \forall \mathbf{s}^*, \mathbf{p}^* \text{ s.t. } (3.1), (3.6) \text{ hold.} \tag{3.8}$$

By using the linear elasticity law (3.4) into the virtual work equation (3.7), and by virtue of (3.2) we get:

$$\mathbf{u}^T\mathbf{G}\mathbf{K}_0^T\mathbf{G}^T(\mathbf{u}^* - \mathbf{u}) - (\mathbf{p} + \mathbf{G}\mathbf{K}_0\mathbf{e}_0)^T(\mathbf{u}^* - \mathbf{u}) = 0,$$
$$\forall \mathbf{u}^* \in V_{ad} = \{\mathbf{u} \in \mathbb{R}^n \mid (3.5) \text{ holds } \}. \tag{3.9}$$

Here $\mathbf{K} = \mathbf{G}\mathbf{K}_0\mathbf{G}^T$ denotes the stiffness matrix of the structure and $\overline{\mathbf{p}} = \mathbf{p} + \mathbf{G}\mathbf{K}_0\mathbf{e}_0$ denotes the nodal equivalent loading vector (including initial deformations' effects).

For a stress based formulation (force method) we use the elasticity law in the form of (3.3) and the boundary conditions (3.6). By analogous reasoning we get from (3.3) the complementary virtual work equality:

$$(\mathbf{e}_0 + \mathbf{s}\mathbf{F}_0)^T(\mathbf{s}^* - \mathbf{s}) = 0, \ \forall \mathbf{s}^* \in \Sigma_{ad} = \{\mathbf{s} \in \mathbb{R}^n \mid (3.6) \text{ holds } \}. \tag{3.10}$$

Note that due to the linearity of the equilibrium and the compatibility equations (3.1) , (3.2) and due to the linearity of the material constitutive law (3.3) (or (3.4)) the previous variational problems involve linear and bilinear terms. Accordingly, both the potential energy optimization problem and the complementary energy optimization problem involve linear and quadratic constituents. The two potential minimization problems read:
Find $\mathbf{u} \in V_{ad}$ such that:

$$\Pi(\mathbf{u}) = \min_{\mathbf{v} \in V_{ad}} \left\{ \Pi(\mathbf{v}) = \frac{1}{2}\mathbf{v}^T\mathbf{K}\mathbf{v} - \overline{\mathbf{p}}^T\mathbf{v} \right\}. \tag{3.11}$$

Find $\mathbf{s} \in \Sigma_{ad}$ such that:

$$\Pi^c(\mathbf{s}) = \min_{\mathbf{t} \in \Sigma_{ad}} \left\{ \Pi^c(\mathbf{t}) = \frac{1}{2}\mathbf{t}^T\mathbf{F}_0\mathbf{t} + \mathbf{e}_0^T\mathbf{t} \right\}. \tag{3.12}$$

One may easily verify that problem (3.7) (resp. (3.10)) actually expresses the minimality conditions for the energy optimization problem (3.11) (resp.

(3.12)). Moreover, by means of the convex analysis duality theory it is possible to show that (3.11) and (3.12) are connected to each other in the sense that the potential energy function Π is the convex conjugate of the complementary energy function Π^c and vice versa (see, e.g., Panagiotopoulos, 1985, Chapt. 62, Zeidler, 1988).

Recall that, due to the physical restriction that a material element subjected to a deformation history cannot produce energy, matrices $\mathbf{K}_0, \mathbf{F}_0$ are symmetric and positive semidefinite. Moreover, the stiffness matrix $\mathbf{K} = \mathbf{G}\mathbf{K}_0\mathbf{G}^T$ has the same properties. Its restriction on the subspace (or affine variety) described by sufficient boundary support conditions (3.5) such as to prevent every (even infinitesimal) rigid body displacement and rotation, i.e., the matrix $\mathbf{K}_{V_{ad}} = \mathbf{V}^T\mathbf{K}\mathbf{V}$, where $\mathbf{V}$ is a matrix composed of base vectors of the subspace V_{ad}, is positive definite (thus nonsingular and invertible). The latter properties guarantee that the optimization problems (3.11)-(3.12) are convex problems (and strictly convex for a sufficiently supported structure without rigid body degrees of freedom).

The convex energy optimization framework can also be used if instead of the linear elasticity law (3.3), (3.4), a general nonlinear monotone elastic constitutive law is adopted. A hyperelastic law is usually adopted which is derived by a local potential relation (instead of (3.3), (3.4)):

$$\mathbf{s} = \frac{\partial w(\mathbf{e})}{\partial \mathbf{e}}, \text{resp. } \mathbf{e} = \frac{\partial w^c(\mathbf{s})}{\partial \mathbf{s}}. \tag{3.13}$$

The total potential energy has in this case the form:

$$\Pi(\mathbf{u}, \mathbf{e}) = W(\mathbf{e}) - \overline{\mathbf{p}}^T\mathbf{u} = \sum_{j=1,m_1} w_j(\mathbf{e}_j) - \overline{\mathbf{p}}^T\mathbf{u}, \tag{3.14}$$

where summation index j runs over all finite elements of the structure. It is obvious that the convexity of $W(\mathbf{e})$ is not affected by the linear kinematic transformation (3.2), thus we finally arrive at a convex potential energy function in terms of the displacement variables:

$$\Pi(\mathbf{u}) = \Pi_{in}(\mathbf{u}) - \overline{\mathbf{p}}^T\mathbf{u} = W(\mathbf{G}^T\mathbf{u}) - \overline{\mathbf{p}}^T\mathbf{u}. \tag{3.15}$$

For the derivation of the variational formulation or of the energy minimization problem, the latter potential can be used, by using the appropriate calculus rules for the partial derivatives with respect to $\mathbf{u}$. Analogous formulations can be derived for the stress based problem.

Details of the valid expressions for the above used discretized versions of the elastostatic analysis problem can be found in classical books on finite element analysis and on computational mechanics, see, e.g., Cook, 1978, Bathe, 1981,

Zienkiewicz and Taylor, 1991, Chen, 1994, Vol. 1. Valid forms for several hyperelastic material laws which obey to the objectivity requirements can be found, among others, in Chen, 1994, Vol. 1.

3.1.2 Unilateral contact problems within the small displacements framework: a convex, inequality constrained, potential energy minimum problem

A model two–dimensional discretized elastic structure with interfaces is considered in the sequel. This framework is general since interfaces may be replaced by boundary conditions (seen as interfaces with a rigid support). Without loss of generality a structure consisting of two parts, Ω_1 (resp. Ω_2), with boundaries Γ_1 (resp. Γ_2) and an interface $\Gamma^{(1.2)}$ connecting them is considered. A right–hand Cartesian orthogonal coordinate system Ox_1x_2 is used throughout. In the framework of a small displacement and deformation theory, a simple, node–to–node, collocation type technique is used to model the interface relative displacement vs. the interface traction mechanical behaviour. In this Section the possible nonlinearities of the problem are restricted in the interfaces and are of a unilateral type.

The mechanical behaviour of each couple of nodes along the interface is considered separately in the normal and in the tangential to the interface direction. Thus, interface laws between the relative normal interface displacements $[u]_N \in \mathbf{R}$ and the normal interface tractions $-S_N \in \mathbf{R}$ and between the tangential interface displacements $[u]_T \in \mathbf{R}$ and the tangential interface tractions $-S_T \in \mathbf{R}$ are considered. Concerning the positive sign conventions, $S_N, [u]_N$ are referred to the outward unit normal to the interface, whereas $S_T, [u]_T$ are perpendicular to the N direction, such as to form a local N,T–right–handed coordinate system.

As in the previous Section, let the structure be discretized by means of m_1 finite elements and let the stress and deformation vectors of the finite element assemblage be denoted by $\mathbf{s}_i, \mathbf{e}_i, i = 1, \ldots, m$. Here m depends on the number of independent stresses (resp. strains) of each finite element of the structure. Let $\mathbf{u}$ be the n–dimensional vector of nodal displacements (the degrees of freedom in the displacement method) and $\mathbf{p}$ be the energy corresponding n–dimensional vector of nodal forces. The discrete interface quantities are assembled in the q–dimensional vectors $\mathbf{S}_N$, $\mathbf{S}_T$ and $[\mathbf{u}]_N$, $[\mathbf{u}]_T$ respectively, where q is the number of couples of nodes which model the interface of the structure. For the whole structure (including the interfaces) the enlarged stress $\bar{\mathbf{s}}$ and deformation $\bar{\mathbf{e}}$ vectors read:

$$\bar{\mathbf{s}} = \begin{bmatrix} \mathbf{s} \\ -\mathbf{S}_N \\ -\mathbf{S}_T \end{bmatrix}, \ \bar{\mathbf{e}} = \begin{bmatrix} \mathbf{e} \\ [\mathbf{u}]_N \\ [\mathbf{u}]_T \end{bmatrix}. \tag{3.16}$$

The static analysis problem is described by the following relations:

- *Stress equilibrium equations*:

$$\overline{\mathbf{G}}\bar{\mathbf{s}} = \begin{bmatrix} \mathbf{G} & \mathbf{G}_N & \mathbf{G}_T \end{bmatrix} \begin{bmatrix} \mathbf{s} \\ -\mathbf{S}_N \\ -\mathbf{S}_T \end{bmatrix} = \mathbf{p} \tag{3.17}$$

where $\mathbf{G}$ is the equilibrium matrix of the discretized structure and $\overline{\mathbf{G}}$ is the enlarged equilibrium matrix such as to take into account the interface tractions $\mathbf{S}_N$ and $\mathbf{S}_T$.

- *Strain–displacements compatibility equations*:

$$\bar{\mathbf{e}} = \overline{\mathbf{G}}^T\mathbf{u} \text{ or explicitly } \begin{bmatrix} \mathbf{e} \\ [\mathbf{u}]_N \\ [\mathbf{u}]_T \end{bmatrix} = \begin{bmatrix} \mathbf{G}^T \\ \mathbf{G}_N^T \\ \mathbf{G}_T^T \end{bmatrix} \mathbf{u}. \tag{3.18}$$

- *Linear material constitutive law* for the structure (outside of the interface):

$$\mathbf{e} = \mathbf{e}_0 + \mathbf{F}_0\mathbf{s}, \tag{3.19}$$

or

$$\mathbf{s} = \mathbf{K}_0(\mathbf{e} - \mathbf{e}_0). \tag{3.20}$$

Here $\mathbf{F}_0$ and $\mathbf{K}_0 = \mathbf{F}_0^{-1}$ are the natural and stiffness flexibility matrices of the unassembled structure and $\mathbf{e}_0$ is the initial deformation vector.

- *Monotone (possibly multivalued) interface laws* (decomposed normally and tangentially to the interface) in the general *subdifferential* form

$$-S_a \in \partial\phi_a([u]_a), \ a = N, T, \tag{3.21}$$

or

$$[u]_a \in \partial\overline{\phi}_a(-S_a), \ a = N, T. \tag{3.22}$$

Here $\phi_a(\cdot)$, $\overline{\phi}_a(\cdot)$, $a = N, T$ are convex, lower semicontinuous (l.s.c.) and proper potential energy functions which "produce" the pointwise interface laws. After integration, for the whole interface, we get the potentials:

$$\Phi_a(\mathbf{u}) = \sum_{i=1}^{q} \phi_a^{(i)}([\mathbf{u}]_a), \ a = N, T, \tag{3.23}$$

and

$$\overline{\Phi}_a(\mathbf{s}) = \sum_{i=1}^{q} \overline{\phi}_a^{(i)}(-\mathbf{S}_a), \; a = N, T. \tag{3.24}$$

For the convenience of the reader we recall here that a function $f(\mathbf{x}) : X \to (-\infty, +\infty]$ is called proper if it is not identically equal to the value $+\infty$, or, equivalently, if its effective domain (i.e., the arguments $\mathbf{x}$ where the function is not equal to $+\infty$) is nonempty.

Moreover, the function f is lower semicontinuous (l.s.c.) if for every $\lambda \in \mathbf{R}$ the set $\{\mathbf{x} \mid \mathbf{x} \in X \; : \; f(\mathbf{x}) \leq \lambda\}$ is closed.

Note here that the same interface law can be equivalently expressed by (3.21) or (3.22), where now $\overline{\phi} = \phi^c$ and ϕ^c is the convex conjugate function of $\phi(\cdot)$.

- Classical *equality boundary conditions* written in the general form:

$$\mathbf{E}\mathbf{u} = \mathbf{u}_0 \tag{3.25}$$

and

$$\mathbf{Z}\mathbf{s} = \mathbf{F} \tag{3.26}$$

where $\mathbf{E}$ and $\mathbf{Z}$ are appropriately defined transformation matrices and $\mathbf{u}_0$, $\mathbf{F}$ denote the known nodal boundary displacements (support) and boundary loading (traction).

For the variational formulations of the problem the *virtual work* equation and the *complementary virtual work* equation are also needed in their discretized form. According to the previous Section both are variational equalities. The virtual work equation involves variations of the displacements and reads:

$$\mathbf{s}^T(\mathbf{e}^* - \mathbf{e}) = \mathbf{p}^T(\mathbf{u}^* - \mathbf{u}) + \mathbf{S}_N^T([\mathbf{u}]_N^* - [\mathbf{u}]_N) + \mathbf{S}_T^T([\mathbf{u}]_T^* - [\mathbf{u}]_T),$$
$$\forall \mathbf{e}^*, \mathbf{u}^*, [\mathbf{u}]_N^*, [\mathbf{u}]_T^* \text{ s.t. } (3.18), (3.25) \text{ hold.} \tag{3.27}$$

The *complementary virtual work* involves variations of the stresses and reads:

$$\mathbf{e}^T(\mathbf{s}^* - \mathbf{s}) = \mathbf{u}^T(\mathbf{p}^* - \mathbf{p}) + [\mathbf{u}]_N^T(\mathbf{S}_N^* - \mathbf{S}_N) + [\mathbf{u}]_T^T(\mathbf{S}_T^* - \mathbf{S}_T),$$
$$\forall \mathbf{p}^*, \mathbf{s}^*, \mathbf{S}_N^*, \mathbf{S}_T^* \text{ s.t. } (3.17), (3.26) \text{ hold.} \tag{3.28}$$

Let us first formulate the previously described elastostatic analysis problem within the framework of the displacements (or direct) analysis method. To this end we first introduce the elasticity law (3.19) into the virtual work equation (3.27), and by using (3.18) we get:

$$\mathbf{u}^T\mathbf{G}\mathbf{K}_0^T\mathbf{G}^T(\mathbf{u}^* - \mathbf{u}) - (\mathbf{p} + \mathbf{G}\mathbf{K}_0\mathbf{e}_0)^T(\mathbf{u}^* - \mathbf{u}) = \mathbf{S}_N^T([\mathbf{u}]_N^* - [\mathbf{u}]_N) + \\ +\mathbf{S}_T^T([\mathbf{u}]_T^* - [\mathbf{u}]_T),\ \forall \mathbf{u}^* \in V_{ad} = \{\mathbf{u} \in \mathbf{R}^n \mid (3.18), (3.25) \text{ hold }\}. \quad (3.29)$$

Here $\mathbf{K} = \mathbf{G}\mathbf{K}_0\mathbf{G}^T$ denotes the stiffness matrix of the structure and $\overline{\mathbf{p}} = \mathbf{p} + \mathbf{G}\mathbf{K}_0\mathbf{e}_0$ denotes the nodal equivalent loading vector (including initial deformations' effects).

At this point we use the inequalities introduced by the interface relations (3.21), (3.23), i.e.

$$-\mathbf{S}_a([\mathbf{u}]_a^* - [\mathbf{u}]_a) \leq \Phi_a(\mathbf{u}^*) - \Phi_a(\mathbf{u}), a = N, T, \quad (3.30)$$

for $\Phi_a(\mathbf{u}^*) < \infty$. Thus from (3.27) we obtain the following *variational inequality* problem in terms of displacements:

Find kinematically admissible displacements $\mathbf{u} \in V_{ad} \cap K$ such that

$$\mathbf{u}^T\mathbf{K}(\mathbf{u}^* - \mathbf{u}) - \overline{\mathbf{p}}^T(\mathbf{u}^* - \mathbf{u}) + \Phi_N(\mathbf{u}^*) - \Phi_N(\mathbf{u}) + \Phi_T(\mathbf{u}^*) - \Phi_T(\mathbf{u}) \geq 0, \\ \forall \mathbf{u}^* \in V_{ad} \cap K \quad (3.31)$$

where $K = \{\mathbf{u}^* \text{ such that } \Phi_N(\mathbf{u}^*) < \infty \text{ and } \Phi_T(\mathbf{u}^*) < \infty\}$.

Note that the fact that $\mathbf{u}$ is sought in K and $\mathbf{u}^* \in K$ permits us to incorporate into this framework conditions where Φ_N or Φ_T is the indicator of a closed convex set as it is the case of the Signorini-Fichera boundary conditions which will be studied further.

For a stress based formulation (force method) we use the elasticity law in the form of (3.20) and the interface relations (3.22), (3.24). By analogous reasoning we get from (3.28) the complementary virtual work equality:

$$\mathbf{e}^T(\mathbf{s}^* - \mathbf{s}) + \mathbf{s}^T\mathbf{F}_0^T(\mathbf{s}^* - \mathbf{s}) = [\mathbf{u}]_N^T(\mathbf{S}_N^* - \mathbf{S}_N) + [\mathbf{u}]_T^T(\mathbf{S}_T^* - \mathbf{S}_T), \\ \forall \overline{\mathbf{s}}^* = [\mathbf{s}^*, -\mathbf{S}_N^*, -\mathbf{S}_T^*] \in \Sigma_{ad} = \{\mathbf{s} \in \mathbf{R}^n \mid (3.17), (3.26) \text{ hold }\}. \quad (3.32)$$

Finally we get the variational inequality:

Find static admissible stresses $\mathbf{s} \in \Sigma_{ad} \cap L$ such as to satisfy:

$$(\mathbf{e}_0 + \mathbf{s}\mathbf{F}_0)^T(\mathbf{s}^* - \mathbf{s}) + \overline{\Phi}_N(\mathbf{s}^*) - \overline{\Phi}_N(\mathbf{s}) + \overline{\Phi}_T(\mathbf{s}^*) - \overline{\Phi}_T(\mathbf{s}) \geq 0, \\ \forall \mathbf{s}^* \in \Sigma_{ad} \cap L \quad (3.33)$$

where $L = \left\{\mathbf{s}^* \text{ such that } \overline{\Phi}_N(\mathbf{s}^*) < \infty \text{ and } \overline{\Phi}_T(\mathbf{s}^*) < \infty\right\}$.

Note that due to the linearity of the equilibrium and the compatibility equations (3.17), (3.18) and due to the linearity of the material constitutive law (3.19), (3.20) the previous variational problems involve linear and bilinear forms of the unknown variables, apart from the nonlinear interface contributions. Accordingly, both the potential and complementary energy optimization problems involve linear and quadratic constituents and the contribution of the interface mechanisms. The two potential minimization problems read:

Find $\mathbf{u} \in V_{ad} \cap K$ such that:

$$\Pi(\mathbf{u}) = \min_{\mathbf{v} \in V_{ad} \cap K} \left\{ \Pi(\mathbf{v}) = \frac{1}{2}\mathbf{v}^T\mathbf{K}\mathbf{v} - \overline{\mathbf{p}}^T\mathbf{v} + \Phi_N(\mathbf{v}) + \Phi_T(\mathbf{v}) \right\}. \tag{3.34}$$

Find $\mathbf{s} \in \Sigma_{ad} \cap L$ such that:

$$\Pi^c(\mathbf{s}) = \min_{\mathbf{t} \in \Sigma_{ad} \cap L} \left\{ \Pi^c(\mathbf{t}) = \frac{1}{2}\mathbf{t}^T\mathbf{F}_0\mathbf{t} + \mathbf{e}_0^T\mathbf{t} + \overline{\Phi}_N(\mathbf{t}) + \overline{\Phi}_T(\mathbf{t}) \right\}. \tag{3.35}$$

One may easily verify that problem (3.31) (resp. (3.33)) actually expresses the minimality conditions for the energy optimization problem (3.34) (resp. (3.35)). Moreover, by means of the convex analysis duality theory it is possible to show that if relations (3.21) and (3.22) describe the same interface law, then $\overline{\phi}_a$ is the convex conjugate of ϕ_a and analogously, function Π is the convex conjugate of Π^c and vice versa.

The previously introduced framework will be followed for the formulation and the study of all small displacement structural analysis problems in this book. In this Chapter only convex problems are considered. Due to the convexity of the quadratic strain energy contribution in the potential energy and the linearity of the external loading contribution, convex problems require that the potentials of the interface laws are convex functions. Accordingly, only monotone, possibly multivalued relations of the type (3.21), (3.22) are considered in this Section. From them, simple inequality relations which describe the frictionless unilateral contact problem are introduced in the remaining of this Section. They lead to inequality constrained convex potential optimization problems. The case of frictional laws which introduce nondifferentiable terms in the potential energy is considered in the next Section.

The Unilateral Contact or Signorini-Fichera Conditions

The pointwise frictionless unilateral contact law reads (Fig. 3.1):

$$-S_N \geq 0, \ [u]_N - g \leq 0, \ -S_N([u]_N - g) = 0, \tag{3.36}$$

where g denotes an initial opening (gap) of the unilateral contact joint. The inequality constraints on the interface tractions (no tensile tractions are permitted), on the relative normal interface displacements (no interpenetration is allowed) and the complementarity relation between them are clearly depicted in Fig.3.1. Fig.3.1 depicts also other types of unilateral contact laws. For example, the solid lines in Figs 3.1a,c correspond to zero initial gap, the dashed lines correspond to the case of an initial interface gap g and the dotted line to a case where a prestressing is also applied on the interface with zero interface gap. Figs 3.1b,d give the superpotentials which correspond to the previous cases.

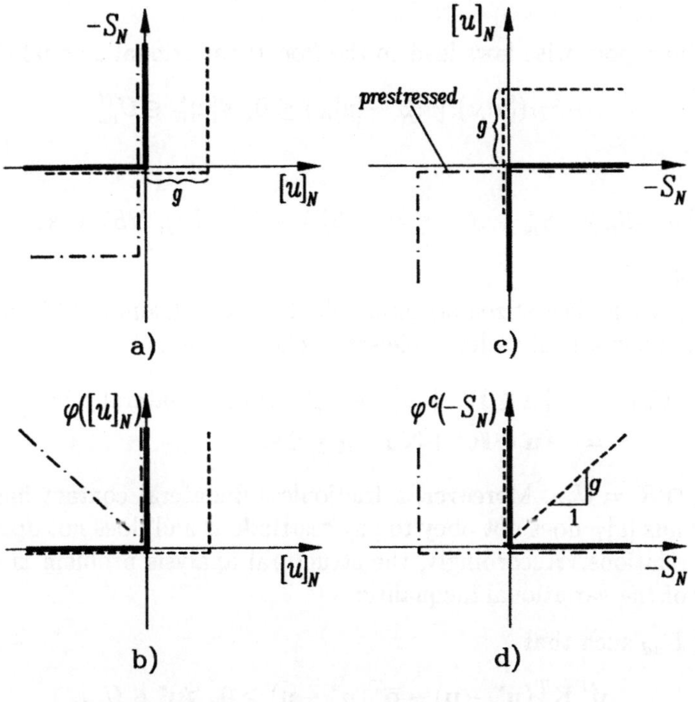

Figure 3.1. Various unilateral contact laws

By introducing the set:

$$U_{ad}^N = \{[u]_N \mid [u]_N - g \leq 0\} \tag{3.37}$$

and by using the notion of the indicator function and the normal cone from the convex analysis, the previous law (3.36) is written in the concise form:

$$-S_N \in \partial \mathbf{I}_{U_{ad}^N}([u]_N) = \mathcal{N}_{U_{ad}^N}([u]_N) \text{ or } -S_N \in \partial\phi([u]_N). \tag{3.38}$$

Equivalently, by using the notion of the support function $\sigma_{U_{ad}^N}$ of the convex set U_{ad}^N we get:

$$\begin{aligned}\phi^c(-S_N) &= \sup_{[u]_N \in \mathbb{R}} \left\{-([u]_N S_N) - \mathbf{I}_{U_{ad}^N}([u]_N)\right\} = \\ &= \sup_{[u]_N \in U_{ad}^N} \{-([u]_N S_N)\}\,. \end{aligned} \tag{3.39}$$

Accordingly, the unilateral contact law is also written as:

$$[u]_N \in \partial\phi^c(-S_N). \tag{3.40}$$

The above pointwise laws lead to the (local) variational inequalities

$$-S_N([u]_N)([u]_N^* - [u]_N) \leq 0, \ \forall [u]_N^* \in U_{ad}^N \tag{3.41}$$

and

$$[u]_N(-S_N)(-S_N^* + S_N) \leq \phi^c(-S_N^*) - \phi^c(-S_N), \ \forall S_N^* \in \mathbb{R} \tag{3.42}$$

respectively.

For the whole discretized structure the local constraints (3.37) are used to define the kinematically admissible set of displacements:

$$\begin{aligned} U_{ad} &= \{\mathbf{u} \in \mathbb{R}^n \mid [u]_N \in U_{ad}^N \text{ for all unilateral joints}\} \\ &= \{\mathbf{u} \in \mathbb{R}^n \mid \mathbf{N}\mathbf{u} - \mathbf{g} \leq \mathbf{0}\}\,. \end{aligned} \tag{3.43}$$

Here $V_{ad} \cap K = U_{ad}$. Moreover, a frictionless unilateral contact has been assumed. Thus $[u]_T$ does not obey to any restrictions and does not appear in the previous relations. Accordingly, the structural analysis problem is written in the form of the variational inequality:

Find $\mathbf{u} \in U_{ad}$ such that

$$\mathbf{u}^T\mathbf{K}^T(\mathbf{u}^* - \mathbf{u}) - \overline{\mathbf{p}}^T(\mathbf{u}^* - \mathbf{u}) \geq 0, \ \forall \mathbf{u}^* \in U_{ad}. \tag{3.44}$$

Classical bilateral support boundary conditions (cf. (3.25)) can be taken into account by changing U_{ad} by $V_{ad} \cap U_{ad}$ in (3.44).

The potential energy minimization problem (cf. (3.34)) is in this case a quadratic, linearly constrained optimization problem and reads:

Find $\mathbf{u} \in U_{ad}$ such that

$$\Pi(\mathbf{u}) = \min_{\mathbf{v} \in U_{ad}} \left\{\frac{1}{2}\mathbf{v}^T\mathbf{K}\mathbf{v} - \overline{\mathbf{p}}^T\mathbf{v}\right\}. \tag{3.45}$$

Furthermore, by following the general mathematical optimization theory outlined in Chapter 2, the Kuhn–Tucker optimality conditions for (3.45) lead after minor manipulations to the Linear Complementarity Problem (L.C.P.):

Find $\mathbf{u} \in \mathbb{R}^n$, $\mathbf{S}_N \in \mathbb{R}^q$ such that

$$\mathbf{K}\mathbf{u} + \mathbf{N}^T \mathbf{S}_N = \mathbf{0},\ \mathbf{N}\mathbf{u} \leq \mathbf{0},\ \mathbf{S}_N \leq \mathbf{0},\ \mathbf{S}_N^T(\mathbf{N}\mathbf{u}) = 0. \tag{3.46}$$

The previous relation constitutes a Linear Complementarity Problem, and in this form can directly be used for the numerical treatment of the problem (see, among others, Panagiotopoulos, 1985, Klarbring, 1986). The case of structures with rigid body displacements and rotations, which leads to semidefinite stiffness matrices $\mathbf{K}$ and thus leads to problems which are only solvable if the Fichera solvability conditions hold true, is treated in Panagiotopoulos, 1985, Stavroulakis et al., 1991, Alart, 1993, He et al., 1996, Pang and Ralph, 1996, Jarušek, 1994, Cheng et al., 1995 (for the general nonconvex case cf. Goeleven and Mentagui, 1995, Goeleven and Théra, 1995, Goeleven et al., 1997, Naniewicz, 1993). The transformation of the L.C.P. (3.46) such as to condense (eliminate) the free variables $\mathbf{u}$ has been discussed in Chapter 2 previously (cf. relation (2.22)).

Besides the general discussions on the solution of the arising mathematical programming problems (3.45), (3.46) given previously, and the discussions of Chapter 6, unilateral contact problems have been treated by the augmented Lagrangian techniques Glowinski et al., 1981, conjugate gradient methods with projection Dostál, 1992, Tamma et al., 1994, see also, among others, Curnier, 1992, Raous et al., 1995.

3.1.3 Friction problems with convex energy potentials

A simplified static Coulomb friction problem, first proposed by Duvaut and Lions, 1972, will be introduced first. Within this model the stick–slip relations of the frictional mechanism are expressed in terms of the toral (static) mechanical variables. Moreover, the (normal) contact mechanism is decoupled from the frictional one by assuming that the normal contact traction is given, i.e., $S_N = C_N$. The governing relations of the friction joint are (Fig. 3.2a):

$$-S_T = \begin{cases} -T_0 & \text{if} \quad [u]_T \leq 0 \\ [-T_0\ ,\ T_0] & \text{if} \quad [u]_T = 0. \\ T_0 & \text{if} \quad [u]_T \geq 0 \end{cases} \tag{3.47}$$

By using the nondifferentiable, friction potential (Fig. 3.2b):

$$\phi_T([u]_T) = T_0 |[u]_T|, \tag{3.48}$$

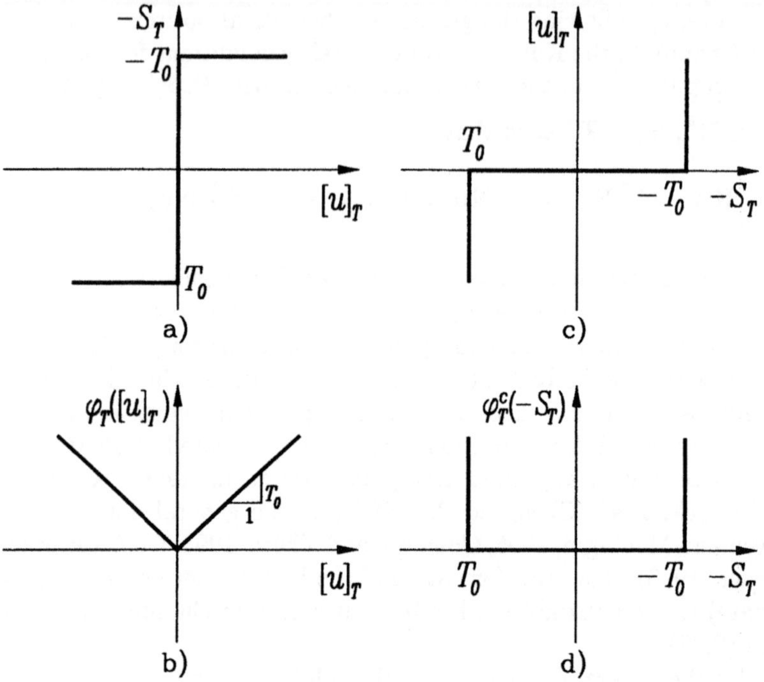

Figure 3.2. The Coulomb friction law and the corresponding superpotentials

where $|.|$ denotes the absolute value, the law can be written in the equivalent subdifferential form:

$$-S_T \in \partial\phi_T([u]_T). \tag{3.49}$$

The inverse to (3.47) relations read (Fig. 3.2c):

$$\begin{array}{llll} \text{If} & -S_T = -T_0 & \text{then} & [u]_T \leq 0, \\ \text{If} & -T_0 \leq -S_T \leq T_0 & \text{then} & [u]_T = 0, \\ \text{If} & -S_T = T_0 & \text{then} & [u]_T \geq 0. \end{array} \tag{3.50}$$

Defining the admissible tractions' set:

$$S_{ad} = \{S_T \mid -T_0 \leq -S_T \leq T_0\}, \tag{3.51}$$

we get the law (Fig. 3.2d):

$$[u]_T \in \partial\mathbf{I}_{S^T_{ad}}(-S_T) \text{ or } [u]_T \in \partial\phi_T^c(-S_T). \tag{3.52}$$

The local variational inequality formulation of the above laws read

$$-S_T([u]_T)([v]_T - [u]_T) \leq \phi_T([v]_T) - \phi_T([u]_T),\ \forall [u]_T \in \mathbb{R}, \tag{3.53}$$

and

$$[u]_T(-S_T)(-S_T^* + S_T) \leq 0,\ \forall S_T^* \in S_{ad}^T. \tag{3.54}$$

Moreover, we can write a variational inequality problem of the (3.31) type with

$$\Phi_T(\mathbf{u}) = \sum \phi_T([u]_T) \tag{3.55}$$

or, in terms of stresses (force method) the variational inequality:
Find $\mathbf{s} \in \Sigma_{ad}$ such that:

$$\mathbf{e}_0^T(\mathbf{s}^* - \mathbf{s}) + \mathbf{s}^T\mathbf{F}_0^T(\mathbf{s}^* - \mathbf{s}) \leq 0,$$
$$\forall \mathbf{s}^* \in \Sigma_{ad} = \{\mathbf{s} \in \mathbb{R}^n \mid (3.17), (3.26) \text{ and } (3.51) \text{ hold}\}. \tag{3.56}$$

Note that $\Phi_T(\mathbf{u}) < \infty$ is fulfilled for the friction boundary conditions.

Variational inequality problems for unilateral (frictionless or frictional) boundary conditions were among the first studied applications of inequality mechanics (see e.g. Duvaut and Lions, 1972, Panagiotopoulos, 1975, Panagiotopoulos, 1985, etc.). Moreover, for this specific case an important remark can be made. Variational inequality formulations concerning smooth potentials and inequality constrained sets of admissible variations can be formulated (see e.g. (3.44), (3.56)) by appropriately using convex duality theory. In the previous examples due to the relatively simple relations, duality merely means the appropriate choice of displacement or stress based formulations for the structural analysis problem.

The previously studied static friction law involves two simplifications which have allowed us to write the variational inequality formulation of the problem, or, equivalently, the potential and complementary energy minimization problems. Namely the dynamic nature of the frictional effects and the implicit connection between normal and tangential mechanical behaviour are not considered in the relations (3.47)-(3.52).

3.1.3.1 Combined frictional contact problem. Concerning the relation between normal and tangential mechanical behaviour one may formulate the following law (with a friction coefficient μ):

$$\begin{array}{llll} \text{If} & -S_T = -\mu|S_N| & \text{then} & [u]_T \leq 0, \\ \text{If} & -\mu|S_N| \leq -S_T \leq \mu|S_N| & \text{then} & [u]_T = 0, \\ \text{If} & -S_T = \mu|S_N| & \text{then} & [u]_T \geq 0. \end{array} \tag{3.57}$$

The coupling between normal and tangential mechanical behaviour is manifested by the dependence of (3.57) on the solution of the unilateral contact problem (through the contact stresses S_N). A friction potential can be written, and it is also a function of S_N:

$$\phi_T([u]_T, S_N) = \mu |S_N| |[u]_T|. \tag{3.58}$$

Analogously the set of admissible tractions reads:

$$S_{ad}(S_N) = \{S_T \mid -\mu|S_N| \leq -S_T \leq \mu|S_N|\}. \tag{3.59}$$

Due to the fact that S_N is a function of u which is not given, any formulation with respect to the displacements leads to implicit variational inequalities, which finally give rise to quasivariational inequality problems (Mosco, 1976, Baiocchi and Capelo, 1984, Telega, 1988, Zavarise et al., 1992, Bisbos, 1995, Outrata and Zowe, 1995) of the following type:

- For the displacement problem we get the implicit variational inequality:

  Find kinematically admissible displacements $\mathbf{u} \in V_{ad} \cap U_{ad}$ such that

$$\mathbf{u}^T \mathbf{K}^T (\mathbf{u}^* - \mathbf{u}) - \overline{\mathbf{p}}^T (\mathbf{u}^* - \mathbf{u}) + \Phi_T(\mathbf{u}^*, \mathbf{S}_N) - \Phi_T(\mathbf{u}, \mathbf{S}_N) \geq 0, \quad \forall \mathbf{u}^* \in V_{ad} \cap U_{ad}. \tag{3.60}$$

- In terms of stresses we get the quasivariational inequality:
  Find $\mathbf{s} \in \Sigma_{ad}(S_N)$ such that:

$$\mathbf{e}_0^T (\mathbf{s}^* - \mathbf{s}) + \mathbf{s}^T \mathbf{F}_0^T (\mathbf{s}^* - \mathbf{s}) \leq 0, \quad \forall \mathbf{s}^* \in \Sigma_{ad}(S_N) = \{\mathbf{s} \in \mathbb{R}^n \mid (3.17), (3.26) \text{ and } (3.59) \text{ hold}\}. \tag{3.61}$$

By using an iterative solution strategy, analogous to the ones proposed in Chapter 6, the above quasivariational inequalities can be approximated by a series of variational inequalities. For the latter subproblems the whole methodology introduced previously can be used. This strategy has first been proposed in Panagiotopoulos, 1975 for the unilateral contact problem with friction, and is now accepted as a powerful solution method for the study of frictional contact problems. This technique was named by Kalker, PANA-algorithm (Kalker, 1988, Kalker, 1990).

3.1.3.2 Direct L.C.P. formulation of the unilateral frictional contact problem. Note here that a direct nonsymmetric L.C.P. formulation of the frictional unilateral contact problem is possible. We follow here the formulation of Kwack and Lee, 1988, Klarbring and Björkman, 1988 (see also Al-Fahed et al., 1991 for the case of three-dimensional problems, where the friction cone is linearized by means of a convex polyhedron). Let the normal forces be assembled in vector $\mathbf{S}_N = \{S_{N1}, \ldots, S_{Nn}\}^T$ (the same with the vector used in the previous Section for the frictionless case). The friction forces are assembled in vector $\mathbf{S}_T$ where

$$\mathbf{S}_T = \{S_{T11}, S_{T12}, S_{T21}, S_{T22}, \ldots, S_{Tn1}, S_{Tn2}\}^T.$$

Coulomb's law of dry friction connects the tangential (frictional) forces with the normal (contact) forces by the relation

$$\gamma_i = \mu|S_{Ni}| - |S_{Ti}|, \; i = 1, \ldots, n, \; \gamma_i \geq 0. \tag{3.62}$$

Here $|*|$ denotes the norm in $\mathbb{R}^3$, μ is the friction coefficient (anisotropic friction may also be considered). The friction mechanism is considered to work in the following way: If $|S_{Ti}| < \mu|S_{Ni}|$ (i.e. $\gamma_i > 0$) the slipping value γ_{iT} must be equal to zero and if $|S_{Ti}| = \mu|S_{Ni}|$ (i.e. $\gamma_i = 0$) then we have slipping in the opposite direction of S_{Ti}. Explicitly we have:

$$\begin{array}{lll} \text{If} \;\; \gamma_i > 0 & \text{then} & y_{Ti} = 0, \\ \text{If} \;\; \gamma_i = 0 & \text{then} & \text{there exists } \sigma > 0 \text{ such that } y_{Ti} = -\sigma S_{Ti}. \end{array} \tag{3.63}$$

In order to achieve a L.C.P. formulation of the above-described frictional contact mechanism we introduce a piecewise linearization of the friction law (3.63) by a polyhedral approximation of the friction cone from the interior. In this piecewise linear approximation, we can write that:

$$\boldsymbol{\gamma} = \mathbf{T}_N^T \mathbf{S}_N + \mathbf{T}_T^T \mathbf{S}_T \tag{3.64}$$

where the matrices $\mathbf{T}_T$ and $\mathbf{T}_N$ of the linearised friction law have the form:

$$\mathbf{T}_T = \text{diag}\left[\mathbf{T}_T^1, \mathbf{T}_T^2, \ldots, \mathbf{T}_T^n\right], \; \mathbf{T}_N = \text{diag}\left[\mathbf{T}_N^1, \mathbf{T}_N^2, \ldots, \mathbf{T}_N^n\right] \tag{3.65}$$

and the submatrices of (3.65) are constructed from the adopted linearization of the friction cone. By taking appropriate projections, the slip value in (3.62), (3.63) is written in the form:

$$\mathbf{y}_T = \mathbf{T}_T \boldsymbol{\lambda}, \; \boldsymbol{\lambda} \geq 0 \tag{3.66}$$

where $\boldsymbol{\lambda}$ is a vector of nonnegative slipping parameters. Then $\boldsymbol{\gamma}$ and $\boldsymbol{\lambda}$ fulfill the following orthogonality condition:

$$\boldsymbol{\gamma}^T \boldsymbol{\lambda} = 0. \tag{3.67}$$

Note that the slipping value $\boldsymbol{\lambda}$ and the tangential displacements $\mathbf{u}_T$, are related by the compatibility relation:

$$\mathbf{T}_T\boldsymbol{\lambda} - \mathbf{u}_T = \mathbf{d}_T \tag{3.68}$$

where $\mathbf{d}_T$ denotes the initial distance in the tangential sense.

A linear elastic behaviour of the structure is assumed now, which, on the assumption that everything outside of the frictional contact interfaces has been condensed out (elimination of the internal d.o.f's), reads:

$$\tilde{\mathbf{u}} = \tilde{\mathbf{F}}\tilde{\mathbf{S}} \tag{3.69}$$

where

$$\tilde{\mathbf{u}} = \begin{bmatrix} \mathbf{u}_N \\ \mathbf{u}_T \end{bmatrix}, \ \tilde{\mathbf{F}} = \begin{bmatrix} \mathbf{F}_{NN} & \mathbf{F}_{NT} \\ \mathbf{F}_{TN} & \mathbf{F}_{TT} \end{bmatrix}, \ \tilde{\mathbf{S}} = \begin{bmatrix} \mathbf{S}_N \\ \mathbf{S}_T \end{bmatrix}. \tag{3.70}$$

Here $\tilde{\mathbf{F}}$ is the symmetric flexibility matrix, $\mathbf{F}_{NN}$ is an $n \times n$ nonsingular matrix with the mechanical meaning of the normal flexibility matrix, $\mathbf{F}_{TT}$ is a $2n \times 2n$ nonsingular matrix (the tangential flexibility) and $\mathbf{F}_{NT}$, $\mathbf{F}_{TN}$ are the corresponding couple flexibility matrices.

Recall here that the unilateral contact relations in the here adopted notation read:

$$\mathbf{y} \geq \mathbf{0}, \ \ \mathbf{S}_N \geq \mathbf{0}, \ \ \mathbf{y}^T\mathbf{S}_N = 0. \tag{3.71}$$

Using the previous relations, the unilateral kinematic relations normally and tangentially to the interface take the form:

$$\begin{aligned} \mathbf{y}_N - \mathbf{F}_{NN}\mathbf{S}_N - \mathbf{F}_{NT}\mathbf{S}_T &= \mathbf{d}_N, \\ \mathbf{T}_T\boldsymbol{\lambda} - \mathbf{F}_{TN}\mathbf{S}_N - \mathbf{F}_{TT}\mathbf{S}_T &= \mathbf{d}_T. \end{aligned} \tag{3.72}$$

A standard L.C.P. formulation is derived by means of the following change of variables. First, from the second relation in (3.72), $\mathbf{S}_T$ is expressed as follows:

$$\mathbf{S}_T = -\mathbf{F}_{TT}^{-1}\mathbf{F}_{TN}\mathbf{S}_N + \mathbf{F}_{TT}^{-1}\mathbf{T}_T\boldsymbol{\lambda} - \mathbf{F}_{TT}^{-1}\mathbf{d}_T. \tag{3.73}$$

Then by eliminating $\mathbf{S}_T$ from equations (3.72), we obtain

$$\begin{aligned} &\mathbf{y}_N - \left(\mathbf{F}_{NN} - \mathbf{F}_{NT}\mathbf{F}_{TT}^{-1}\mathbf{F}_{TN}\right)\mathbf{S}_N - \mathbf{F}_{NT}\mathbf{F}_{TT}^{-1}\mathbf{T}_T\boldsymbol{\lambda} \\ &= \mathbf{d}_N - \mathbf{F}_{NT}\mathbf{F}_{TT}^{-1}\mathbf{d}_T\boldsymbol{\gamma} + \left(\mathbf{T}_T^T\mathbf{F}_{TT}^{-1}\mathbf{F}_{TN} - \mathbf{T}_N^T\right)\mathbf{S}_N + \\ &\quad -\mathbf{T}_T^T\mathbf{F}_{TT}^{-1}\mathbf{T}_T\boldsymbol{\lambda} = -\mathbf{T}_T^T\mathbf{F}_{TT}^{-1}\mathbf{d}_T. \end{aligned} \tag{3.74}$$

Finally a standard L.C.P. is obtained from equations (3.74):

$$\begin{gathered} \mathbf{w} - \mathbf{M}\mathbf{z} = \mathbf{b} \\ \mathbf{w} \geq \mathbf{0}, \ \mathbf{z} \geq \mathbf{0}, \ \mathbf{w}^T\mathbf{z} = 0 \end{gathered} \tag{3.75}$$

with

$$\mathbf{w} = \begin{bmatrix} \mathbf{y}_N \\ \gamma \end{bmatrix}, \; \mathbf{z} = \begin{bmatrix} \mathbf{S}_N \\ \lambda \end{bmatrix}, \; \mathbf{b} = \begin{bmatrix} \mathbf{d}_N - \mathbf{F}_{NT}\mathbf{F}_{TT}^{-1}\mathbf{d}_T \\ -\mathbf{T}_T^T\mathbf{F}_{TT}^{-1}\mathbf{d}_T \end{bmatrix},$$

$$\mathbf{M} = \begin{bmatrix} (\mathbf{F}_{NN} - \mathbf{F}_{NT}\mathbf{F}_{TT}^{-1}\mathbf{F}_{TN}) & \mathbf{F}_{NT}\mathbf{F}_{TT}^{-1}\mathbf{T}_T \\ -(\mathbf{T}_T^T\mathbf{F}_{TT}^{-1}\mathbf{F}_{TN} - \mathbf{T}_N^T) & \mathbf{T}_T^T\mathbf{F}_{TT}^{-1}\mathbf{T}_T \end{bmatrix}.$$

L.C.P. (3.75) is defined on $\mathbb{R}^{n+m_2}$ (n is the number of discrete unilateral joints, $m_2 = n \times l$ and l is the number of faces in the polyhedral approximation of the friction cone). Note that here we get a nonsymmetric L.C.P., due to the nonsymmetric (but positive semidefinite) matrix $\mathbf{M}$ in (3.75) (a well-known fact from analogous cases of nonassociated laws in friction, plasticity etc., see e.g. Panagiotopoulos, 1985, Panagiotopoulos, 1993).

Bibliographical Remarks

The L.C.P. formulation to the above unilateral contact and friction problems has been used in various works. For numerical solution methods based on this formulation see, among others, Al-Fahed et al., 1991, Stavroulakis et al., 1991, Liolios, 1986. The flexibility matrices in (3.75) can be formulated by either the substructure technique of the Finite Element Method (e.g., Panagiotopoulos, 1985, Stavroulakis et al., 1991) or directly by the Boundary Element Method (e.g, Antes and Panagiotopoulos, 1992). In some cases equivalent quadratic programming problems have been formulated and solved (see, e.g., Lee, 1994b, Baniotopoulos et al., 1994, Refaat and Meguid, 1994, Simunovič and Saigal, 1995, Tzaferopoulos, 1993, Refaat and Meguid, 1996).

In view of the complexity of the inequality constraints and the complementarity relations, augmented Lagrangian techniques or mixed penalty Lagrangian techniques have also been used for the treatment of unilateral frictional problems (see the discussion on Lagrangian techniques in optimization in Chapter 2 and, among others, Curnier, 1984, Oden and Kikuchi, 1988, Simo et al., 1985, Alart and Curnier, 1991, DelPiero and Maceri, 1991, Klarbring, 1992, Simo and Laursen, 1992, Perić and Owen, 1992, Cescotto and Charlier, 1993, Laursen and Oancea, 1994, Cescotto and Charlier, 1994, Bille et al., 1995). See also the review article Zhong and Mackerle, 1992.

It should be mentioned here that the existence of a solution for a unilateral contact problem with Coulomb friction has been proved only on the assumption that the friction coefficient is "small" enough (see, among others, Nečas et al., 1980, Jarušek, 1983, Jarušek, 1984, Cocu, 1984, Panagiotopoulos, 1985, Martins and Oden, 1987, Klarbring et al., 1991a, Klarbring et al., 1991b, Doudoumis et al., 1995, Telega, 1995).

For various models of anisotropic friction effects the reader is referred to Michalowski and Mróz, 1978, Zmitrowicz, 1989, Lee et al., 1994 among others.

Note here that some friction effects can be modelled by appropriate modification of plasticity laws, which are reviewed later on in this book. Moreover, all results cited in the next Chapter for large displacement and deformation problems can also be used in this case as well.

3.1.3.3 Dynamic friction problem. The dynamic nature of the problem is captured by writing the friction law in terms of velocities, i.e.

$$\begin{array}{llll} \text{If} & -S_T = -\mu|S_N| & \text{then} & [\dot{u}]_T \leq 0, \\ \text{If} & -\mu|S_N| \leq -S_T \leq \mu|S_N| & \text{then} & [\dot{u}]_T = 0, \\ \text{If} & -S_T = \mu|S_N| & \text{then} & [\dot{u}]_T \geq 0. \end{array} \tag{3.76}$$

After time discretization by using implicit or explicit time marshing schemes, the above relation can be transformed to stepwise, static–like relations which again can be treated by the techniques introduced previously.

More details on dynamical frictional contact problems can be found in Oden and Pires, 1983, Oden and Martins, 1985, Lötstedt, 1981, Mitsopoulou, 1983, Chaudhary and Bathe, 1986, Moreau, 1988, Klarbring, 1990, Marques, 1992, Glocker and Pfeiffer, 1992, Taylor and Papadopoulos, 1993, Zhong, 1993, Zhong and Mackerle, 1994, Lee, 1994a, Cocu et al., 1995, Pang and Trinkle, 1996, Pfeiffer and Glocker, 1996, Pfeiffer, 1996a, Pfeiffer, 1996b, Andrews et al., 1996, Fritz and Pfeiffer, 1997, Trinkle et al., 1997. Some material on time discretization techniques is also given later in the discussion of elastoplasticity problems.

Static and especially dynamic friction problems have significant importance for engineering mechanics applications (see, among others, the recent contributions Pfeiffer, 1991, Glocker and Pfeiffer, 1995, Pfeiffer, 1992, Srnik and Pfeiffer, 1996, Pfeiffer, 1996b, Pfeiffer, 1997, Beitelschmidtand and Pfeiffer, 1997, Wösle and Pfeiffer, 1997).

3.1.3.4 General monotone material laws. It should be mentioned here that the previously presented approach can be easily extended to cover general monotone material constitutive laws of the subdifferential type (Panagiotopoulos, 1985). In this case, a nonlinear hyperelastic material law is considered which may have complete vertical branches. This law is produced through generalized differentiation, by means of the subdifferential operator of convex analysis, from a convex, possibly nondifferentiable in the classical sense, but subdifferentiable strain energy density function. It is l.s.c. and proper. These laws are written in the following general form:

$$\mathbf{s}_i \in \partial w_i(\mathbf{e}_i), \text{ or } \mathbf{e}_i \in \partial \overline{w}_i(\mathbf{s}_i), \; i = 1, \ldots, m, \tag{3.77}$$

where $w_i(\cdot)$ and $\overline{w}_i(\cdot)$ are appropriately defined convex and possibly nondifferentiable potentials.

Let us outline here the formulation of the problem by assuming the general framework of the previous Section and the general subdifferential interface laws introduced therein.

From (3.77a,b) one gets the relations:

$$\mathbf{s}_i^T(\mathbf{e}_i^* - \mathbf{e}_i) \leq w_i(\mathbf{e}_i^*) - w_i(\mathbf{e}_i),\ i = 1, \dots, m \tag{3.78}$$

and

$$\mathbf{e}_i^T(\mathbf{s}_i^* - \mathbf{s}_i) \leq \overline{w}_i(\mathbf{s}_i^*) - \overline{w}_i(\mathbf{s}_i),\ i = 1, \dots, m. \tag{3.79}$$

In the following we denote by K_1 the set

$$K_1 = \{\mathbf{u}^* \mid w_i(\mathbf{e}_i^*) < \infty, i = 1, 2, \dots, m\} \tag{3.80}$$

and by L_1 the set

$$L_1 = \{\mathbf{s}^* \mid \overline{w}_i(\mathbf{s}_i^*) < \infty, i = 1, 2, \dots, m\}. \tag{3.81}$$

By using the variational inequalities (3.78), (3.79) and the discrete virtual work (resp. complementary virtual work) equation (3.27) (resp. (3.28)), we formulate the following variational inequality problems:

Find $\mathbf{u} \in \mathbf{V}_{ad} \cap K \cap K_1$ such that:

$$W(\mathbf{e}^*) - W(\mathbf{e}) + \Phi_N(\mathbf{u}^*) - \Phi_N(\mathbf{u}) + \Phi_T(\mathbf{u}^*) - \Phi_T(\mathbf{u}) - \overline{\mathbf{p}}^T(\mathbf{u}^* - \mathbf{u}) \geq 0,$$
$$\forall \mathbf{u}^* \in V_{ad} \cap K \cap K_1, \mathbf{e}^* \text{ s.t. (3.18) is satisfied.} \tag{3.82}$$

Find $\mathbf{s} \in \Sigma_{ad} \cap L \cap L_1$ such that:

$$\overline{W}(\mathbf{s}^*) - \overline{W}(\mathbf{s}) + \overline{\Phi}_N(\mathbf{s}^*) - \overline{\Phi}_N(\mathbf{s}) + \overline{\Phi}_T(\mathbf{s}^*) - \overline{\Phi}_T(\mathbf{s}) \geq 0, \tag{3.83}$$
$$\forall \mathbf{s}^* \in \Sigma_{ad} \cap L \cap L_1. \tag{3.84}$$

Here W (resp. $\overline{W}$) is the sum of the local potentials w_i (resp. $\overline{w}_i$) over all finite elements and K, L are the sets defined in Section 3.1.2.

Equivalent potential energy and complementary potential energy, convex minimization problems can be written.

Since the arising problems have certain similarities with the stepwise holonomic problems which arise in the theory of elastoplasticity, the details about their treatment will be given in the next Section.

3.1.4 Uniaxial holonomic elastic plastic relations

In the following we present some examples of uniaxial holonomic elastic plastic laws. This kind of uniaxial material laws appears in several engineering problems and its theoretical and numerical treatment is of great importance.

3.1.4.1 Elastic perfectly plastic spring. The uniaxial mechanical behaviour of a holonomic elastic, perfectly plastic one–dimensional element reads (Fig. 3.3a):

$$\begin{aligned} \varepsilon &= \frac{\sigma}{E} && \text{for } \sigma < \sigma_0 \\ \varepsilon &= \frac{\sigma_0}{E} + \lambda && \text{for } \sigma = \sigma_0,\ \lambda \geq 0. \end{aligned} \tag{3.85}$$

Relation (3.85) is written in compact form

$$\varepsilon \in \partial\phi^c(\sigma), \tag{3.86}$$

by means of the convex superpotential (Fig. 3.3b)

$$\phi^c(\sigma) = \frac{1}{2}E^{-1}\sigma^2 + \mathbf{I}_{(-\infty,\sigma_0]}(\sigma). \tag{3.87}$$

The inverse to (3.85) relation reads (Fig. 3.3c):

$$\begin{aligned} \sigma &= E\varepsilon \text{ for } \varepsilon < \varepsilon_0 = \frac{\sigma_0}{E} \\ \sigma &= E\varepsilon_0 = \sigma_0 \text{ for } \varepsilon > \varepsilon_0. \end{aligned} \tag{3.88}$$

Relation (3.88) is written in the potential form:

$$\sigma \in \partial\phi(\varepsilon) = \nabla\phi(\varepsilon), \tag{3.89}$$

by means of the smooth and convex (not strictly) potential (Fig. 3.3d):

$$\phi(\varepsilon) = \frac{1}{2}E\varepsilon^2 - \frac{1}{2}E(\varepsilon - \varepsilon_0)_+^2. \tag{3.90}$$

One may easily show the relation between the convex, conjugate potentials $\phi(\varepsilon)$ and $\phi^c(\sigma)$, i.e.

$$\phi(\varepsilon) = \max_{\sigma}\{\sigma\varepsilon - \phi^c(\sigma)\} \tag{3.91}$$

and

$$\phi^c(\sigma) = \max_{\varepsilon}\{\sigma\varepsilon - \phi(\varepsilon)\}. \tag{3.92}$$

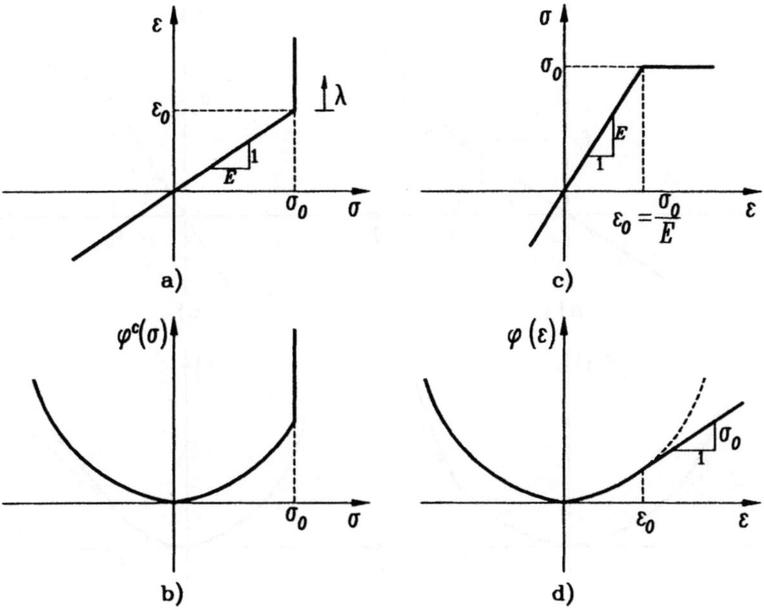

Figure 3.3. Elastic perfectly plastic spring and the corresponding superpotentials

3.1.4.2 Elastic linear hardening spring. The uniaxial mechanical behaviour of an elastic linear hardening spring, as a generalization of the one for the holonomic elastic, perfectly plastic element (3.85), reads (Fig. 3.4a):

$$\begin{aligned} \varepsilon &= \frac{\sigma}{E} && \text{for } \sigma \leq \sigma_0 \\ \varepsilon &= \frac{\sigma_0}{E} + \frac{\sigma - \sigma_0}{E_t} && \text{for } \sigma > \sigma_0. \end{aligned} \tag{3.93}$$

In compact form relation (3.93) reads:

$$\varepsilon \in \partial \phi^c(\sigma), \tag{3.94}$$

by means of the convex, differentiable superpotential (Fig. 3.4b)

$$\begin{aligned} \phi^c(\sigma) &= \begin{cases} \frac{1}{2}E^{-1}\sigma^2 & \text{for } \sigma < \sigma_0, \\ \dfrac{\sigma_0}{E}\sigma + \dfrac{1}{2}E_t^{-1}(\sigma - \sigma_0)^2, & \text{for } \sigma > \sigma_0 \end{cases} \\ &= \frac{1}{2}E^{-1}\sigma^2 + \frac{1}{2}(E_t^{-1} - E^{-1})(\sigma - \sigma_0)_+^2. \end{aligned} \tag{3.95}$$

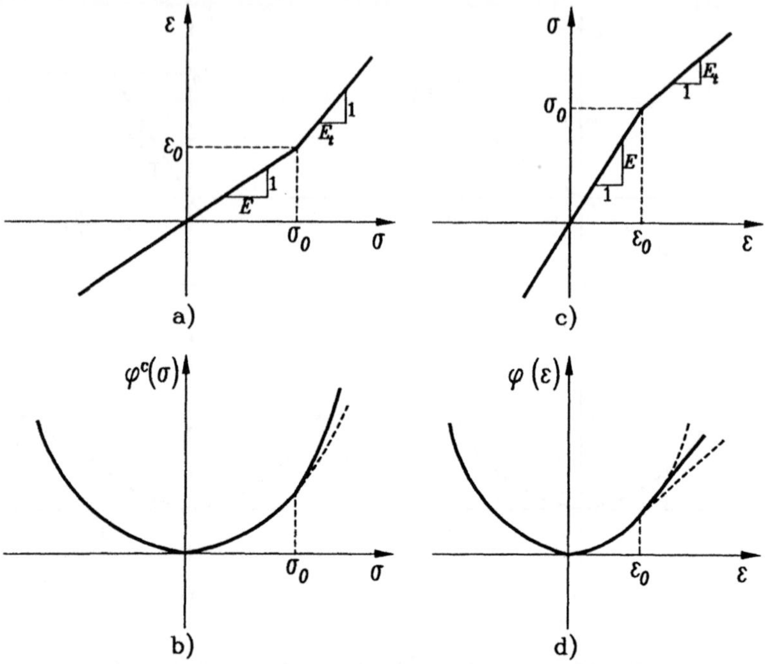

Figure 3.4. Elastic linear hardening spring and the corresponding superpotentials

The inverse to (3.93) relation reads (Fig. 3.4c)

$$\begin{aligned} \sigma &= E\varepsilon \quad \text{for } \varepsilon \leq \varepsilon_0 \\ \sigma &= \sigma_0 + E_t(\varepsilon - \varepsilon_0), \quad \text{for } \varepsilon \geq \varepsilon_0. \end{aligned} \tag{3.96}$$

Relation (3.96) is written in the potential form:

$$\sigma \in \partial\phi(\varepsilon), \tag{3.97}$$

by means of the smooth and convex (not strictly) potential (Fig. 3.4d):

$$\begin{aligned} \phi(\varepsilon) &= \begin{cases} \frac{1}{2}E\varepsilon^2 & \text{for } \varepsilon \leq \varepsilon_0, \\ \sigma_0\varepsilon + \frac{1}{2}E_t(\varepsilon - \varepsilon_0)^2, & \text{for } \varepsilon \geq \varepsilon_0 \end{cases} \\ &= \frac{1}{2}E\varepsilon^2 + \frac{1}{2}(E_t - E)(\varepsilon - \varepsilon_0)_+^2. \end{aligned} \tag{3.98}$$

3.1.4.3 Elastic locking spring. The uniaxial mechanical behaviour of an elastic locking spring, reads (Fig. 3.5a):

$$\varepsilon = \frac{\sigma}{E} \qquad \text{for } \sigma_2 \leq \sigma \leq \sigma_1$$

$$\varepsilon = \varepsilon_1 \qquad \text{for } \sigma \geq \sigma_1,$$
$$\varepsilon = \varepsilon_2 \qquad \text{for } \sigma \leq \sigma_2, \tag{3.99}$$

where

$$\sigma_1 = E\varepsilon_1, \ \sigma_2 = E\varepsilon_2.$$

In compact form relation (3.99) reads:

$$\varepsilon \in \partial\phi^c(\sigma), \tag{3.100}$$

by means of the convex, differentiable superpotential (Fig. 3.5b):

$$\phi^c(\sigma) = \frac{1}{2}E^{-1}\sigma^2 - \frac{1}{2}E^{-1}(\sigma - \sigma_1)_+^2 - \frac{1}{2}E^{-1}(\sigma_2 - \sigma)_+^2. \tag{3.101}$$

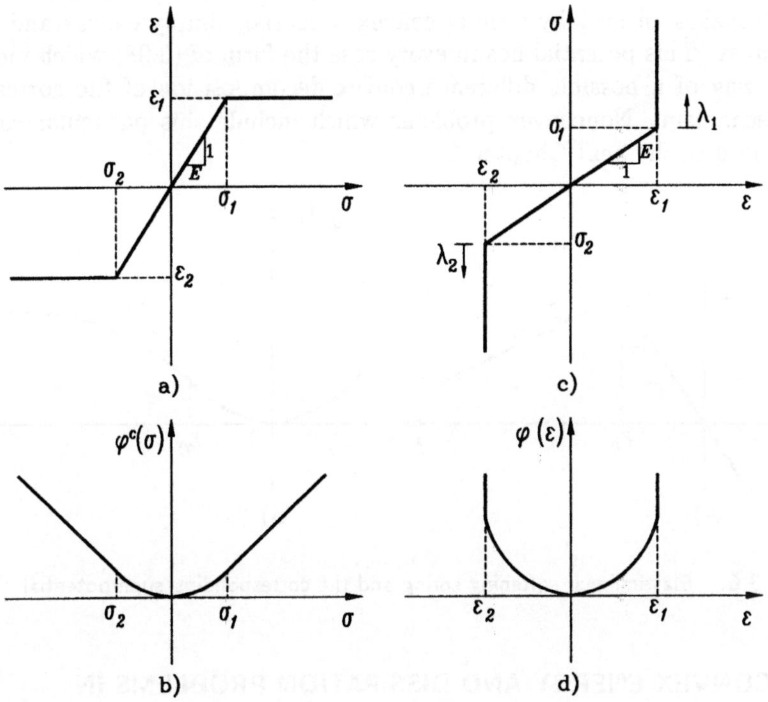

Figure 3.5. Elastic locking spring and the corresponding superpotentials

The inverse to (3.99) relation has the form (Fig. 3.5c):

$$\sigma = E\varepsilon \qquad \text{for } \varepsilon_2 \leq \varepsilon \leq \varepsilon_1,$$
$$\sigma = \sigma_1 + \lambda_1, \quad \text{for } \varepsilon = \varepsilon_1, \ \lambda_1 \geq 0,$$
$$\sigma = \sigma_2 - \lambda_2, \quad \text{for } \varepsilon = \varepsilon_2, \ \lambda_2 \geq 0. \tag{3.102}$$

Relation (3.102) is written in the potential form:

$$\sigma \in \partial\phi(\varepsilon), \tag{3.103}$$

by means of the convex potential (Fig. 3.5d):

$$\phi(\varepsilon) = \frac{1}{2}E\varepsilon^2 + \mathbf{I}_{[\varepsilon_2,\varepsilon_1]}(\varepsilon). \tag{3.104}$$

3.1.4.4 Elastic linear softening spring. An elastic linear softening spring and its corresponding superpotential is depicted in Fig. 3.6. We observe that in this case relation (3.96) holds true with $E_t < 0$. The transition from $E_t > 0$ (hardening behaviour) to $E_t = 0$ (perfect plastic behaviour) and finally to $E_t < 0$ makes an initially strictly convex potential, simply convex and finally nonconvex. This potential has in every case the form of (3.98) which indicates also a way of a possible difference convex decomposition of the corresponding mechanism. Nonconvex problems which include this particular case are introduced in the next Chapter.

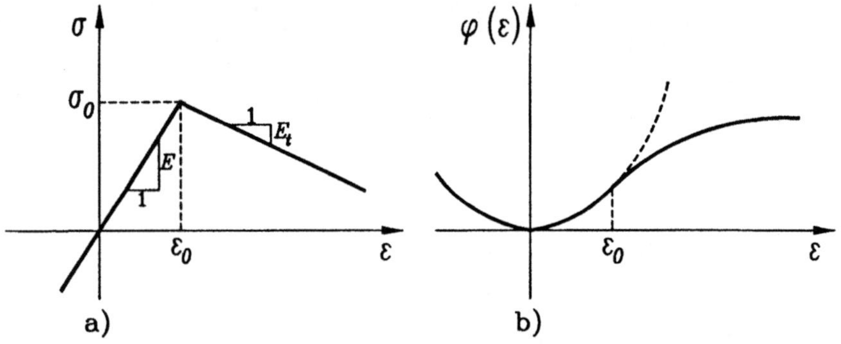

Figure 3.6. Elastic linear softening spring and the corresponding superpotential

3.2 CONVEX ENERGY AND DISSIPATION PROBLEMS IN STANDARD GENERALIZED ELASTOPLASTICITY

In this Section variational inequality problems are formulated and studied for convex, standard generalized elastoplasticity. Methods of convex analysis and finite element discretization techniques are used, by following the pioneering approach of J.J. Moreau and G. Maier (cf. Moreau, 1968, Moreau, 1974, Moreau, 1975, Cohn and Maier, 1979, Maier, 1970). In particular, the standard generalized elastoplasticity models provide the unified framework that includes various classical plasticity theories and linear and nonlinear hardening and softening laws. The constitutive laws which include the plastic evolution relations are in turn discretized in time to produce step-wise holonomic problems. The latter problems are actually problems analogous to the deformation theory of plasticity, which in fact are equivalent to nonlinear and possibly nonsmooth elastostatic problems (i.e., without the time derivatives of the strains etc.), if the internal variables (e.g. the plastic strains and stresses) are appropriately interpreted as initial strains or stresses. Thus, nonsmooth optimization problems arise, which give a concrete interpretation for existing elastoplastic structural analysis algorithms, permit the derivation of new ones and allow for convergence and error estimate analysis. This Section is restricted to the treatment of convex problems; nonconvex generalizations are considered in Chapter 4.

In the classical, stress–based formulation of plasticity, the stresses are assumed to lie within a convex closed set, which, in turn, is defined by a smooth yield surface in the stress space. If this surface is attained (i.e., the constraint is satisfied as an equality) an additional plastic strain arises (cf., the Lagrangian multiplier of an inequality constrained optimization problem). In the framework of small strains and small displacements theory and for certain (linear) hardening laws, the problem is described by linear equalities, inequalities and complementarity conditions (the loading - unloading condition of plasticity). As expected, quadratic optimization problems arise in this case. Certain formulations of this type can be extended to cover nonlinear hardening and softening problems, which lead to nonquadratic, general optimization problems. A yield surface with corners can be described by a finite number of smooth, intersecting yield surfaces. This is the case of the Tresca criterion, of the multi–surface plasticity models proposed, among others, for the description of polycrystalline bodies, or for certain cup models of concrete or soils. This approach has been initiated by Moreau, 1963, Moreau, 1966, Moreau, 1968, Moreau, 1974, Moreau, 1975, Maier, 1970a and has been followed by a number of researchers Comi et al., 1992, Eve et al., 1990, Panagiotopoulos, 1985, Simo et al., 1988, Simo, 1993, Lubliner, 1990.

In addition, internal variable formulations of elastoplasticity has provided a sound basis for the development of the mathematical theory of plasticity and the link with mathematical programming and optimization techniques (Germain, 1973, Maier, 1970a, Halphen and Son, 1975, Martin, 1975, Martin et al., 1987, Martin and Nappi, 1990). The technique of introducing internal (or hidden) variables in a mechanical model, which count for the (from the past time or loading history) nonlinear response, can be traced back to Ziegler, 1963. The introduction of a pseudopotential allows for the automatic satisfaction of the thermodynamical principles. Convex analysis techniques and the introduction of a couple of convex conjugate pseudo-potentials in plasticity are attributed to Moreau and for the discretized problems especially in connection with linearized models and quadratic optimization techniques to Maier (for more details the reader is referred to Lemaitre and Chaboche, 1985, Lubliner, 1990, Ladèveze, 1995 among others).

In the sequel, after a short introduction to elastoplasticity for holonomic and rate problems, model elastoplasticity problems are briefly introduced. Time discretization leads to a series of stepwise holonomic elastoplastic analysis problems, which can be formulated as mathematical optimization problems. This approach permits us to apply well-tested optimization algorithms for the treatment of the elastoplastic analysis problem and at the same time provides a framework for the study of convergence and error approximation questions (see, e.g., Comi et al., 1992, Martin and Caddemi, 1994, Simo et al., 1988, Reddy and Martin, 1991, Romano et al., 1993a, Romano et al., 1993b, Simo, 1993). This Section discusses the link between convex analysis and optimization from the one side and elastoplasticity (including computational issues) from the other side. It is also referred to in the next Chapter with respect to the nonconvex problems which are studied there. Details can be found in the cited literature.

3.2.1 Holonomic (Hencky–type) elastoplasticity

A holonomic elastoplasticity problem is formulated first. It is actually a nonlinear elasticity model which predicts a stress which lies within a set defined by a yield function $\sigma \to F(\sigma)$ and the corresponding plasticity mechanisms as Lagrangian based additional plastic deformations. Convex and nonconvex yield functions can be used and they require different mathematical tools for their treatment, especially in Section 4.2 dealing with nonconvex problems.

An additive decomposition of the strain vector into elastic and plastic contributions, which is a valid assumption for a small displacements theory, is first introduced (Temam, 1983, p.66, Panagiotopoulos, 1985):

$$\epsilon = \epsilon^e + \epsilon^p, \tag{3.105}$$

where ϵ^e denotes the elastic and ϵ^p the plastic strains of the three–dimensional elastoplastic body.

The complementary virtual work for the whole structure takes the following form:

$$(\epsilon^e, \tau - \sigma) + (\epsilon^p, \tau - \sigma) = < l, \tau - \sigma >, \ \forall \tau \in \Sigma_{ad}. \tag{3.106}$$

In (3.106) and in the sequel, the following energy and work expressions appear respectively:

$$(\epsilon, \sigma) = \int_\Omega \epsilon_{ij} \sigma_{ij} d\Omega \tag{3.107}$$

and

$$< l, \sigma > = \int_{\Gamma_U} U_i S_i d\Gamma. \tag{3.108}$$

Moreover, the set of statically admissible stresses Σ_{ad} reads :

$$\Sigma_{ad} = \{\tau | \tau_{ij,j} + f_i = 0 \text{ on } \Omega, T_i = F_i \text{ on } \Gamma_F, \ i,j = 1,2,3\}. \tag{3.109}$$

Here we assume that the body on a part Γ_U of its boundary has given displacements, i.e. $u_i = U_i$ on Γ_U and that on the rest of its boundary $\Gamma_F = \Gamma - \Gamma_U$ the boundary tractions are given, i.e. $S_i = F_i$ on Γ_F.

In general we assume that the material of the structure is hyperelastic and such that we can write $(\epsilon^e, \tau - \sigma) \le w'_m(\sigma, \tau - \sigma)$, $\forall \tau \in \mathbf{R}^6$, where w_m denotes the superpotential of the constitutive hyperelastic law and $w'_m(x, y)$ is the directional derivative of w_m at the point x in the direction given by y.

The plasticity law is defined in the following form:

$$\epsilon^p \in \mathcal{N}_P(\sigma). \tag{3.110}$$

Recall here that $\mathcal{N}_P(\sigma)$ denotes the normal cone to the convex set P at the point σ. Here σ is the actual stress state and the yield locus is defined by the yield function $F(\sigma)$ as:

$$P = \{\sigma \in \Sigma_{ad} : F(\sigma) \le 0\}. \tag{3.111}$$

By means of the polar cone $\mathcal{N}_P^0(\sigma)$ to the previously introduced normal cone (cf. (2.31), (2.32)) and of the indicator function $\mathcal{I}_P(\sigma)$ of the set P one gets the equivalent variational formulations:

$$(\epsilon^p, \tau - \sigma) \le 0, \ \forall \tau - \sigma \in \mathcal{N}_P^0(\sigma), \tag{3.112}$$

and

$$(\epsilon^p, \tau - \sigma) \le \mathcal{I}_P(\tau) - \mathcal{I}_P(\sigma), \ \forall \tau \in \Sigma_{ad}. \tag{3.113}$$

In this Section we assume that the elastic specific energy function w_m is a convex function and the yield surface P is a convex set. Thus we have, for $\sigma \in \Sigma_{ad}$

$$(\epsilon^e, \tau - \sigma) \le w_m(\tau) - w_m(\sigma), \ \forall \tau \in \Sigma_{ad}, \tag{3.114}$$

and (cf. (3.112)), for $\sigma \in P$,

$$(\epsilon^p, \tau - \sigma) \le 0, \ \forall \tau \in P. \tag{3.115}$$

Finally the holonomic elastoplasticity problem takes the following variational inequality formulation (cf. (3.106)):
Find $\sigma \in P$ such that

$$w_m(\tau) - w_m(\sigma) - < l, \tau - \sigma > \geq 0, \ \forall \tau \in P. \tag{3.116}$$

Moreover, for linear elastic material, $w_m(\sigma)$ is a quadratic form.

Concrete forms of the yield locus (3.111) will be discussed in the next Section with respect to the more realistic rate elastoplasticity formulation. In general, the volume, the shape and the position of the yield surface P may change during the loading of the structure and may depend on additional internal parameters of the system. In the case of locally expandable yield surfaces we talk about isotropic, anisotropic and kinematic hardening respectively. More details are given later (see also Section 3.2.2.2).

3.2.2 Rate elastoplasticity

3.2.2.1 Perfect elastoplasticity. The rate formulation of the elastoplasticity problem is outlined here. The additive decomposition adopted for the strain rates (cf. (3.105)) reads:

$$\dot{\epsilon} = \dot{\epsilon}^e + \dot{\epsilon}^p, \tag{3.117}$$

where $\dot{x}$ denotes the time derivative of the variable x. The flow rule (cf. (3.110)) is written for the rate of the plastic deformation as:

$$\dot{\epsilon}^p \in \mathcal{N}_P(\sigma), \tag{3.118}$$

or, if σ does not lie on a corner of P,

$$\begin{aligned} \dot{\epsilon}^p &= \lambda \nabla F(\sigma) \ \text{ with} \\ \lambda &\geq 0, \ F(\sigma) \leq 0, \ \lambda F(\sigma) = 0. \end{aligned} \tag{3.119}$$

If σ lies on a corner of P then we have the case of multimodal plasticity which is described by the relations (3.122) to (3.124) given in the sequel. The latter Lagrangian formulation can be justified mechanically by means of the following arguments. Let us consider the dissipative plastic work, i.e., the dissipated energy along the elastoplastic deformation:

$$dW^p = (\sigma, \dot{\epsilon}^p). \tag{3.120}$$

By assuming that for the real process the above dissipative work, considered as a linear function of σ, is maximized under the inequality restriction posed by

(3.111), relation (3.119) results as the optimality condition of this mathematical programming problem.

Note that in the previous formulation the yield surface $F(\sigma)$ is taken to be the plastic potential for the derivation of the first relation in (3.119). It is called an associated plasticity model. More general models, where the plastic potential is different from the yield surface function have also been proposed in the framework of nonassociated elastoplasticity theories. Nevertheless, the latter problems are not directly connected with the mathematical optimization theory and will not be discussed further (cf. in this respect, the coupled unilateral contact and Coulomb friction problem discussed in Section 3.1.3.1).

Recall that the flow rule (3.118) is written also in the variational form (cf. (3.115)):

$$(\dot{\epsilon}^P, \tau - \sigma) \leq 0, \ \forall \tau \in P, \tag{3.121}$$

where σ is the actual stress tensor and $\tau - \sigma$ is a virtual variation of the stress variable. From the engineering point of view the latter inequality is the Drucker's stability postulate (Drucker, 1959).

Drucker's material stability postulate provides the mechanical justification for the assumption that the yield surface P (see (3.111)) is a convex set. In this framework all proposed yield surfaces must qualify this criterion (see, among others, Chen, 1994, Betten, 1993). The case where the elastic properties change along the plastic deformation path is an exception and allows for nonconvex yield surfaces which nevertheless comply with the Drucker's stability postulate (see Panagiotopoulos, 1985 p. 153 and Section 4.2.1).

Within the general framework of the thermodynamics of irreversible processes, Drucker's stability postulate is accompanied by the requirement that the dissipative plastic work is nonnegative: $dW^p = (\sigma, \dot{\epsilon}^P) \geq 0$. Moreover, the principle of maximal plastic dissipation can be obtained from the more general principle of the maximum specific entropy production, written for the specific case of adiabatic and isothermal processes (see, e.g., Betten, 1993, p. 133, Ziegler, 1961, Lubliner, 1990).

For the rate elastoplasticity problem the virtual work principle (3.106) is written for the time derivatives of the variables involved and it is combined with the flow rule (3.118) to produce a variational inequality problem. Details and model time-discretization techniques are discussed later on.

A multimodal plasticity model involves several yield functions (modes) $F_i(\sigma)$. The yield set is defined as the intersection of all single mode sets (3.111), i.e.,

$$\begin{aligned} P &= \{\sigma \in \Sigma_{ad} : F_i(\sigma) \leq 0, \ i = 1, \ldots, m\} = \\ &= \cap_{i=1}^m P_i, \text{ with } P_i = \{\sigma \in \Sigma_{ad} : F_i(\sigma) \leq 0\}. \end{aligned} \tag{3.122}$$

The latter representation is equivalent to

$$P = \left\{ \sigma \in \Sigma_{ad} \ : \ \max_{i \in \{1,\dots,m\}} \{F_i(\sigma)\} \leq 0 \right\}. \qquad (3.123)$$

The normality rule (3.118) reads in this case:

$$\begin{aligned} \dot{\epsilon}^p &= \sum_{i \in I} \lambda_i \nabla F_i(\sigma) \text{ with} \\ \lambda_i &\geq 0, \ F_i(\sigma) \leq 0, \ \lambda_i F_i(\sigma) = 0, \end{aligned} \qquad (3.124)$$

and $I = \{i \in \{1,\dots,m\} : F_i(\sigma) = 0\}$ is the set of active modes at point σ. Obviously the sum sign in relation (3.124) holds at corner points σ where the set I has more than one elements. The Lagrangian multipliers λ_i are also referred as the plastic consistency parameters.

Example 1: Von Mises criterion is written with

$$F(\sigma) = \|dev\sigma\|_\sigma - \sqrt{\frac{2}{3}}\sigma_Y,$$

where $dev\sigma = \sigma - \frac{1}{3}(tr\sigma)I$ is the deviatoric part of the stress tensor σ ; $\|.\|_\sigma$ is the norm of the tensor and σ_Y is the yield stress of the material.

Example 2: Tresca multi-mode criterion is written with

$$F(\sigma) = \max_{i,j=\{1,2,3\}} \{\sigma_i - \sigma_j - \sigma_Y\},$$

where σ_i are the principal values of the deviatoric stress component of the stress tensor σ.

3.2.2.2 Elastoplasticity with hardening. The shape, the position and the orientation of the yield surface (3.111) in the stress space may change as a result of the loading history and the nonlinear changes experienced by the material of the structure during the previous loading. This way the Bauschinger effect can be modelled, among others. By introducing the scalar hardening parameter κ, the stress-type (internal variables) tensorial hardening parameters α_{ij} and the hardening function $k(\kappa)$, a general yield function can be expressed as:

$$F(\sigma, \alpha, \kappa) = F_0(\sigma - \alpha) - k(\kappa). \qquad (3.125)$$

Isotropic hardening, i.e., the change of the size of the yield locus is included by the function $k(\kappa)$ in (3.125). The position of the center of the yield surface changes by the kinematic hardening hypothesis which is included in (3.125) by means of the variable α.

Appropriate flow rules must be added to the elastoplastic analysis problem in order to account for the change of the yield locus by means of the hardening parameters κ, α in this case. Note that the assumption of a hardening behaviour accounts for only local expansion of the yield surface near the actual stress point (if this bounds on the boundary of the yield locus) under the action of the loading. This behaviour guarantees that the arising model remains convex. Thus, certain restrictions must be posed on the admissible forms of the functions $F_0(\sigma - \alpha)$ and $k(\kappa)$. For instance, linear isotropic hardening with $k(\kappa) = k_0\kappa$ or a nonlinear isotropic hardening where $k(\kappa)$ is a monotonously increasing function of κ are included in this convex framework.

A linear hardening elastoplasticity problem is formulated in Section 6.3.1. Various multimode elastoplasticity models, both associated and nonassociated, have been proposed. See, among others, Ottosen and Ristinmaa, 1996.

3.2.2.3 Generalized standard elastoplasticity model. In the framework of small displacements and strains the additive decomposition of strains is a valid assumption for the formulation of elastoplasticity problems. A relatively simple model problem involves the plastic strains e^p as the only internal variables. Moreover, let the elastic region in stress space be a closed convex subset P in $\mathbb{R}^n$. The perfect elastoplasticity problem introduced previously is formulated here within the generalized standard elastoplasticity framework. More complicated models with several internal variables are outlined in the next Section.

For the studied problem, the existence of a convex free energy potential is assumed:

$$W(e - e^p) = \frac{1}{2}(e - e^p)^T K_0 (e - e^p). \tag{3.126}$$

Moreover, the existence of an internal dissipation potential, which is a convex, nonnegative, sublinear and positively homogeneous of order one function is assumed:

$$\tilde{\mathcal{D}} = I_P^c(\dot{e}^p) \equiv \sup_{\sigma \in P}\{\sigma^T \dot{e}^p\}. \tag{3.127}$$

Here K_0 is the natural elastic stiffness matrix and I_P^c is the support function of the convex set P.

The governing relations of a perfect elastoplastic analysis problem are derived by means of differentiation of (3.126) (resp. (3.127)) with respect to e (resp. $\dot{e}^p$), i.e.

$$\sigma = K_0(e - e^p), \tag{3.128}$$

$$\sigma = \partial I_P^c(\dot{e}^p). \tag{3.129}$$

Note that the inverse of (3.129) is the flow rule (3.124).

Remark: Recall that for plasticity relations defined by means of convex sets C the transition between the formulations (3.129) and (3.124) can be performed automatically. In fact by using the results of Minkowski duality, the function (3.127) is equal to the gauge function $|.|_{P^0}$ of the polar set of P, denoted by P^0, thus

$$\tilde{\mathcal{D}} = I_P^c(\dot{e}^p) = |\dot{e}^p|_{P^0}. \tag{3.130}$$

Here the gauge function is defined by:

$$|\dot{e}^p|_{P^0} \equiv \min\{\lambda > 0 \mid \dot{e}^p \in \lambda P^0\} \tag{3.131}$$

and the polar set C^0 is a convex closed set, including the zero element such that

$$P^0 = \{\dot{e}^p \mid \sigma^T \dot{e}^p \leq 1, \ \forall \sigma \in P\}. \tag{3.132}$$

3.2.2.4 Hardening and internal variables. We follow here the approach of Panagiotopoulos, 1985, pp. 103, 157, 350, Simo et al., 1988, Simo, 1993 for a description of a general elastoplasticity model with hardening. Let the yield function $F(\sigma, q) = F(\sigma, \alpha, \kappa)$ (see (3.125)) define the elasticity area by (cf. (3.111)):

$$P = \{\sigma, q \ : \ F(\sigma, q) \leq 0\}. \tag{3.133}$$

Here, for notational simplicity, all hardening parameters (α, κ) are put together into the supervector q with appropriate dimension. The number and the mechanical meaning of the elements of vector q are dictated by the concrete application on hand. The additive decomposition (3.105) is also assumed. Let first the following special form of the complementary Helmholtz free energy be assumed (cf. (3.126)):

$$\Xi(\sigma, q) = \frac{1}{2}\sigma^T \mathbf{K}_0^{-1} \sigma + \frac{1}{2} q^T \mathbf{H}^{-1} q, \tag{3.134}$$

where $\mathbf{H}$ is, for the present, a given symmetric positive definite matrix of plastic hardening moduli. The dissipation function is defined as the difference between the total stress power and the rate of change in the free energy, i.e.,

$$\begin{aligned} \mathcal{D} \ &= \ (\sigma, \dot{\epsilon}) - \frac{d}{dt}\Xi(\sigma, q) = \\ = (\sigma, \dot{\epsilon} - \mathbf{K}_0^{-1}\, \dot{\sigma} + (q, -\mathbf{H}^{-1}\, \dot{q}) \ &= \ (\sigma, \dot{\epsilon}^p) + (q, \mathbf{H}^{-1}\, \dot{q}). \end{aligned} \tag{3.135}$$

Assuming that for a given increment $\dot{\epsilon^p}$, $\dot{q}$ the linear dissipation function $\mathcal{D}(\sigma, q)$ is maximized over the convex set defined by P in (3.133), we get, by writing the optimality conditions of the latter optimization problem, the flow rules for $\dot{\epsilon^p}$ and $\dot{q}$:

$$\dot{\epsilon^p} = \dot{\epsilon} - \mathbf{K}_0^{-1}\,\dot{\sigma} = \lambda^T \nabla_\sigma F(\sigma, q), \tag{3.136}$$

$$-\mathbf{H}^{-1}\,\dot{q} = \lambda^T \nabla_q F(\sigma, q), \tag{3.137}$$

with

$$\lambda \geq 0,\ F(\sigma, q) \leq 0,\ \lambda^T F(\sigma, q) = 0. \tag{3.138}$$

The algorithmic analysis of this problem follows the methodology that will be discussed later on in this Section and can be found in Simo, 1993.

Note here that the flow relations (3.136), (3.137) can be introduced directly, without any reference to a yield surface (3.133). For a standard generalized elastoplasticity formulation they are supposed to be produced by subdifferentiation of a convex, positive, possibly nondifferentiable pseudopotential of dissipation $\phi^*(\sigma, q)$ (with $\phi^*(0,0) = 0$), as follows:

$$\begin{bmatrix} \dot{\epsilon^p} \\ -\dot{q} \end{bmatrix} \in \begin{bmatrix} \partial_\sigma \phi^*(\sigma, q) \\ \partial_q \phi^*(\sigma, q) \end{bmatrix}. \tag{3.139}$$

More details on this approach can be found in Ladèveze, 1995, among others.

As with relation (3.126) it is possible to derive the elastoplastic analysis problem with internal variables by starting from a free energy potential $W(e - e^p, \overline{q})$ which includes the effect of a set of kinematic internal variables $\overline{q}$. These internal variables describe the hardening process. Stresses σ and force-type hardening parameters q are produced by subdifferentiation from the free energy potential:

$$[\sigma, -q] \in \partial W(e, \overline{q}). \tag{3.140}$$

Details of this approach can be found in Matthies, 1991, Comi et al., 1992, Romano et al., 1993a, Romano et al., 1993b. For a holonomic problem see, among others, Corradi and Genna, 1990.

For instance, in the example of the holonomic elastic one-dimensional spring with hardening of Section 3.1.4.2, the plastic deformation $e^p = \overline{q} = (\epsilon - \epsilon_0)_+$ is the only internal variable while the yield locus, including hardening, is given by $\sigma - \sigma_0 - \overline{q}E_T \leq 0$. Thus, the stress-type hardening variable is in this case equal to $q = \overline{q}E_T = E_T(\epsilon - \epsilon_0)_+$. The free energy potential is given in (3.98). Simple one-dimensional examples can also be found in Martin and Nappi, 1990.

3.2.2.5 Stepwise holonomic problems by time discretization. Time discretization is used to resolve the elastoplasticity problem, written initially in terms of velocities, into a series of holonomic elastoplastic analysis problems, i.e., a series of Hencky type elastoplasticity problems or of nonlinear elasticity problems. The simple perfect elastoplasticity problem is again assumed for notational simplicity. Within a finite time step $\Delta t = t_{n+1} - t_n > 0$ the assumption is done that no plastic unloading is permitted. Time discretization of the velocity variables is performed by an appropriate numerical scheme. Let us consider the implicit Euler scheme (see e.g. LeTallec, 1990). Let at the time step $(n+1)$, the plastic flow relation be discretized:

$$\frac{e^p_{n+1} - e^p_n}{\Delta t} \in \partial I_P(\sigma_{n+1}) \Rightarrow e^p_{n+1} - e^p_n \in \partial I_P(\sigma_{n+1}). \tag{3.141}$$

Here the cone property of $\partial I_P = \mathcal{N}_P$ has been used to eliminate Δt from (3.141). Moreover (3.128) is written for the time step $(n+1)$ as (with $F_0 = K_0^{-1}$)

$$e_{n+1} - e^p_{n+1} = F_0 \sigma_{n+1}. \tag{3.142}$$

Relations (3.141) and (3.142) lead to the incremental relation

$$0 \in F_0 \sigma_{n+1} - e_{n+1} + e^p_n + \partial I_P(\sigma_{n+1}) \tag{3.143}$$

or equivalently to the minimization (stepwise holonomic) problem

$$\min_{\sigma_{n+1} \in P} J_{n+1}(\sigma_{n+1}) \tag{3.144}$$

with

$$J_{n+1}(\sigma_{n+1}) = \frac{1}{2} \sigma^T_{n+1} F_0 \sigma_{n+1} - e^T_{n+1} \sigma_{n+1} + e^{p\,T}_n \sigma_{n+1}. \tag{3.145}$$

Solutions of problem (3.144) are also solutions of the differential inclusion (multivalued equation)

$$0 \in \partial_{\sigma_{n+1}} f(\sigma_{n+1}) + \partial I_P(\sigma_{n+1}), \tag{3.146}$$

or

$$\partial_{\sigma_{n+1}} f(\sigma_{n+1}) \cap \{ -\mathcal{N}_P(\sigma_{n+1}) \} \neq \emptyset, \tag{3.147}$$

where $f(\sigma_{n+1})$ is defined by:

$$f(\sigma_{n+1}) = \left\{ \frac{1}{2} \sigma^T_{n+1} F_0 \sigma_{n+1} - e^T_{n+1} \sigma_{n+1} + e^{p\,T}_n \sigma_{n+1} \right\}. \tag{3.148}$$

An equivalent convex variational inequality problem can be written: Find $\sigma_{n+1} \in \mathbb{R}^n$ such that

$$\nabla f(\sigma_{n+1})(\tau - \sigma_{n+1}) + I_P(\tau) - I_P(\sigma_{n+1}) \geq 0, \ \forall \tau \in \mathbb{R}^n. \tag{3.149}$$

Equivalently from (3.129) we get (recall that $I_P^c(.)$ is positively homogeneous of degree one) that

$$\sigma_{n+1} \in \partial I_P^c\left(\frac{e_{n+1}^p - e_n^p}{\Delta t}\right) = \frac{1}{\Delta t}\partial I_P^c(e_{n+1}^p - e_n^p). \tag{3.150}$$

Thus (3.128) written for the time step $(n+1)$ gives rise by means of (3.148) to the inclusion

$$0 \in K_0(e_{n+1}^p - e_{n+1}) + \frac{1}{\Delta t}\partial I_P^c(e_{n+1}^p - e_n^p) \tag{3.151}$$

or equivalently to the minimization problem

$$\min_{e_{n+1}^p} H_{n+1}(e_{n+1}^p) \tag{3.152}$$

with

$$H_{n+1}(e_{n+1}^p) = \frac{1}{2}e_{n+1}^{pT}K_0 e_{n+1}^p + \frac{1}{\Delta t}I_C^c(e_{n+1}^p - e_n^p) - e_{n+1}^T K_0 e_{n+1}^p. \tag{3.153}$$

Details of numerical algorithms for concrete elastoplastic analysis problems based on this general approach can be found in Simo et al., 1988, Glowinski and LeTallec, 1989, Reddy and Martin, 1991, LeTallec, 1990, among others.

The case of von Mises linear kinematic and isotropic hardening elastoplasticity in connection with backward difference, midpoint and trapezoidal integration rules is studied in Caddemi, 1994. Generalized and mid-point integration schemes have been considered in Simo and Govindjee, 1990, Reddy and Martin, 1991, Corigliano and Perego, 1993, Corigliano, 1994, Martin and Caddemi, 1994, Chaboche and Gailletaud, 1996, Han et al., 1997 among others. Various holonomic problems are studied in Reddy, 1992, Comi and Perego, 1995, de Sciarra and Rosati, 1995.

References

Al-Fahed, A. M., Stavroulakis, G. E., and Panagiotopoulos, P. D. (1991). Hard and soft fingered robot grippers. *Zeitschrift fuer Angew. Mathematik und Mechanik (ZAMM)*, 71(7/8):257–266.

Alart, P. (1993). Critères d' injectivité et de surjectivité pour certaines applications de $\mathbb{R}^n$ dans lui même. Application a la mécanique du contact. *RAIRO Mod. Math. et An. Num.*, 27:203–222.

Alart, P. and Curnier, A. (1991). A mixed formulation for frictional contact problems prone to Newton like solution methods. *Computer Methods in Applied Mechanics and Engineering*, 92:353–375.

Andrews, K. T., Shillor, M., and Wright, S. (1996). On the dynamic vibrations of an elastic beam in frictional contact with a rigid obstacle. *Journal of elasticity*, 42(1):1–30.

Antes, H. and Panagiotopoulos, P. D. (1992). *The boundary integral approach to static and dynamic contact problems. Equality and inequality methods.* Birkhäuser, Basel-Boston-Berlin.

Argyris, J. H. (1965). Continua and discontinua. In *Proc. 1st Conf. Matrix Meth. Struct. Mech.*, pages 66–80, Dayton, Ohio. Wright Petterson Air Force Base. AFFDL TR.

Baiocchi, C. and Capelo, A. (1984). *Variational and quasivariational inequalities. Applications to free boundary problems.* J. Wiley and Sons, Chichester.

Baniotopoulos, C. C., Abdalla, K. M., and Panagiotopoulos, P. D. (1994). A variational inequality and quadratic programming approach to the separation problem of steel bolted brackets. *Computers and Structures*, 53(4):983–991.

Bathe, K. J. (1981). *Finite element procedures in engineering analysis.* Prentice–Hall, New Jersey.

Beitelschmidtand, M. and Pfeiffer, F. (1997). Impacts with friction and normal and tangential reversibility in multibody systems. *Zeitschrift für Angewandte Mathematik und Mechanik (ZAMM)*, 77(1):S29.

Betten, J. (1993). *Kontinuumsmechanik. Elasto-, Plasto- und Kriechmechanik.* Springer Verlag, Berlin - Heidelberg.

Bille, J. P., Cescotto, S., Habraken, A. M., and Charlier, R. (1995). Numerical approach for contact using an augmented Lagrangian method. In Raous, M., Jean, M., and Moreau, J. J., editors, *Contact mechanics*, pages 243–246, New York-London. Plenum Press.

Bisbos, C. D. (1995). A competitive game algorithm with five players for unilateral contact problems including the rotational and the thermal degrees of freedom. In Raous, M., Jean, M., and Moreau, J. J., editors, *Contact mechanics*, pages 251–258, New York-London. Plenum Press.

Caddemi, S. (1994). Computational aspects of the integration of the von Mises linear hardening constitutive laws. *International Journal of Plasticity*, 10(8): 935–956.

Cescotto, S. and Charlier, R. (1993). Frictional contact finite elements based on mixed variational principles. *International Journal of Numerical Methods in Engineering*, 36:1681–1701.

Cescotto, S. and Charlier, R. (1994). Frictional contact finite elements based on mixed variational principles. *International Journal for Numerical Methods in Engineering*, 36:1681–1701.

Chaboche, J. L. and Gailletaud, G. (1996). Integration methods for complex plastic constitutive equations. *Computer Methods in Applied Mechanics and Engineering*, 133:125–155.

Chaudhary, A. B. and Bathe, K. J. (1986). A solution method for static and dynamic analysis of three-dimensional contact problems with friction. *Computers and Structures*, 24:855–873.

Chen, W. F. (1994). *Constitutive equations for engineering materials. Vol. 1: Elasticity and modeling. Vol. 2: Plasticity and modeling.* Elsevier, Holland.

Cheng, Y.-C., Walker, I. D., and Cheatham, J. B. (1995). Visualization of force-closure grasps for objects through contact force decomposition. *The Int. Journal of Robotics Research*, 14(1):37–75.

Cocu, M. (1984). Existence of solutions of Signorini problems with friction. *International Journal of Engineering Sciences*, 22:567.

Cocu, M., Pratt, E., and Raous, M. (1995). Analysis of an incremental formulation for frictional contact problems. In Raous, M., Jean, M., and Moreau, J. J., editors, *Contact mechanics*, pages 13–21, New York-London. Plenum Press.

Cohn, M. Z. and Maier, G., editors (1979). *Engineering plasticity by mathematical programming.* Pergamon Press, Oxford.

Comi, C., Maier, G., and Perego, U. (1992). Generalized variable finite element modelling and extremum theorems in stepwise holonomic elastoplasticity with internal variables. *Computer Methods in Applied Mechanics and Engineering*, 96:213–237.

Comi, C. and Perego, U. (1995). A unified approach for variationally consistent finite elements in elastoplasticity. *Computer Methods in Applied Mechanics and Engineering*, 121:323–344.

Cook, R. D. (1978). *Finite element method. Concepts and applications.* J. Wiley, New York.

Corigliano, A. (1994). Numerical analysis of discretized elastoplastic systems using the generalized mid-point integration. *Engineering Computations*, 11:389–411.

Corigliano, A. and Perego, U. (1993). Generalized mid-point finite element dynamic analysis of elastoplastic systems. *International Journal for Numerical Methods in Engineering*, 35:361–383.

Corradi, L. and Genna, F. (1990). Kinematic extremum theorems for holonomic plasticity. *International Journal of Plasticity*, 6:63–82.

Curnier, A. (1984). A theory of friction. *International Journal of Solids and Structures*, 20(7):637–647.

Curnier, A., editor (1992). *Contact mechanics*. Presses Polytechniques et Universitaires Romandes, Lausanne.

de Sciarra, F. M. and Rosati, L. (1995). Extremum theorems and a computational algorithm for an internal variable model of elastoplasticity. *European Journal of Mechanics A/Solids*, 14(6):843–871.

DelPiero, G. and Maceri, F. (1991). *Unilateral problems in structural analysis IV : Proc. Meeting Capri, June 14-16, 1989*. Birkhäuser Verlag, Basel Boston.

Dostál, Z. (1992). Conjugate projection preconditioning for the solution of contact problems. *International Journal for Numerical Methods in Engineering*, 34:271–277.

Doudoumis, I., Mitsopoulou, E., and Charalambakis, N. (1995). The influence of the friction coefficients on the uniqueness of the solution of the unilateral contact problem. In Raous, M., Jean, M., and Moreau, J., editors, *Contact mechanics*, pages 79–86, New York-London. Plenum Press.

Drucker, D. C. (1959). A definition of stable plastic materials. *Journal of Applied Mechanics*, 26:101–106.

Duvaut, G. and Lions, J. L. (1972). *Les inéquations en méchanique et en physique*. Dunod, Paris.

Eve, R., Reddy, B., and Rockafellar, R. (1990). An internal variable theory of plasticity based on the maximum plastic work inequality. *Quarterly of Applied Mathematics*, 68(1):59–83.

Fritz, P. and Pfeiffer, F. (1997). Dynamics of highspeed roller chain drives. *Zeitschrift für Angewandte Mathematik und Mechanik (ZAMM)*, 76(S5):151–152.

Germain, P. (1973). *Cours de mecanique des mileux continus*. Masson, Paris.

Glocker, C. and Pfeiffer, F. (1992). Dynamical systems with unilateral contacts. *Nonlinear Dynamics*, 3:245–259.

Glocker, C. and Pfeiffer, F. (1995). Multiple impacts with friction in rigid multibody systems. *Nonlinear Dynamics*, 7(4):471–498.

Glowinski, R. and LeTallec, P. (1989). *Augmented Lagrangian and operator–splitting methods for nonlinear mechanics*. SIAM, Philadelphia.

Glowinski, R., Lions, J. L., and Trémolières, R. (1981). *Numerical analysis of variational inequalities*. Studies in Mathematics and its Applications, Vol. 8. Elsevier, Amsterdam-New York.

Goeleven, D. and Mentagui, D. (1995). Well-posed hemivariational inequalities. *Numerical Functional Analysis and Optimization*, 16(7–8):909–921.

Goeleven, D., Stavroulakis, G. E., Salmon, G., and Panagiotopoulos, P. D. (1997). Solvability theory and projection methods for a class of singular variational inequalities. Elastostatic unilateral contact applications. *Journal of Optimization Theory and Applications*, (to appear).

Goeleven, D. and Théra, M. (1995). Semicoercive variational hemivariational inequalities. *Journal of Global Optimization*, 6:367–381.

Halphen, B. and Son, N. Q. (1975). Sur les matériaux standards généralisés. *J. de Mecanique*, 14:39–61.

Han, W., Reddy, B. D., and Schroeder, G. C. (1997). Qualitative and numerical analysis of quasi-static problems in elastoplasticity. *SIAM Journal of Numerical Analysis*, 34(1):143–177.

He, Q. C., Telega, J. J., and Curnier, A. (1996). Unilateral contact of two solids subject to large deformations: Formulation and existence results. *Proceedings of the Royal Society of London / A*, 452(1955):2691–2718.

Jarušek, J. (1983). Contact problems with bounded friction. Coercive case. *Czech. Math. Journal*, 33:237.

Jarušek, J. (1984). Contact problems with bounded friction. Semicoercive case. *Czech. Math. Journal*, 34:619.

Jarušek, J. (1994). Solvability of the variational inequality for a drum with a memory vibrating in the presence of an obstacle. *Bullettino della Unione Matematica Italiana / A*, (1):113–122.

Kalker, J. J. (1988). Contact mechanical algorithms. *Communications in Applied Numerical Methods*, 4:25–32.

Kalker, J. J. (1990). *Three–dimensional elastic bodies in rolling contact*. Kluwer Academic, Dordrecht Boston.

Klarbring, A. (1986). A quadratic program in frictionless contact problems. *International Journal of Engineering Sciences*, 24:459–479.

Klarbring, A. (1990). Derivation and analysis of rate boundary-value problems. *European Journal of Mechanics A/Solids*, 9(1):53–86.

Klarbring, A. (1992). Mathematical programming and augmented Lagrangian methods for frictional contact problems. In Curnier, A., editor, *Proc. Contact Mechanics Int. Symp.*, pages 409–422, Lausanne. Presses Polytechniques et Universitaires Romandes.

Klarbring, A. and Björkman, G. (1988). A mathematical programming approach to contact problem with friction and varying contact surface. *Computers and Structures*, 30:1185–1198.

Klarbring, A., Mikelič, A., and Shillor, M. (1991a). A global existence result for the quasistatic frictional contact problem with normal compliance. In *International Series of Numerical Mathematics Vol. 101*, page 85, Basel Boston. Birkhäuser Verlag.

Klarbring, A., Mikelič, A., and Shillor, M. (1991b). A global existence result for the quasistatic frictional contact problem with normal compliance. In DelPiero, G. and Maceri, F., editors, *Unilateral Problems in Structural Mechanics IV*, pages 85–111, Basel Boston. Birkhäuser Verlag.

Kwack, B. M. and Lee, S. S. (1988). A complementarity problem formulation for two–dimensional frictional contact problems. *Computers and Structures*, 28:469–480.

Ladèveze, P. (1995). *Mecanique des structures nonlineaires*. Hérmes, Paris.

Laursen, T. A. and Oancea, V. G. (1994). Automation and assessment of augmented Lagrangian algorithms for frictional contact problems. *ASME Journal of Applied Mechanics*, 61:956–963.

Lee, K. (1994a). A numerical solution for dynamic contact problems satisfying the velocity and acceleration compatibilities on the contact surface. *Computational Mechanics*, 15:189–200.

Lee, S., Kwak, B., and Kwon, O. (1994). Analysis of incipient sliding contact by three-dimensional LCP formulation. *Computers and Structures*, 53(3):695–705.

Lee, S.-S. (1994b). A computational method for frictional contact problem using finite element method. *Int. Journal for Numerical Methods in Engineering*, 37:217–228.

Lemaitre, J. and Chaboche, J. L. (1985). *Mécanique des matériaux solides*. Dunod, Paris. English Translation Cambridge Univ. Press 1994.

LeTallec, P. (1990). *Numerical analysis of viscoelastic problems*. Masson, Springer, Paris, Berlin.

Liolios, A. A. (1986). A linear complementarity approach for the Signorini problem with friction. *Zeitschrift für Angewandte Mathematik und Mechanik (ZAMM)*, 66:349–352.

Lötstedt, P. (1981). Coulomb friction in two-dimensional rigid body systems. *Zeitschrift für Angewandte Mathematik und Mechanik (ZAMM)*, 61:605–616.

Lubliner, L. (1990). *Plasticity theory*. Macmillan Publ., New York, London.

Maier, G. (1970a). A matrix structural analysis theory of piecewise-linear plasticity with interacting yield planes. *Meccanica*, 5:55–66.

Maier, G. (1970b). A matrix structural theory of piecewise linear elastoplasticity with interacting yield planes. *Meccanica*, March:54–66.

Marques, M. (1992). *Differential inclusions in nonsmooth mechanical problems: shocks and dry friction.* Birkhäuser Verlag, Basel - Boston.

Martin, J. B. (1975). *Plasticity: foundamentals and general results.* MIT Press, Cambridge.

Martin, J. B. and Caddemi, S. (1994). Sufficient conditions for convergence of the Newton–Raphson iterative algorithms in incremental elastic–plastic analysis. *European Journal of Mechanics, A/Solids*, 13(3):351–365.

Martin, J. B. and Nappi, A. (1990). An internal variable formulation for perfectly plastic and linear hardening relations in plasticity. *European Journal of Mechanics A. Solids*, 9(2):107–131.

Martin, J. B., Reddy, B. D., Griffin, T. B., and Bird, W. W. (1987). Applications of mathematical programming concepts to incremental elastic-plastic analysis. *Engineering Structures*, 9:171–176.

Martins, J. A. C. and Oden, J. T. (1987). Existence and uniqueness results for dynamic contact problems with nonlinear normal and friction interface laws. *Nonlinear Analysis Theory Methods and Applications*, 11(3):407–428.

Matthies, H. G. (1991). Computation of constitutive response. In Wriggers, P. and Wagner, W., editors, *Nonlinear computational mechanics*, pages 573–585. Springer Verlag, Berlin Heidelberg.

Michalowski, R. and Mróz, Z. (1978). Associated and non-associated sliding rules in contact friction problems. *Arch. Mech*, 30:259–276.

Mitsopoulou, E. (1983). Unilateral contact, dynamic analysis of beams by a time-stepping quadratic programming procedure. *Meccanica*, 18:254–265.

Moreau, J. J. (1963). Fonctionnelles sous - différentiables. *C.R. Acad. Sc. Paris*, 257A:4117 – 4119.

Moreau, J. J. (1966). *Fonctionnelles convexes. Séminaire sur les équations aux dérivées partielles.* College de France, Paris.

Moreau, J. J. (1968). La notion de sur-potentiel et les liaisons unilatérales en élastostatique. *C.R. Acad. Sc. Paris*, 267A:954 – 957.

Moreau, J. J. (1974). On unilateral constraints, friction and plasticity. In Capriz, G. and Stampacchia, G., editors, *New variational techniques in mathematical physics*, pages 175–322, Roma. Edizioni Cremonese.

Moreau, J. J. (1975). Application of convex analysis to the treatment of elastoplastic systems. In Germain, P. and Nayroles, B., editors, *Application of methods of functional analysis to problems in mechanics*, pages 56–89, Berlin. Springer.

Moreau, J. J. (1988). Unilateral contact and dry friction in finite freedom dynamics. In Moreau, J. J. and Panagiotopoulos, P. D., editors, *Nonsmooth mechanics and applications*, CISM Lect. Notes. Vol. 302, pages 1–82. Springer Verlag, Wien - New York.

Mosco, V. (1976). Implicit variational problems and quasi-variational inequalities. In *Nonlinear operators and the calculus of variations*, pages 83–156, Berlin. Springer Verlag. Lect Notes in Math. 543.

Naniewicz, Z. (1993). On the existence of solutions to the continuum model of delamination. *Nonlinear analysis*, 20(5):481–508.

Nečas, J., Jarusek, J., and Haslinger, J. (1980). On the solution of the variational inequality to the Signorini problem with small friction. *Bulletino U.M.I.*, 17B:796–811.

Oden, J. T. and Kikuchi, N. (1988). *Contact problems in elasticity: a study of variational inequalities and finite element methods.* SIAM, Philadelphia.

Oden, J. T. and Martins, J. A. C. (1985). Models and computational methods for dynamic friction phenomena. *Computer Methods in Applied Mechanics and Engineering*, 52:527–635.

Oden, J. T. and Pires, E. (1983). Nonlocal and nonlinear friction laws and variational principles for contact problems in elasticity. *Journal of Applied Mechanics*, 50:67–76.

Ottosen, N. S. and Ristinmaa, M. (1996). Corners in plasticity. Koiter's theory revisited. *International Journal of Solids and Structures*, 33(25):3697–3721.

Outrata, J. V. and Zowe, J. (1995). A Newton method for a class of quasi-variational inequalities. *Computational Optimization and Applications*, 4:5–21.

Panagiotopoulos, P. D. (1975). A nonlinear programming approach to the unilateral contact and friction boundary value problem in the theory of elasticity. *Ing. Archiv*, 44:421–432.

Panagiotopoulos, P. D. (1985). *Inequality problems in mechanics and applications. Convex and nonconvex energy functions.* Birkhäuser, Basel - Boston - Stuttgart. Russian translation, MIR Publ., Moscow 1988.

Panagiotopoulos, P. D. (1993). *Hemivariational inequalities. Applications in mechanics and engineering.* Springer, Berlin - Heidelberg - New York.

Pang, J. S. and Ralph, D. (1996). Piecewise smoothness, local invertibility, and parametric analysis of normal maps. *Mathematics of operations research*, 21(2):401–426.

Pang, J. S. and Trinkle, J. C. (1996). Complementarity formulations and existence of solutions of dynamic multi-rigid-body contact problems with Coulomb friction. *Mathematical Programming*, 73:199–226.

Perić, G. and Owen, D. R. J. (1992). Computational model for 3D-contact problems with friction based on the penalty method. *International Journal for Numerical Methods in Engineering*, 35:1289–1309.

Pfeiffer, F. (1991). Dynamical systems with time-varying or unsteady structure. *Zeitschrift für Angewandte Mathematik und Mechanik (ZAMM)*, 71(4):6–22.

Pfeiffer, F. (1992). Stick-slip motion of turbine blade dampers. *Phil. Trans. R. Soc. London A*, 338:503–517.

Pfeiffer, F. (1996a). Complementarity problems of stick-slip vibrations. *Journal of vibration and acoustics*, 118(2):177–183.

Pfeiffer, F. (1996b). Robotics in theory and practice. In Kirchgässner, K., Mahrenholtz, O., and Memmicken, R., editors, *ICIAM 95*, pages 315–400, Berlin. Akademie Verlag.

Pfeiffer, F. (1997). Assembly processes with robotics systems. *Robotics and Autonomous Systems*, 19(2):151–166.

Pfeiffer, F. and Glocker, C. (1996). *Multibody dynamics with unilateral contacts.* John Wiley, New York.

Raous, M., Jean, M., and Moreau, J., editors (1995). *Contact mechanics.* Plenum Press, New York-London.

Reddy, B. D. (1992). Mixed variational inequalities arising in elastoplasticity. *Nonlinear Analysis Theory Methods and Applications*, 19(11):1074–1089.

Reddy, B. D. and Martin, J. B. (1991). Algorithms for the solution of internal variable problems in plasticity. *Computer Methods in Applied Mechanics and Engineering*, 93:253–273.

Refaat, M. H. and Meguid, S. A. (1994). On the elastic solution of frictional contact problems using variational inequalities. *Int. J. of Mechanical Science*, 36(4):329–342.

Refaat, M. H. and Meguid, S. A. (1996). A novel finite element approach to frictional contact problems. *International Journal for Numerical Methods in Engineering*, 39:3889–3902.

Romano, G., Rosati, L., and de Sciara, F. M. (1993a). A variational theory for finite–step elasto–plastic problems. *Intern. J. of Solids and Structures*, 30(17):2317–2334.

Romano, G., Rosati, L., and de Sciarra, F. M. (1993b). Variational principles for a class of finite step elastoplastic problems with non–linear mixed hardening. *Computer Methods in Applied Mechanics and Engineering*, 109:293–314.

Simo, J., Wriggers, P., and Taylor, R. (1985). A perturbed Lagrangian formulation for the finite element solution of contact problems. *Computer Methods in Applied Mechanics and Engineering*, 50:163–180.

Simo, J. C. (1993). Recent developments in the numerical analysis of plasticity. In Stein, E., editor, *Progress in computational analysis of inelastic structures*, volume 321 of *CISM Courses and Lectures*, pages 115–174. Springer, Wien - New York.

Simo, J. C. and Govindjee, S. (1990). Nonlinear B-stability and symmetry preserving return mapping algorithms for plasticity and viscoplasticity. *International Journal for Numerical Methods in Engineering*, 31:151–176.

Simo, J. C., Kennedy, J. G., and Govindjee, S. (1988). Nonsmooth multisurface plasticity and viscoplasticity. Loading / unloading conditions and numerical analysis. *Int. J. Num. Meth. Engng.*, 26:2161–2185.

Simo, J. C. and Laursen, T. (1992). An augmented Lagrangian treatment of contact problems involving friction. *Computers and Structures*, 42(1):97–116.

Simunovič, S. and Saigal, S. (1995). Quadratic programming contact formulation for elastic bodies using boundary element method. *AIAA Journal*, 33(2):325–331.

Srnik, J. and Pfeiffer, F. (1996). Dynamics of frictional chain drives. *Zeitschrift für Angewandte Mathematik und Mechanik (ZAMM)*, 76(5):495–496.

Stavroulakis, G. E., Panagiotopoulos, P. D., and Al-Fahed, A. M. (1991). On the rigid body displacements and rotations in unilateral contact problems and applications. *Computers and Structures*, 40:599–614.

Tamma, K. K., Li, M., and Sha, D. (1994). Conjugate gradient based projection: a new explicit computational methodology for frictional contact problems. *Communications in Numerical Methods in Engineering*, 10:633–648.

Taylor, R. L. and Papadopoulos, P. (1993). On a finite element method for dynamic contact/impact problems. *International Journal for Numerical Methods in Engineering*, 36:2123–2140.

Telega, J. J. (1988). Topics on unilateral contact problems in elasticity and inelasticity. In Moreau, J. J. and Panagiotopoulos, P. D., editors, *Nonsmooth Mechanics and Applications, CISM Lect. Notes. 302*, pages 341–462. Springer, Wien New York.

Telega, J. J. (1995). Quasi-static Signorini's contact problem with friction and duality. In Raous, M., Jean, M., and Moreau, J., editors, *Contact mechanics*, pages 199–214, New York-London. Plenum Press.

Temam, R. (1983). *Problémes mathématiques en plasticité*. Gauthier–Villars, Paris.

Trinkle, J. C., Pang, J. S., Sudarsky, S., and Lo, G. (1997). On dynamic multi-rigid-body contact problems with coulomb friction. *Zeitschrift für Angewandte Mathematik und Mechanik (ZAMM)*, 77(4):267–280.

Tzaferopoulos, M. A. (1993). On an efficient new numerical method for the frictional contact problem of structures with convex energy density. *Computers and Structures*, 48(1):87–106.

Wösle, M. and Pfeiffer, F. (1997). Dynamics of multibody systems with unilateral constraints. *Zeitschrift für Angewandte Mathematik und Mechanik (ZAMM)*, 77(1):S377.

Zavarise, G., Wriggers, P., Stein, E., and Shreffer, B. A. (1992). Real contact mechanisms and finite element formulation - a coupled thermomechanical approach. *International Journal of Numerical Methods in Engineering*, 35:767–785.

Zeidler, E. (1988). *Nonlinear functional analysis and its applications. IV: Applications to mathematical physics.* Springer Verlag, New York - Heidelberg.

Zhong, Z. H. (1993). *Finite element procedures for contact-impact problems.* Oxford University Press, New York.

Zhong, Z. H. and Mackerle, J. (1992). Static contact problems - a review. *Engineering Computations*, 9:3–37.

Zhong, Z. H. and Mackerle, J. (1994). Contact-impact problems: a review with bibliography. *ASME Applied Mechanics Review*, 47(2):55–76.

Ziegler, H. (1961). Zwei Extremalprinzipien der irreversiblen Thermodynamik. *Ingenieur Archiv*, 30:410–416.

Ziegler, H. (1963). Some extremum principles in irreversible thermodynamics with application to continuum mechanics. In Sneddon, I. and Hill, R., editors, *Progress in Solid Mechanics IV.* North Holland. Chap. 2.

Zienkiewicz, O. C. and Taylor, R. L. (1991). *The finite element method. Vol. II: Solid and fluid mechanics, dynamics and non–linearity.* McGraw-Hill.

Zmitrowicz, Z. (1989). Mathematical descriptions of anisotropic friction. *International Journal of Solids and Structures*, 25:837–862.

4 NONCONVEX SUPERPOTENTIAL PROBLEMS. VARIATIONAL AND HEMIVARIATIONAL INEQUALITIES.

4.1 ORIGIN AND TREATMENT OF NONCONVEXITY IN MECHANICS

Linearity of the kinematics and of the constitutive relations combined with fairly general material stability assumptions guarantee the convexity of a structural analysis problem in either a potential energy or in a complementary energy formulation, as it has been discussed in details in the previous Chapter. In real life applications some of these assumptions may be violated: kinematic nonlinearity which is indispensable for the description of buckling effects, decohesion, damage and fracture problems which introduce material or interface instabilities and softening behaviour in elastoplasticity are some of the applications which lead to nonconvex problems in mechanics. This is due to the fact that most of the materials used are composite, e.g., concrete with steel, fibre reinforced materials etc. Moreover, the composite nature of the materials may appear also at the micromechanical level, e.g., concrete itself is a composition of stone aggregates and cement paste, etc.

A systematic but brief introduction to this area is made in this Chapter. First, within linear kinematics, material instabilities in the form of nonmonotone interface or boundary relations are studied. The effect of nonlinear kinematics is introduced in the sequel. Then some problems in nonconvex elastoplasticity and damage mechanics are discussed.

Most of the proposed algorithms of Chapter 6 are based on appropriate decompositions of the nonconvex problem into a number of convex subproblems and, subsequently, in the approximation of the solution by means of a series composed of the solution of the convex subproblems. In smooth (differentiable) problems this is a linearization technique, where linear subproblems are solved in each step. We have made an effort to adapt the theoretical discussion in this Chapter to the algorithmic approach to these problems. Therefore an attempt is made to transform the nonconvex problems into a series of convex subproblems such as the ones already studied in the previous Chapter.

4.1.1 Nonmonotone contact (adhesive) laws

Let us consider the elastostatic analysis problem of a linear elastic structure in the small displacements and deformations setting. Moreover, the structure contains interfaces or boundaries where nonmonotone contact (adhesive) laws are assumed to hold. Thus the governing relations of the problem are the ones given in Section 3.1 except that in the place of the monotone interface laws used in Section 3.1, nonmonotone, possibly multivalued interface or boundary relations are considered here. Nonmonotone laws can be derived by nonconvex superpotentials. Continuous laws can be derived by smooth (at least one–time continuously differentiable functions) while laws with complete vertical branches are derived by nondifferentiable superpotentials.

4.1.1.1 Hemivariational inequality formulation. An elastic structure with both classical, linearly elastic and degrading elements is considered. The static analysis problem is described by the following relations:

- *Stress equilibrium equations*:

$$\overline{\mathbf{G}}\overline{\mathbf{s}} = \begin{bmatrix} \mathbf{G} & \mathbf{G}_N \end{bmatrix} \begin{bmatrix} \mathbf{s} \\ -\mathbf{S}_N \end{bmatrix} = \mathbf{p} \tag{4.1}$$

  where $\overline{\mathbf{G}}$ is the equilibrium matrix of the discretized structure which takes into account the stress contribution of the linear s and nonlinear q elements.

- *Strain–displacements compatibility equations*:

$$\overline{\mathbf{e}} = \overline{\mathbf{G}}^T \mathbf{u} \text{ or explicitly } \begin{bmatrix} \mathbf{e} \\ [\mathbf{u}]_N \end{bmatrix} = \begin{bmatrix} \mathbf{G}^T \\ \mathbf{G}_N^T \end{bmatrix} \mathbf{u}. \tag{4.2}$$

- *Linear material constitutive law* for the structure :

$$\mathbf{s} = \mathbf{K}_0(\mathbf{e} - \mathbf{e}_0) \tag{4.3}$$

where $\mathbf{K}_0$ is the stiffness matrix and $\mathbf{e}_0$ is the initial deformation.

- *Nonmonotone, superpotential constitutive laws* of the nonlinear elements:

$$-S_N \in \bar{\partial}\widetilde{\phi}_N([u]_N) \text{ or } [u]_N \in \bar{\partial}\overline{\overline{\phi}}_N(-S_N). \tag{4.4}$$

Here $\widetilde{\phi}_N(\cdot)$, , $\overline{\overline{\phi}}_N(\cdot)$, are general nonconvex and nondifferentiable potentials which produce the laws (4.4) by means of an appropriate generalized differential, set–valued operator $\bar{\partial}$. Summation over all nonlinear elements gives the total strain energy contribution of them as:

$$\widetilde{\Phi}_N(\mathbf{u}) = \sum_{i=1}^{q} \widetilde{\phi}_N^{(i)}([\mathbf{u}]_N) \text{ or } \overline{\overline{\Phi}}_N(\mathbf{s}) = \sum_{i=1}^{q} \overline{\overline{\phi}}_N^{(i)}(-\mathbf{S}_N). \tag{4.5}$$

- Classical *support boundary conditions* which are either of the displacement form:

$$\mathbf{E}\mathbf{u} = \mathbf{u}_0 \tag{4.6}$$

or of the boundary traction form:

$$\mathbf{Z}\mathbf{s} = \mathbf{F}. \tag{4.7}$$

For notational simplicity a model interface which is modelled by nonlinear one-dimensional elements in the normal to the interface direction is formulated and studied in this Section. This case is an extension of the frictionless unilateral contact problem (resp. the unilateral contact problem with given frictional tractions) of Section 3.1.2, where the remaining interface relations, i.e. the zero (resp. the given) tangential traction S_T, can be considered in the set of classical boundary conditions (4.6). More general nonlinear tangential interface behaviour can be considered by means of relations between tangential tractions S_T and tangential relative displacements $[u]_T$ (analogously to the relations (4.4) and (4.5)). Then the formulation follows the lines of Section 3.1.2.

Note also that in this Section the generalized gradient notion of Clarke-Rockafellar has been used for the formulation of the nonmonotone nonlinear interface laws (4.4). Therefore $\bar{\partial}$ denotes the generalized gradient operator in the sense of Clarke (see relation (2.79) of Section 2.4.3, and Clarke, 1983,

Rockafellar, 1979, Panagiotopoulos, 1985, p. 143, Dem'yanov et al., 1996, pp. 46, 75).

For the variational formulations of the problem the *virtual work* equation is first formulated in a discretized form:

$$\mathbf{s}^T(\mathbf{e}^* - \mathbf{e}) = \mathbf{p}^T(\mathbf{u}^* - \mathbf{u}) + \mathbf{S}_N^T([\mathbf{u}]_N^* - [\mathbf{u}]_N), \ \forall \mathbf{e}^*, \mathbf{u}^*, [\mathbf{u}]_N^* \text{ s.t. } (4.2),\ (4.6), \text{hold.} \tag{4.8}$$

Introducing the elasticity law (4.3) into the virtual work equation (4.8), and by using (4.2) we get:

$$\mathbf{u}^T\mathbf{G}\mathbf{K}_0^T\mathbf{G}^T(\mathbf{u}^* - \mathbf{u}) - (\mathbf{p} + \mathbf{G}\mathbf{K}_0\mathbf{e}_0)^T(\mathbf{u}^* - \mathbf{u}) = \mathbf{S}_N^T([\mathbf{u}]_N^* - [\mathbf{u}]_N),$$
$$\forall \mathbf{u}^* \in V_{ad} = \{\mathbf{u} \in \mathbf{R}^n \mid (4.2),\ (4.6) \text{ hold } \}, \tag{4.9}$$

where $\mathbf{K} = \mathbf{G}\mathbf{K}_0\mathbf{G}^T$ denotes the stiffness matrix of the structure and $\overline{\mathbf{p}} = \mathbf{p} + \mathbf{G}\mathbf{K}_0\mathbf{e}_0$ denotes the nodal equivalent loading vector.

At this point we use the nonlinear laws (4.4) in the following form

$$-S_N^T([u]_N^* - [u]_N) \leq \widetilde{\phi}_N^o([u]_N^* - [u]_N), \ \forall [u]_N^* \tag{4.10}$$

where $\widetilde{\phi}_N^o([u]_N^* - [u]_N)$ is the directional derivative in the sense of Clarke of the potential $\widetilde{\phi}_N$ or, in terms of mechanics the virtual work of the nonlinear structural elements for a small deformation equal to $[u]_N^* - [u]_N$. Thus we obtain the following *hemivariational inequality*:
Find kinematically admissible displacements $\mathbf{u} \in V_{ad}$ such that

$$\mathbf{u}^T\mathbf{K}(\mathbf{u}^* - \mathbf{u}) - \overline{\mathbf{p}}^T(\mathbf{u}^* - \mathbf{u}) + \widetilde{\Phi}_N^o(\mathbf{u}^* - \mathbf{u}) \geq 0, \ \forall \mathbf{u}^* \in V_{ad}, \tag{4.11}$$

where $\widetilde{\Phi}_N^o$ results from the summation of $\widetilde{\phi}_N^o$ over all the nonlinear elements. The corresponding substationarity problems read (cf. Chapter 1):
Find $\mathbf{u} \in V_{ad}$ such that:

$$\Pi(\mathbf{u}) = \underset{\mathbf{v} \in V_{ad}}{\text{substat}} \left\{ \Pi(\mathbf{v}) = \frac{1}{2}\mathbf{v}^T\mathbf{K}\mathbf{v} - \overline{\mathbf{p}}^T\mathbf{v} + \widetilde{\Phi}_N(\mathbf{v}) \right\}. \tag{4.12}$$

In the same manner, by using the nonmonotone laws (4.4b) and by introducing them in the complementary virtual work expression, we obtain a hemivariational inequality in terms of the stresses and the corresponding complementary energy substationarity problem:
Find $\mathbf{s} \in \Sigma_{ad}$ such that:

$$\Pi^c(\mathbf{s}) = \underset{\mathbf{t} \in \Sigma_{ad}}{\text{substat}} \left\{ \Pi^c(\mathbf{t}) = \frac{1}{2}\mathbf{t}^T\mathbf{F}_0\mathbf{t} - \mathbf{e}_0^T\mathbf{t} + \overline{\widetilde{\Phi}}_N(\mathbf{t}) \right\}. \tag{4.13}$$

Here the set of statically admissible stresses is defined by:

$$\Sigma_{ad} = \{\bar{s} \in \mathbf{R}^m \quad \text{such that } (4.1), (4.7) \text{ hold true}\}. \tag{4.14}$$

It should be mentioned here that the previously presented approach can be specialized to lead to classical, nonlinear minimization problems and variational equalities for differentiable potentials, to variational inequality problems for convex nondifferentiable potentials (as in the previous Chapter) and to systems of variational inequalities for difference convex potentials (see Panagiotopoulos, 1985, Panagiotopoulos, 1993, Dem'yanov et al., 1996). Thus all types of nonlinear relations, even with vertical branches (ascending, e.g. in locking effects, and descending ones) can be considered.

4.1.1.2 Difference convex optimization approach. In addition to the previous Section, let us consider that the interface or boundary adhesive contact relations (4.4) can be derived by a nonconvex superpotential $\widetilde{\phi}_N([u]_N)$, which can be expressed as a difference of convex constituent functions, i.e. it is a difference of convex functions $\phi_{N.1}$ and $\phi_{N.2}$. The formulation of the elastostatic analysis problem with adhesive laws obtained by difference convex (d.c.) superpotentials will first be derived by using the critical point theory for d.c. functions in the sense of Toland, 1979, Stuart and Toland, 1980, Auchmuty, 1989 (see also the mathematical preliminaries in Chapter 2). In the following the normal contact stresses S_N are decomposed by means of the assumed difference convex structure of the potential into a difference of two components. Each component corresponds to a monotone law without concrete physical meaning which, in turn, can be treated separately by the convex analysis tools presented in the previous Chapter.

Following Auchmuty, 1989 let us denote the set of all possible tangential directions of the function $\widetilde{\phi}$ at $[u]$ by

$$\Theta\widetilde{\phi}([u]_N) = \left\{ -S_N \;\middle|\; \begin{array}{l} -S_N([u]_N) = w_1([u]_N) - w_2([u]_N), \\ w_1([u]_N) \in \partial\phi_1([u]_N), \\ w_2([u]_N) \in \partial\phi_2([u]_N) \end{array} \right\}. \tag{4.15}$$

The above interface law is concisely written as $-S_N \in \Theta\widetilde{\phi}([u]_N)$, where Θ denotes a generally nonconvex set which describes exactly the directional derivative information supplied by the function $\widetilde{\phi}$ at $[u]_N$ (see also Section 2.3.1). Moreover, the convex analysis subdifferential ∂ is used in (4.15) to allow for, in general, nondifferentiable convex constituents in the d.c. decomposition of $\widetilde{\phi}(\cdot)$. It is clear that classical differentiation is sufficient for smooth functions.

By using the decomposition of the interface law given in (4.15), and by considering the inverse relation concerning the second constituent, the following equivalent form of the law can be written:

$$-S_N \in \partial\phi_1([u]_N) - w_2 \text{ for } w_2 \in \partial\phi_2([u]_N) \text{ or equiv. } [u]_N \in \phi_2^c(w_2). \quad (4.16)$$

Intuitively, variable w_2 is considered as an artificial (hidden) variable, which has the mechanical meaning of a correction interface traction, and which enforces the fulfillment of the adopted nonmonotone law by using monotone constituents. Moreover, w_2^c denotes the convex conjugate (Moreau–Fenchel conjugate, see e.g. Rockafellar, 1970 and (2.48)) of the function ϕ_2. Expressions (4.15), (4.16) give rise to the following local system of variational inequalities

$$\begin{aligned}
\phi_1([v]_N) - \phi_1([u]_N) &\geq -S^T([v]_N - [u]_N) + w_2^T([v]_N - [u]_N), \ \forall [v]_N \in \mathbb{R}, \\
\phi_2([v]_N) - \phi_2([u]_N) &\geq w_2^T([v]_N - [u]_N), \ \forall [v]_N \in \mathbb{R}. \quad (4.17)
\end{aligned}$$

The previous inequalities can be written for each discrete nonlinear interface or boundary element and, after integration, lead to global interface action inequalities. Note that a given d.c. decomposition can be used for the construction of the global one, i.e., (cf. (4.5))

$$\widetilde{\Phi}_N(\mathbf{u}) = \Phi_{1.N}(\mathbf{u}) - \Phi_{2.N}(\mathbf{u}),$$

by simply summing up the two convex constituents separately. Then, by using the discrete virtual work equation (4.8) we get the following system of variational inequalities:

Find the $\mathbf{u} \in V_{ad}$ and $\mathbf{w}_2 \in \mathbb{R}^q$ such as to satisfy the following variational inequality

$$\begin{aligned}
\mathbf{u}^T\mathbf{K}(\mathbf{u}^* - \mathbf{u}) - \bar{\mathbf{p}}^T(\mathbf{u}^* - \mathbf{u}) + \Phi_{1.N}(\mathbf{u}^*) - \Phi_{1.N}(\mathbf{u}) - \\
-\mathbf{w}_2^T(\mathbf{u}^* - \mathbf{u}) \geq 0, \ \forall \mathbf{u}^* \in V_{ad} \quad (4.18)
\end{aligned}$$

and a $\mathbf{w}_2 \in \mathbb{R}^q$ such that

$$\Phi_{2.N}(\mathbf{u}^*) - \Phi_{2.N}(\mathbf{u}) \geq \mathbf{w}_2^T(\mathbf{u}^* - \mathbf{u}), \ \forall \mathbf{u}^* \in V_{ad}. \quad (4.19)$$

Further let us consider the d.c. potential energy

$$\begin{aligned}
\Pi(\mathbf{u}) &= \frac{1}{2}\mathbf{u}^T\mathbf{K}\mathbf{u} - \bar{\mathbf{p}}^T\mathbf{u} + \Phi_{1.N}(\mathbf{u}) - \Phi_{2.N}(\mathbf{u}) = \\
&= \{\Pi_{in}(\mathbf{e}) + \Phi_1(\mathbf{u}) - \Phi_2(\mathbf{u}) - \mathbf{p}^T\mathbf{u} \mid \mathbf{u} \in \mathbb{R}^n\} = \\
&= \underbrace{\Pi_{in}(\mathbf{u}) + \Phi_1(\mathbf{u}) - \mathbf{p}^T\mathbf{u}}_{\Pi_1} - \underbrace{\Phi_2(\mathbf{u})}_{\Pi_2} = \\
&= \Pi_1(\mathbf{u}) - \Pi_2(\mathbf{u}). \quad (4.20)
\end{aligned}$$

Here $\Pi(\mathbf{u})$ is the total potential energy of the structure, $\Pi_{in}(\mathbf{e})$ is the internal elastic energy, $\Phi_1(\mathbf{u})$, $\Phi_2(\mathbf{u})$ are the convex and concave parts of the potential energy that corresponds to the nonlinear elements and, finally, $\Pi_1(\mathbf{u})$ (resp. $\Pi_2(\mathbf{u})$) is the convex (resp. the concave) constituent of $\Pi(\mathbf{u})$.

For completeness one should mention here that a d.c. function belongs to the more general class of quasidifferentiable functions (see Section 2.3.2 and Dem'yanov et al., 1996).

It can easily be shown that the above derived variational problem is equivalent to the calculation of a critical point for the d.c. superpotential. In fact, the necessary optimality condition for this problem reads:
Find $\mathbf{u} \in \mathbb{R}^n$ such that

$$\partial\overline{\Pi}_2(\mathbf{u}) \subset \partial\overline{\Pi}_1(\mathbf{u}). \tag{4.21}$$

The latter differential inclusion leads to the following system of variational inequalities:
Find $\mathbf{u} \in \mathbb{R}^n$ and a field of fictitious stresses $\mathbf{w}_2 \in W$ such as to satisfy:

$$\Pi_1(\mathbf{u}^*) - \Pi_1(\mathbf{u}) - \langle \mathbf{w}_2, \mathbf{u}^* - \mathbf{u} \rangle \geq 0, \ \forall \mathbf{u}^* \in \mathbb{R}^n \tag{4.22}$$

and for each $\mathbf{w}_2 \in W$ such that

$$\Pi_2(\mathbf{u}^*) - \Pi_2(\mathbf{u}) \geq \langle \mathbf{w}_2, \mathbf{u}^* - \mathbf{u} \rangle, \ \forall \mathbf{u}^* \in \mathbb{R}^n. \tag{4.23}$$

The space W is in general a subspace of $\mathbb{R}^n$ and is defined by the effective domain of the conjugate function of Π_2.

It should be mentioned here that the d.c. decomposition of a given function is not uniquely defined, since it holds:

$$\begin{aligned}\phi(x) &= \phi_1(x) - \phi_2(x) = \{\phi_1(x) + h(x)\} - \{\phi_2(x) + h(x)\} = \\ &= \overline{\phi}_1(x) - \overline{\phi}_2(x).\end{aligned} \tag{4.24}$$

Here $h(x)$ is a given convex function and $\{\phi_1(x) - \phi_2(x)\}, \{\overline{\phi}_1(x) - \overline{\phi}_2(x)\}$ are two possible d.c. decompositions of $\phi(x)$. For the numerical analysis of the problem this freedom allows us to consider a more favourable representation. Moreover, an iterative redefinition of a local d.c. representation within an iterative solution scheme is possible.

Analogously to the previous case one may consider interface laws of the form

$$[\mathbf{u}]_N \in \Theta\overline{\widetilde{\phi}}(-\mathbf{S}_N) \tag{4.25}$$

where $\widetilde{\phi}(-\mathbf{S}_N)$ is a nonconvex superpotential, which is assumed to be a difference convex function with convex constituents denoted by $\overline{\phi}_1(\cdot)$ and $\overline{\phi}_2(\cdot)$.

Recall that the same simplification, that only normal to the interface mechanical behaviour is assumed to be of the previous nonlinear type, has been made throughout this Section.

By counting on all interface points we get for the whole interface the global interface potentials

$$\overline{\overline{\Phi}}_{j.N}(\mathbf{s}) = \sum_{i=1}^{q} \overline{\overline{\phi}}_N^{(i)}(-\mathbf{S}_N^{(i)}),\ j = 1, 2, \tag{4.26}$$

which are also d.c. functions. Following analogous reasoning as before, we get the system of variational inequalities:

Find $\mathbf{s} \in \Sigma_{ad}$ and $\overline{\mathbf{w}}_2 \in \mathbf{R}^q$ such as to satisfy the following system of variational inequalities:

$$\begin{aligned} \mathbf{e}^T(\mathbf{s}^* - \mathbf{s}) + \mathbf{e}^T\mathbf{F}_0^T(\mathbf{s}^* - \mathbf{s}) \ &+\ \overline{\Phi}_{1.N}(\mathbf{s}^*) - \overline{\Phi}_{1.N}(\mathbf{s}) - \\ &-\ \overline{\mathbf{w}}_2^T(\mathbf{s}^* - \mathbf{s}) \geq 0,\ \forall \mathbf{s}^* \in \Sigma_{ad} \end{aligned} \tag{4.27}$$

and for $\overline{\mathbf{w}}_2 \in \mathbf{R}^q$ such that

$$\overline{\Phi}_{2.N}(\mathbf{s}^*) - \overline{\Phi}_{2.N}(\mathbf{s}) \ \geq\ \overline{\mathbf{w}}_2^T(\mathbf{s}^* - \mathbf{s}),\ \forall \mathbf{s}^* \in \Sigma_{ad}. \tag{4.28}$$

As previously, the complementary energy function for problem (4.27), (4.28) has the form of the d.c. superpotential

$$\tilde{\Pi}(\overline{\mathbf{s}}) = \frac{1}{2}\mathbf{s}^T\mathbf{F}_0\mathbf{s} + \mathbf{e}_0^T\mathbf{s} + \overline{\Phi}_{1.N}(\mathbf{s}) - \overline{\Phi}_{2.N}(\mathbf{s}). \tag{4.29}$$

One should note here that, in contrast to what happens in the convex case, there is no direct connection, through a conjugacy relation, between the laws described by (4.25) and (4.4). Thus, the two above formulated problems are different although they are equivalent formulations of the same mechanical problem.

4.1.1.3 Heuristic nonconvex optimization approach. Another approach for the solution of problem (4.12) which works even in the cases of complex adhesive laws, where a d.c. decomposition of the superpotential is not possible, is based on the heuristic nonconvex optimization algorithm presented in Chapter 2. According to this method, the problem is reduced to the solution of an appropriate sequence of convex optimization problems.

Let us suppose that instead of the nonmonotone law (4.4) the monotone law

$$-S_N \in \partial\phi_N([u]_N) \tag{4.30}$$

describes the behaviour of the nonlinear elements. Then, as it was demonstrated in Chapter 3, the equilibrium configuration of the structure can be obtained by solving the convex minimization problem of the potential or the complementary energy of the structure.

We notice now, that it is possible to write the nonmonotone law problem in terms of the monotone law problem, i.e. the superpotential of (4.12) can be written in the form

$$\Pi(\mathbf{v}) = \Pi_c^{(i)}(\mathbf{v}) + \left[\widetilde{\Phi}_N(\mathbf{v}) - \Phi_N^{(i)}(\mathbf{v})\right] \tag{4.31}$$

where

$$\Pi_c^{(i)}(\mathbf{v}) = \frac{1}{2}\mathbf{v}^T\mathbf{K}\mathbf{v} - \overline{\mathbf{p}}^T\mathbf{v} + \Phi_N^{(i)}(\mathbf{v}). \tag{4.32}$$

Then, according to the heuristic nonconvex optimization algorithm presented in Section 2.4.3, the following approximation scheme is constructed:

$$\begin{aligned}
\min\left\{\Pi^{(1)}(\mathbf{v})\right\} &= \min\left\{\Pi_c^{(1)}(\mathbf{v}) + \left[\widetilde{\Phi}_N(\mathbf{v}^{(0)}) - \Phi_N^{(0)}(\mathbf{v}^{(0)})\right]\right\} \\
&= \min\left\{\Pi_c^{(1)}(\mathbf{v})\right\} + C^{(1)} \\
&\ \vdots \\
\min\left\{\Pi^{(i)}(\mathbf{v})\right\} &= \min\left\{\Pi_c^{(i)}(\mathbf{v}) + \left[\widetilde{\Phi}_N(\mathbf{v}^{(i-1)}) - \Phi_N^{(i-1)}(\mathbf{v}^{(i-1)})\right]\right\} \\
&= \min\left\{\Pi_c^{(i)}(\mathbf{v})\right\} + C^{(i)} \\
&\ \vdots \\
\min\left\{\Pi^{(n)}(\mathbf{v})\right\} &= \min\left\{\Pi_c^{(n)}(\mathbf{v}) + \left[\widetilde{\Phi}_N(\mathbf{v}^{(n-1)}) - \Phi_N^{(n-1)}(\mathbf{v}^{(n-1)})\right]\right\} \\
&= \min\left\{\Pi_c^{(n)}(\mathbf{v})\right\} + C^{(n)}
\end{aligned} \tag{4.33}$$

where the convex superpotentials $\Phi_N^{(n)}$ are selected such that the following relation is fulfilled

$$\partial\Phi_N^{(i)}(\mathbf{v}^{(i-1)}) = \bar{\partial}\widetilde{\Phi}_N(\mathbf{v}^{(i-1)}). \tag{4.34}$$

Concerning the convergence of the previous iterative scheme the reader is referred to the relative discussion of Section 2.4.3.

Using the above iterative scheme, a solution of (4.12) can be found by solving only the convex minimization problems that correspond to the functions $\Pi_c^{(i)}$. We notice here, that the initial superpotential $\Pi(\mathbf{v})$ and the last one $\Pi_c^{(n)}(\mathbf{v})$ are minimized in the same point $\mathbf{v}^{(n)}$. Of course, the minimum is not the same at the two problems but their difference is the given term $C^{(n)}$. As the solution of (4.12) is not unique, the obtained solution depends on the initial selection of $\mathbf{v}^{(0)}$. Different starting points $\mathbf{v}^{(0)}$ may lead to different solutions $\mathbf{v}^{(n)}$ as it is demonstrated in the numerical applications of Chapter 7.

Of course, by means of the previous relations it is easily verified that a solution of the initial problem (4.12) is obtained using the proposed iterative scheme, but the full proof of convergence remains still an open problem. However, in the various numerical applications presented in Chapter 7, convergence was always achieved.

We have to notice here that the relation (4.34) for the approximation of the nonconvex superpotential $\widetilde{\Phi}_N$, gives to the method a great flexibility. Indeed, there is usually an infinite number of convex superpotentials satisfying (4.34). In general, the convex superpotentials that approximate the nonconvex ones are selected in such a way that the computational effort for the solution of the arising convex minimization subproblems will be minimum. This task depends on the particular nonconvex superpotentials to be approximated. Moreover, in the case of one-dimensional superpotentials the situation is much simpler. In this case, the nonconvex (resp. convex) superpotential gives rise to a one-dimensional nonmonotone law $g([v]_N)$ (resp. monotone law $h([v]_N)$) that relates the normal traction S_N and the relative normal displacements $[v]_N$. Indeed, in this case the differential relation (4.34) reduces to a simple relation between the monotone and the nonmonotone interface laws, i.e. instead of (4.34) we have

$$h^{(i)}(\mathbf{v}^{(i-1)}) = g(\mathbf{v}^{(i-1)}). \tag{4.35}$$

Fig. 4.1 explains this procedure and gives some possible monotone laws fulfilling (4.35). In this figure, A is the solution point provided by the solution of the problem involving the superpotential $\Pi_c^{(i-1)}$. Analogous iterative schemes can also be formulated for the complementary energy substationarity problem. For more details about the algorithmic treatment the reader is referred to Chapter 6 and to Mistakidis and Panagiotopoulos, 1994, Mistakidis et al., 1995, Tzaferopoulos et al., 1995.

Bibliographical Remarks The application of nonconvex potentials, the formulation of nonconvex variational inequality problems and the hemivariational inequalities (after Panagiotopoulos) for the treatment of the corresponding problems, have been studied in several recent publications. The reader is referred to the references given in Chapter 1, the monographies Panagiotopoulos, 1985, Moreau et al., 1988, Moreau and Panagiotopoulos, 1988, Panagiotopoulos, 1993, Naniewicz and Panagiotopoulos, 1995, Dem'yanov et al., 1996 and, among others, to the following publications: Panagiotopoulos, 1983, Baniotopoulos and Panagiotopoulos, 1987, Naniewicz, 1989, Motreanu and Panagiotopoulos, 1993, Miettinen, 1993, Naniewicz, 1994a, Naniewicz, 1994b, Goeleven,

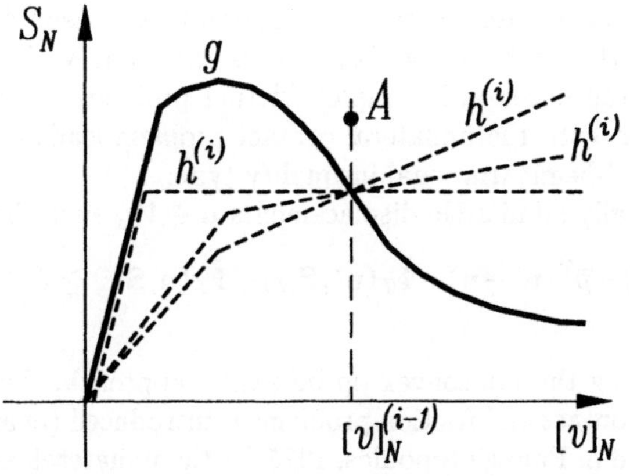

Figure 4.1. Different monotone laws $h^{(i)}$ approximating the nonmonotone law g

1995, Miettinen, 1995, Motreanu and Panagiotopoulos, 1995, Naniewicz, 1995, Goeleven, 1997.

4.1.2 Friction Problems with nonconvex energy potential

By following the methodology of the previous Section and by using a nonconvex friction potential, one may introduce a nonconvex friction potential law (cf. (3.49)):

$$-S_T \in \widetilde{\phi}_T([u]_T).$$

Here, in analogy to the model studied in the previous Section, we first consider only the tangential mechanical behaviour of the interface or boundary. The corresponding mechanical behaviour in the normal direction is assumed to be of a more classical nature (for instance, perfect contact between the two interfaces, i.e., a zero normal relative displacement $[u]_N$ which can be modelled by more classical means).

Moreover, for a difference convex optimization approach this potential is assumed to be written as the difference of two convex functions:

$$\widetilde{\phi}_T([u]_T) = \phi_{1.T}([u]_T) - \phi_{2.T}([u]_T). \tag{4.36}$$

The variational formulation of the friction law with nonconvex friction potential, for a nonmonotone version of the static Coulomb friction models discussed in Section 3.1.3, has the same form with the problems (4.11), (4.15)–(4.20) of the previous Section. In this case $\widetilde{\phi}_T$ takes the place of $\widetilde{\phi}_N$, S_T of S_N, etc.

The same method is used for the formulation of combined unilateral contact and friction problems as in Section 3.1.3. A nonconvex friction potential

which depends on the normal contact traction is assumed: $\widetilde{\phi}_T([u]_T, S_N)$. The formulation of the problem for fixed S_N is the same with the one outlined previously for the decoupled (simple) friction problem. The coupling of the friction problem with the unilateral contact problem leads to problems of the quasi-variational-hemivariational inequality type:
Find kinematically admissible displacements $\mathbf{u} \in V_{ad}$ such that

$$\mathbf{u}^T\mathbf{K}^T(\mathbf{u}^* - \mathbf{u}) - \overline{\mathbf{p}}^T(\mathbf{u}^* - \mathbf{u}) + \widetilde{\Phi}_T(\mathbf{u}^*, \mathbf{S}_N) - \widetilde{\Phi}_T(\mathbf{u}, \mathbf{S}_N) \geq 0, \ \forall \mathbf{u}^* \in V_{ad}. \quad (4.37)$$

Here, following the nonconvex optimization approach, the iterative decoupling between contact and friction problems is introduced (in analogy to the approach proposed in Panagiotopoulos, 1975 for the unilateral, frictional problem with classical Coulomb friction and convex potentials). Within this approach the unilateral contact subproblem presents no difficulty and can be studied by the convex methods of Section 3.1. For the nonconvex friction subproblem the method outlined in the beginning of this Section will be used. More details on this algorithmic approach are given in Chapter 6 and in Mistakidis and Panagiotopoulos, 1994, Mistakidis and Panagiotopoulos, 1993, Mistakidis and Panagiotopoulos, 1997, Stavroulakis and Mistakidis, 1995, Koltsakis et al., 1995.

If both the normal and tangential to the interface tractions admit a nonconvex superpotential, a more general problem can be formulated. In this case the two subproblems arising in the normal and tangential directions are coupled, i.e. the superpotential arising in the tangential direction takes the form $\widetilde{\phi}_T([u]_T, S_N)$ and the superpotential arising in the normal to the interface direction the form $\widetilde{\phi}_N([u]_N, S_T)$. In this case the application of the principle of virtual work to the above problem leads to the quasi-hemivariational inequality:

Find kinematically admissible displacements $\mathbf{u} \in V_{ad}$ such that

$$\begin{aligned}\mathbf{u}^T\mathbf{K}^T(\mathbf{u}^* - \mathbf{u}) - \overline{\mathbf{p}}^T(\mathbf{u}^* - \mathbf{u}) + \widetilde{\Phi}_N(\mathbf{u}^*, \mathbf{S}_T) - \widetilde{\Phi}_N(\mathbf{u}, \mathbf{S}_T) + \\ + \widetilde{\Phi}_T(\mathbf{u}^*, \mathbf{S}_N) - \widetilde{\Phi}_T(\mathbf{u}, \mathbf{S}_N) \geq 0, \ \forall \mathbf{u}^* \in V_{ad}. \quad (4.38)\end{aligned}$$

The above problem is very common in the study of structures connected with adhesive materials, e.g. laminated structure, and its algorithmic treatment will be given in detail in Chapter 6.

It should be mentioned here that nonmonotone friction laws can be used for the modelling of stick-slip phenomena which take into account the difference between the static and the dynamic friction coefficients (the so-called sticktion effect), see, e.g., Rabinowicz, 1959, Ionescu and Paumier, 1994, Visintin, 1994, p. 49, Glocker, 1995.

4.1.3 *General nonmonotone material laws*

In this Section, the treatment presented in Section 3.1.3.4 for the case of general monotone material laws will be extended to cover general nonmonotone material constitutive laws. In this case a nonmonotone material law is considered which may have complete vertical branches. This law is produced through generalized differentiation from a nonconvex, possibly nondifferentiable strain energy density function. These laws are written in the following general form:

$$\mathbf{s}_i \in \bar{\partial}\widetilde{w}_i(\mathbf{e}_i), \text{ or } \mathbf{e}_i \in \bar{\partial}\overline{\overline{w}}_i(\mathbf{s}_i), \; i = 1, \ldots, m, \tag{4.39}$$

where $\widetilde{w}_i(\cdot)$ and $\overline{\overline{w}}_i(\cdot)$ are appropriately defined nonconvex and possibly nondifferentiable potentials.

In order to have a more general description of the problem we consider it as an extension of the interface problems presented in the previous Section, i.e. in the variational formulation presented in the previous Section we introduce a general nonmonotone material law.

Relations (4.39) imply the hemivariational inequalities:

$$\mathbf{s}_i^T(\mathbf{e}_i^* - \mathbf{e}_i) \leq \widetilde{w}_i^0(\mathbf{e}_i^*) - \widetilde{w}_i^0(\mathbf{e}_i), \; i = 1, \ldots, m \tag{4.40}$$

and

$$\mathbf{e}_i^T(\mathbf{s}_i^* - \mathbf{s}_i) \leq \overline{\overline{w}}_i^0(\mathbf{s}_i^*) - \overline{\overline{w}}_i^0(\mathbf{s}_i), \; i = 1, \ldots, m \tag{4.41}$$

where $\widetilde{w}_i^0(.)$ and $\overline{\overline{w}}_i^0(.)$ are the directional derivatives of the superpotentials $\widetilde{w}_i(.)$ and $\overline{\overline{w}}_i(.)$ respectively. Combining the above relations with the discrete virtual or complementary virtual work equations, we formulate the following hemivariational inequality problems:

Find $\mathbf{e}, \mathbf{u} \in V_{ad}$ such that:

$$\widetilde{W}(\mathbf{e}^*) - \widetilde{W}(\mathbf{e}) + \widetilde{\Phi}_N(\mathbf{u}^*) - \widetilde{\Phi}_N(\mathbf{u}) + \widetilde{\Phi}_T(\mathbf{u}^*) - \widetilde{\Phi}_T(\mathbf{u}) - \overline{\mathbf{p}}^T(\mathbf{u}^* - \mathbf{u}) \geq 0,$$
$$\forall \mathbf{u}^* \in V_{ad}, \mathbf{e}^* \text{ s.t. (4.2) is satisfied.} \tag{4.42}$$

Find $\mathbf{s} \in \Sigma_{ad}$ such that:

$$\overline{\overline{W}}(\mathbf{s}^*) - \overline{\overline{W}}(\mathbf{s}) + \overline{\overline{\Phi}}_N(\mathbf{s}^*) - \overline{\overline{\Phi}}_N(\mathbf{s}) + \overline{\overline{\Phi}}_T(\mathbf{s}^*) - \overline{\overline{\Phi}}_T(\mathbf{s}) \geq 0, \; \forall \mathbf{s}^* \in \Sigma_{ad}. \tag{4.43}$$

Here $\widetilde{W}$ (resp. $\overline{\overline{W}}$) is the sum over all finite elements $i = 1, \ldots, m$ of the local potentials $\widetilde{w}_i$ (resp. $\overline{\overline{w}}_i$). Moreover, the sets V_{ad} and Σ_{ad} are defined as in relations (4.9) and (4.14) previously.

Following the previous approach, nonconvex substationarity problems can be written for the potential and the complementary energy of the structure.

4.1.4 Large displacement and deformation problems: nonconvex optimization problems

Besides the nonlinearity which is caused by the material laws or the boundary and interface conditions, as it has been considered till now, nonlinear kinematics is another cause of nonlinearity in structural analysis applications. The point here is that the displacements or / and the deformations of a given system are no longer small, so that their nonlinearity cannot be neglected in the formulation of the governing relations of the system. Accordingly, for large deformations, the strain-displacement compatibility relation (cf. (3.2)) is in general a nonlinear mapping. Moreover, in order to account for the effect of the large displacements into the equilibrium, the equilibrium equations (cf. (3.1)) must be written in the deformed configuration of the structure. To this end, for instance, a Lagrangian coordinate system that follows the material points of the structure in their current, deformed configuration is employed.

Considering the virtual work expression of the structural analysis problem (cf. (3.7), (3.9)), even if the virtual variations of the involved quantities are assumed to be small, the resulting terms are not linear (or bilinear) if one considers nonlinear kinematic transformations.

On the potential energy level one should first note that not all large displacement problems admit a potential energy description. For instance, the noncommutativity of the large rotations effect leads to structural operators which do not admit a potential. Here, for simplicity, problems which do admit a potential energy will be discussed. Nevertheless, even with this simplification, this potential can not be in general convex and for some applications is nondifferentiable as well.

To fix ideas, consider the potential energy optimization problem. Roughly speaking and without going into the details in this presentation, let a stored elastic energy potential exist and let it be a function of the deformation vector (cf. the first term in (3.11)). Even if this function is convex, as it will be expected to be for every appropriately modelled structure composed of stable materials, the nonlinear, due to the large deformations, strain-displacement compatibility operator makes the composite function in general nonconvex in terms of the displacement variables.

In this Section a small number of techniques will be outlined for the treatment of the previously mentioned problems. These methods can be applied directly either on the virtual work expression or on the potential energy minimization formulation of the problem. They are based on the well-known techniques of iterative linearization of the involved nonlinear quantities. This technique can be enlarged to cope with inequality constraints expressed by nonlinear functions after appropriate linearization of these latter functions. Moreover, from the potential energy minimization point of view, the iterative linearization corresponds to a sequential quadratic programming technique for the treatment of the general energy optimization problem. Accordingly, one may write iterative techniques where the arising subproblems correspond to convex, not necessarily differentiable potential functions. Thus, general sequential convex programming techniques can be proposed. In this sense, the whole procedure is qualified as an iterative or multilevel optimization approach for the treatment of the initially nonconvex problem and is connected with the methods and algorithms studied in this book.

We should mention here that the discussion of all theoretical and computational aspects connected with the modelling of large displacement and deformation problems lies beyond the scope of the present book. The short discussion which follows is mainly based on the approach presented in Zeidler, 1988, Curnier, 1993. More material on the theoretical and computational aspects of nonlinear elasticity can be found, among others, in: Cook, 1978, Bathe, 1981, Marsden and Hughes, 1983, Ciarlet, 1983, Stein et al., 1989, Björkman, 1991, Crisfield, 1991, Zienkiewicz and Taylor, 1991, Betten, 1993, Altenbach and Altenbach, 1994, Pilkey and Wunderlich, 1994, Björkman et al., 1995, Chung, 1996.

4.1.4.1 Formulation of the problem. Let a fixed Cartesian coordinate system be assumed. Let for each material point $x \in \Omega$ the initial (undeformed) position be denoted by x and the displacements of the point x be denoted by $u(x)$. Thus, the final (deformed) position of the sought material point is:

$$y(x) = x + u(x). \tag{4.44}$$

Here a description of the system in terms of Lagrangian (x) or Eulerian (y) coordinates can be used. A Lagrangian description where everything is referred to the undeformed configuration is adopted.

The symmetric Green-Lagrange material deformation tensor is used for the description of the deformation of the structure. It is defined by:

$$E(x) = \frac{1}{2}(C - I) = \frac{1}{2}\left(H + H^T + H^T H\right), \tag{4.45}$$

where $C(x) = F^T F$ is the left Cauchy-Green material metric tensor, $F(x) = \frac{\partial y(x)}{\partial x} = \nabla y$ is the nonsymmetric gradient of deformation matrix and $H = \nabla u = F - I$ is the gradient of the displacements matrix.

At the material level the true, second Piola-Kirchhoff stress tensor S is used to express the stresses of the system. This stress measure is energetically equivalent (dual) to the above defined material deformation tensor E. Nevertheless, for the writing of the structural analysis problem in the initial (undeformed) configuration, the nominal first Piola-Kirchhoff stress tensor $P(x)$ is also used with: $S = F^{-1}P(x)$. Note that the nominal stress is referred to the undeformed configuration of the structure and, although it is an artificial quantity, is used for the formulation of the problem in the undeformed configuration. A more detailed discussion on other possibilities lies beyond the scope of this work.

In the above outlined framework the elastostatic structural analysis problem reads (cf. e.g. Zeidler, 1988, p. 187):

- *kinematic deformation field* $u(x) \in \Omega$ such that (4.44) holds,
- *equilibrium equations* in the undeformed configuration are satisfied, i.e.:

$$div P(x) \; + p(x) = 0, \text{ on } \Omega, \tag{4.46}$$

- *material constitutive law (nominal)*

$$P(x) = \mathcal{A}(x, u(x), u'(x)) = \mathcal{A}(x, u(x), F(x)), \text{ on } \Omega, \tag{4.47}$$

- and *boundary conditions*, e.g.,

$$u(x) = u_0, \text{ on } \Gamma_U \subset \Gamma, \tag{4.48}$$

$$P(x)n = p_N, \text{ on } \Gamma_F \subset \Gamma. \tag{4.49}$$

By assuming arbitrary small virtual displacements $v \in \Omega$, multiplying (4.46) by v, assuming sufficient differentiability of the involved quantities such as to use the integration by parts, one gets the variational formulation of the problem (see, e.g., Curnier, 1993, p.237):

$$\int_\Omega (\nabla v, P(x, u(x), F(x)) \, d\Omega = \int_\Omega v p(x) d\Omega + \int_\Gamma v p_N(x) d\Gamma, \; \forall \; v. \tag{4.50}$$

First, one observes that the term $P(x, ., .)$ in (4.50) is nonlinear due to the assumption of large, nonlinear deformations. Moreover, the nominal material constitutive law of (4.47) does not have a clear mechanical meaning and is not always objective. Thus, it cannot be used in this form for the formulation of an engineering problem. Recall here that objectivity requires that the same relation holds true irrespective of the coordinate system in which it is expressed and that, especially for constitutive laws, rigid body displacements and rotations do not affect the introduced law. Finally, mechanical restrictions require that a nonobjective nominal material law (4.47) must necessarily be produced by a nonconvex potential (see Curnier, 1993, p. 252).

For all the above outlined reasons, one introduces material constitutive laws expressed in terms of the real variables S, E, i.e., for example a hyperelastic law of the form:

$$S = \frac{dW(E)}{dE}. \tag{4.51}$$

Examples of potentials $W(E)$ or directly material laws $S = S(E)$ which can be used in (4.51) and which comply with the required material restrictions can be found in the specialized literature and will not be given here (see, e.g., Malvern, 1969, Zeidler, 1988, Betten, 1993, Curnier, 1993, Chen, 1994, Vol. I). For instance, under appropriate differentiability assumptions, a general form of the potential energy can be obtained by taking a Taylor series expansion around a given E_0, say $E_0 = 0$ (see Chung, 1996, p. 102):

$$W(E) = W_0 + K_{ij}E_{ij} + \frac{1}{2}K_{ijkm}E_{ij}E_{km} + \frac{1}{3}K_{ijkmnp}E_{ij}E_{km}E_{np} + \ldots, \tag{4.52}$$

where W_0 is a constant, $K_{ij}\ldots$ are tensorial variables, the second term stands for the effect of residual stresses and the nonlinear behaviour is described by the third and eventually higher order terms.

One must only observe that even if (4.51) leads to a linear material relation, for instance to a Hookean law of the form

$$S(E) = \lambda tr(E)I + 2\mu E, \tag{4.53}$$

the nonlinear transformations which are applied to bring this relation back into the nominal material relation (4.47) destroy the linearity (thus, also every convexity assumption concerning the potential).

So far we have discussed the nonlinearity induced by a material relation. Another cause of nonlinearity comes from the kinematical relation (4.45).

In some cases the previously described nonlinear elasticity problem can directly be derived by writing down the Euler critical point conditions for an appropriately defined potential energy function (see, e.g., Zeidler, 1988, p. 190):

$$\Pi(u) = \int_\Omega L(x, u'(x))dx - \int_\Omega p(x)u(x)dx - \int_{\Gamma_u} p_N(x)u(x)dx, \tag{4.54}$$

restricted by the kinematic boundary conditions (4.48). Obviously, $\Pi(u)$, in view of the previous discussion, cannot be a convex function anymore due to the presence of the first term in the r.h.s. of (4.54).

The method of iterative linearization is used for the linearization of all the above nonlinear relations around a given point. Thus, on the assumption of sufficient differentiability of all involved quantities, the initial nonlinear problem is replaced by a number of iteratively linearized problems. To this end, all quantities are linearized around a given point by using the chain rule of differentiation. This way a stepwise linear problem arises where in (4.50) only linear and bilinear terms appear. Respectively, in the potential energy (4.54) only linear and quadratic terms arise (the approximate problem in the terminology of Zeidler, 1988, or a sequential quadratic approximation of the nonlinear optimization problem in the terminology of mathematical programming). In turn, linear elastic solvers are used for the solution of the linearized subproblems and the procedure is repeated for all required iterations till convergence. The readers are referred to the nonlinear computational literature for more details (Zienkiewicz and Taylor, 1991, Curnier, 1993).

This procedure has been extended to include inequality constraints which describe unilateral contact effects (see, e.g., Björkman, 1991, Björkman et al., 1995, Givoli and Doukhovni, 1996, Sun and Natori, 1996).

One should first mention here that, to the authors best knowledge, no attempt has been made to include convexity and concavity information in the above outlined linearization procedure. Moreover, the experience with nondifferentiable relations is restricted.

An extension of the bi-modular elasticity relations into the range of large displacement elastostatics has recently been proposed in Curnier et al., 1995. In this paper piecewise differentiable potential functions have been introduced for the description of the elastic material behaviour and this approach has been integrated into the large displacements regime. Nevertheless, only continuous potentials with continuous first derivatives have been discussed there, which is sufficient for the bi-modular elasticity relations where the $S - E$ material law is continuous. This restriction, in view of the formulations discussed in this book (see also Dem'yanov et al., 1996) seems unnecessary and can be removed to cover, for instance, locking behaviour or cracking and crushing effects.

A linearization technique appropriate for nonsmooth but convex, large displacements and small strains problems which uses the notion of duality gap due to large displacements and leads to linear dual (approximate) problems has been proposed and studied for several applications in Gao, 1992, Gao and Strang, 1989a, Gao and Strang, 1989b, Gao, 1996. In this direction one may also consult the works of Galka and Telega, 1995a, Galka and Telega, 1995b.

If the involved potentials belong to the class of the quasidifferentiable functions (which class includes the difference convex functions), then the classical chain rules of differentiation can be replaced by the quasidifferential calculus rules (Dem'yanov et al., 1996) for the formulation of the variational problem.

In principle, this approach can treat simultaneously the nonconvexity and the nondifferentiability problems.

Another approach for the treatment of nonconvex potential optimization problems arising in mechanics follows parametric optimization or path-following techniques. In this case, for instance, the potential energy function (4.52) is parametrized by means of the variables of interest, e.g., the loading, and the corresponding structural analysis problem is studied as a parametrized potential energy optimization (or substationarity) problem. This approach includes the load incrementation techniques of computational mechanics and can be extended to cover certain classes of nondifferentiable problems. For some first results in this direction the reader is referred to Rohde and Stavroulakis, 1995, Rohde and Stavroulakis, 1997.

Finally a first attempt to introduce piecewise smooth continuous selections of smooth functions in potential energy problems of mechanics has been proposed in Stavroulakis and Rohde, 1996.

In conclusion one should say that this area of research remains at the present open for further investigations.

4.1.4.2 Large displacement unilateral contact problems. In the following an iterative procedure is presented for the solution of a unilateral contact problem within the geometrical nonlinearity framework. In order to improve the comprehension of the formulation, each quantity is provided with a left superscript and a left subscript. The left superscript indicates the configuration to which the quantity occurs and the left subscript indicates the configuration with respect to which the quantity is measured.

We consider now a three-dimensional linear elastic body Ω with a boundary Γ. The boundary is decomposed into three mutually disjoint parts, Γ_U, Γ_F and Γ_N. It is assumed that on Γ_U (resp. Γ_F) the displacements (resp. the tractions) are prescribed. Moreover, Γ_N denotes the region of the boundary on which unilateral contact conditions hold with zero tangential forces, i.e.

$$-S_N \geq 0\ ,\ [u]_N - g \leq 0\ ,\ -S_N([u]_N - g) = 0 \qquad (4.55)$$

Since the body can undergo large displacements and large strains, the direct solution of the problem is not possible. However, an approximate solution can be obtained by referring all variables to a previously known equilibrium configuration and linearizing the resulting relations. This solution can by improved by iteration. In order to develop the governing relations for the approximate solution obtained by linearization, we assume that the solutions for times 0, Δt, $2\Delta t$, ..., t have already been calculated. Thus, any one of the already calculated equilibrium configuration can be used. In the following, in order to simplify the calculations, we prefer to follow the total Lagrangian formulation, where all the quantities are referred to the initial configuration at time 0. Our aim is the calculation of all the quantities for the time $t + \Delta t$. The boundary conditions for the above problem imply the variational inequality

$$\int_{{}^0\Omega} {}^{t+\Delta t}{}_0S_{ij}\delta_0^{t+\Delta t}E_{ij}{}^0 d\Omega - \int_{{}^{t+\Delta t}\Gamma_F} {}^{t+\Delta t}F_i \delta u_i{}^{t+\Delta t} d\Gamma_F -$$

$$- \int_{{}^{t+\Delta t}\Omega} {}^{t+\Delta t}f_i \delta u_i{}^{t+\Delta t} d\Omega \geq 0 \quad \forall\, u^* = u + \delta u \text{ s.t. (4.55) holds on } \Gamma_N$$

$$\text{and } u_i^* = 0 \text{ on } \Gamma_U. \quad (4.56)$$

Here f_i, F_i are the externally applied body and surface force vectors and δ denotes the kinematically admissible variations, i.e. $\delta u = u^* - u$ where $u^* = 0$ on Γ_U. The previous relation is a variational inequality expressing the principle of virtual work in total Lagrange form. After geometric linearization according to Bathe, 1981 and discretization, relation (4.56) yields the following minimization problem:

$$\min\left\{\frac{1}{2}\Delta\mathbf{u}^{(k)T}({}_0^t\mathbf{K}_L + {}_0^t\mathbf{K}_{NL})\Delta\mathbf{u}^{(k)} - ({}^{t+\Delta t}\mathbf{R} - {}^{t+\Delta t}{}_0\mathbf{F}^{(k-1)})^T\Delta\mathbf{u}^{(k)}\right.$$

$$\left. \mid \mathbf{A}\Delta\mathbf{u}^{(k)} \leq \mathbf{g}^{(k)}\right\} \quad (4.57)$$

with

$$\mathbf{g}^{(k)} = \mathbf{g}^{(k-1)} - \Delta\mathbf{u}^{(k-1)}, \quad {}^{t+\Delta t}\mathbf{u}^{(k)} = {}^{t+\Delta t}\mathbf{u}^{(k-1)} - \Delta\mathbf{u}^{(k)}, \quad {}^{t+\Delta t}\mathbf{u}_i^{(0)} = {}^t\mathbf{u}_i. \quad (4.58)$$

Here ${}_0^t\mathbf{K}_L$ (resp. ${}_0^t\mathbf{K}_{NL}$) is the linear (resp. nonlinear) strain incremental stiffness matrix, ${}^{t+\Delta t}\mathbf{R}$ is the vector of externally applied nodal point load at the time $t+\Delta t$, ${}^{t+\Delta t}{}_0\mathbf{F}^{(k-1)}$ is the vector of nodal point forces equivalent to element stresses at the time $t+\Delta t$ in the $(k-1)$ iteration (internal forces) and finally $\mathbf{A}$ and $\mathbf{g}^{(k)}$ are appropriately defined matrix and vector introducing the inequality constrains. The solution of every step of the proposed iterative scheme can be obtained by means of any quadratic programming (Q.P.) algorithm.

The previous optimization problem corresponds to a modified Newton iteration scheme, which is repeated for $k = 1, 2, \ldots$ until the out of balance vector ${}^{t+\Delta t}\mathbf{R} - {}^{t+\Delta t}{}_0\mathbf{F}^{(k-1)}$ becomes negligible. Numerical applications of the above formulation can be found in Panagiotopoulos et al., 1993.

Bibliographical Remarks

Unilateral contact problems for large displacement and deformation problems have been extensively studied in the last years in the computational mechanics literature. The interest on these problems comes, among others, from the need of simulation of metal forming processes. For more details in this area the reader is referred, among others, to Johnson and Quigley, 1989, Wriggers et al., 1990, Ghosh, 1992, Zhu and Jin, 1994, Klarbring, 1995, Björkman et al., 1995, Lochegnies and Oudin, 1995.

4.2 ENERGY AND DISSIPATION PROBLEMS: THE CASE OF STANDARD GENERALIZED ELASTOPLASTICITY

4.2.1 *Nonconvexity in elastoplasticity*

As it has been discussed in the previous Chapter, classical elastoplasticity models are based on the small displacements and deformations assumption with linear or linearized hardening laws. Thus, the holonomic elastoplasticity problem (and the stepwise holonomic one, arising in a time marshing scheme) are equivalent to convex optimization problems. If some of these assumptions are not valid then the arising problems are nonconvex. The most usual causes of nonconvexity come from the adoption of a nonconvex yield surface and from the assumption of strain-softening behaviour. In view of the discussions undertaken in the previous Sections the second assumption is equivalent to the adoption of stress-strain relations with falling (descending) branches. Since several contributions have been made concerning the mechanical meaning of such an assumption and of its validity, this part starts with a small review of the relevant bibliography and with a summary of the proposed arguments. Model problems of nonconvex elastoplasticity are developed in the next Sections.

Let us first concentrate our attention on the assumption that the yield set is a convex set. From the mechanical point of view, convexity of the yield surface and the normality associated flow rule for standard elastoplastic materials are consequences of the Drucker's maximum dissipation postulate. This assumption has been adopted until now, in a large extend, without investigation of the range of its validity. The same holds true for Ilyushin's postulate (see e.g. Lubliner, 1990, p.116, Bažant and Cedolin, 1991, pp. 687, 686). In a more critical interpretation it has been observed that the violation of Drucker's stability postulate at a particular point of a structure does not violate any thermodynamical principle and accordingly, the postulate should be considered for the whole structure. In fact, even if Drucker's postulate is adopted for the whole structure, its validity at every point of the structure is assured only for the highly idealized case of a structure which is subjected to a uniform stress and strain field (see also Malvern, 1969, p.360). On the other hand Ilyushin's postulate relaxes in some sense the convexity requirements for the yield surface. Nevertheless, as with the Drucker's postulate, its validity for a structural system is subjected to the same requirement as previously. In other words convexity, which often is adopted in elastoplasticity, is a consequence of assumptions which may be relaxed without violation of any basic principle of mechanics and thermodynamics. Other arguments for the permission of nonconvex yield surfaces in elastoplasticity can be found in Green and Naghdi, 1965, Justusson and Phillips, 1966, Hill and Rice, 1973, Findley and Michno, 1987, Hill, 1987, Panagiotopoulos, 1985, p.154, Kim and Oden, 1985, Kuczma and Stein, 1994, Nemat-Nasser, 1992 where also references to experiments are given.

In a more general context, star–shaped nonconvex yield functions have been justified in Nadai, 1963 based on thermodynamical arguments. For instance, in the holonomic (Hencky type) elastoplasticity model described in Section 3.2.1 the yield function $\sigma \to F(\sigma)$ can be nonconvex but star–shaped. This means that $0 \in \text{int } F(\sigma)$ and every line of the form $\{\lambda\sigma | \lambda \geq 0, \ \sigma \in F(\sigma)\}$ intersects the boundary of $F(\sigma)$ at most in one point, (Green and Naghdi, 1965 , Salençon and Tristan-Lopez, 1980, p.523).

Moreover, the adoption of a generalized standard elastoplasticity framework does not require the explicit definition of a yield surface: the dissipation potential (cf., (3.127), (3.120)) can directly be defined such as to comply with the thermodynamic requirements (no production of energy) and this potential can be nonconvex (and can be related, in some cases with star-shaped sets, see Rubinov and Yagubov, 1986, Dem'yanov et al., 1996). A generalized nonconvex potential function which produces both plastic strain rates and rates of internal variables has been proposed and used in Kim and Oden, 1984, Kim and Oden, 1985.

It should be noted that if elastic-plastic coupling exists, i.e., if the elasticity tensor of the material changes with the plastic deformation (this case may arise, for instance, in large displacements and deformations problems), then, even if Drucker's stability postulate is adopted, the convexity of the yield surface can not be guaranteed (see, for instance, Panagiotopoulos, 1985, p. 153, Khan and Juang, 1995, p. 155, Fosdick and Volkmann, 1993).

Finally, several criteria introduced to describe, e.g., anisotropic plasticity, can be nonconvex for some values of their parameters (see, among others, the works Hill, 1979 and Zhu et al., 1987 referred also in Kuczma and Stein, 1994, the bounding surface model for cohesive solid proposed by Argyris and et. al., 1974 and discussed in Lin and Bažant, 1986 and Chen, 1994, Vol. II, p. 1050).

In this Section model elastoplasticity problems with yield surfaces which have the form of general nonconvex sets, possibly with reentrant corners, or with softening effects, are considered. For the case of nonconvex yield surface we assume that it is a structured set, i.e., it is a set described by differences of convex functions (sets) or that it is a quasidifferentiable set. Using this structure and an iterative convex decomposition, the nonconvex elastoplasticity problem is written as a sequence of classical, convex subproblems. Finally, the complications introduced by softening effects are discussed, with reference to the plasticity models of the previous Chapter.

4.2.2 *Holonomic, Hencky–type problems*

The quasidifferentiability concept and the application of the theory of star–shaped sets in nonsmooth optimization (see Rubinov and Yagubov, 1986) have been used for the systematic derivation of variational inequalities in a class of discrete nonconvex elastoplasticity problems (Panagiotopoulos and Stavroulakis, 1992, Stavroulakis, 1995b, Stavroulakis, 1995a, Dem'yanov et al., 1996). By using these notions from nonsmooth analysis a variational formulation is obtained for nonconvex problems by extending the classical convex analysis approach to elastoplasticity.

Let us extend the model of Section 3.2.1 by assuming first a classical Hencky-type elastoplasticity problem for a hyperelastic material described by a non-convex and nonsmooth but quasidifferentiable internal energy function $\tilde{w}_m$.

Recall here that a quasidifferentiable superpotential $\tilde{w}_m$ produces the constitutive law of the hyperelastic material in the following way: there exists a couple of convex and compact subsets $\underline{\partial}\tilde{w}_m$, $\overline{\partial}\tilde{w}_m$, such that

$$\tilde{w}'_m(\sigma, \tau - \sigma) = \max_{w_1^e \in \underline{\partial}\tilde{w}_m} (w_1^e, \tau - \sigma) + \min_{w_2^e \in \overline{\partial}\tilde{w}_m} (w_1^e, \tau - \sigma). \tag{4.59}$$

The r.h.s. of (4.59) is the virtual complementary work of the elastic elements for the virtual stress increment $\tau - \sigma$ at the point σ.

From the relations (3.106), (4.59) and (3.115) we get the following variational problems:

a) Find $\sigma \in P$ such that

$$\max_{w_1^e \in \underline{\partial}\tilde{w}_m(\sigma)} (w_1^e, \tau - \sigma) + \min_{w_2^e \in \overline{\partial}\tilde{w}_m(\sigma)} (w_2^e, \tau - \sigma) - < l, \tau - \sigma > \geq 0,$$

$$\forall \tau \in P. \tag{4.60}$$

b) Find $\sigma \in P$ such that $\forall w_2^e \in \overline{\partial}\tilde{w}_m(\sigma)$ the following inequality holds:

$$\max_{w_1^e \in \underline{\partial}\tilde{w}_m(\sigma)} (w_1^e, \tau - \sigma) + (w_2^e, \tau - \sigma) - < l, \tau - \sigma > \geq 0, \ \forall \tau \in P. \tag{4.61}$$

c) Find $\sigma \in P$ such that there exists $w_1^e \in \underline{\partial}\tilde{w}_m(\sigma)$ for which $\forall w_2^e \in \overline{\partial}\tilde{w}_m(\sigma)$ the following inequality holds:

$$(w_1^e, \tau - \sigma) + (w_2^e, \tau - \sigma) - < l, \tau - \sigma > \geq 0, \ \forall \tau \in P. \tag{4.62}$$

By analogous arguments a nonconvex yield surface which is described by a quasidifferentiable function can be assumed.

Let us introduce the generally nonconvex yield surface $P \subset \Sigma_{ad}$ and let the plastic strains be defined to be normals to the boundary of P at the point σ. The yield surface is defined by means of the generally nonconvex but quasidifferentiable function $F(\sigma)$, i.e.

$$P = \{\sigma \in \Sigma_{ad} \mid F(\sigma) \leq 0\}, \tag{4.63}$$

and thus the set P is quasidifferentiable. With respect to the set P of (4.63) the following cones of admissible variations may be defined at a point σ:

$$\gamma_1(\sigma) = \{g \in \Sigma_{ad} \mid F'(\sigma, g) \leq 0\}, \tag{4.64}$$

$$\gamma_0(\sigma) = \{g \in \Sigma_{ad} \mid F'(\sigma, g) = 0\}, \tag{4.65}$$

$$\gamma(\sigma) = \{g \in \Sigma_{ad} \mid F'(\sigma, g) < 0\}. \tag{4.66}$$

As it has been discussed in Chapt. 2 of Dem'yanov et al., 1996, if $F(\sigma)$ is quasidifferentiable, the cones (4.64)–(4.66) can be expressed through the sets $\underline{\partial}F$ and $\overline{\partial}F$ in the following form (see also Dem'yanov and Vasiliev, 1985, pp. 242–251):

$$\gamma_1(\sigma) = \bigcup_{w \in \overline{\partial}F(\sigma)} [K^0(\underline{\partial}F(\sigma) + w)], \tag{4.67}$$

$$\gamma_0(\sigma) = \bigcup_{v \in \underline{\partial}F(\sigma) w \in \overline{\partial}F(\sigma)} [-K^0(\overline{\partial}F(\sigma) + v) \cap K^0(\underline{\partial}F(\sigma) + w)], \tag{4.68}$$

$$\gamma(\sigma) = \Sigma_{ad} \backslash \bigcup_{v \in \underline{\partial}F(\sigma)} [-K^0(\overline{\partial}F(\sigma) + v)]. \tag{4.69}$$

Here K^0 denotes the polar cone with respect to the cone K. Let moreover the regularity condition

$$\overline{\gamma}(\sigma) = \gamma_1(\sigma), \ \sigma \in P \tag{4.70}$$

holds at σ, where $\overline{\gamma}$ denotes the closure of γ in the Euclidean space of the plastic strain tensors. It is obvious that $\gamma_1(\sigma)$ written in the form of (4.67) constitutes a systematic convex partitioning of the general nonconvex cone of feasible directions of the set P at the point σ.

To see this, let us write the cone of feasible directions (Bouligand cone) to a set defined by (4.63) explicitly by means of the directional derivatives of $F(.)$ as

$$\Gamma_1(\sigma) = \{g \in \mathbb{R}^n \mid F'(\sigma, g) \leq 0\}. \tag{4.71}$$

For a quasidifferentiable function F with $\mathcal{D}F(\sigma) = [\underline{\partial}F(\sigma), \overline{\partial}F(\sigma)]$ we have:

$$\Gamma_1(\sigma) = \left\{ g \in \mathbb{R}^n \mid \max_{v \in \underline{\partial}F(\sigma)} < v, g > + \min_{w \in \overline{\partial}F(\sigma)} < w, g > \leq 0 \right\}. \tag{4.72}$$

By means of (4.72), Γ_1 is equivalently expressed as

$$\begin{aligned} \Gamma_1(\sigma) &= \{g \in \mathbb{R}^n \mid < v, g > + < w, g > \leq 0, \\ \forall v &\in \underline{\partial}F(\sigma), \ \text{for some } w \in \overline{\partial}F(\sigma)\} \end{aligned} \tag{4.73}$$

or

$$\begin{aligned} \Gamma_1(\sigma) &= \bigcup_{v \in \overline{\partial}F(\sigma)} \ \bigcap_{w \in \underline{\partial}F(\sigma)} T_{v+w}(\sigma) \\ &= \bigcup_{v \in \overline{\partial}F(\sigma)} T_{\underline{\partial}F(\sigma)+w}(\sigma) \end{aligned} \tag{4.74}$$

where the following abbreviations have been used:

$$T_{v+w}(\sigma) = \{g \in \mathbb{R}^n \mid < v, g > + < w, g > \leq 0 \} \tag{4.75}$$

and

$$\begin{aligned} T_{\underline{\partial}F(\sigma)+w}(\sigma) &= \{g \in \mathbb{R}^n \mid < v, g > + < w, g > \leq 0, \\ \forall v &\in \underline{\partial}F(\sigma) \}. \end{aligned} \tag{4.76}$$

Consequently, every cone of the form:

$$\mathcal{N}_1(\sigma) = K(\underline{\partial}F(\sigma) + w), \ w \in \overline{\partial}F(\sigma) \tag{4.77}$$

is called a quasi–normal cone to the set P, while the convex normal cone to P is given by

$$\mathcal{N}_P(\sigma) = \bigcap_{w \in \overline{\partial}F(\sigma)} K(\underline{\partial}F(\sigma) + w). \tag{4.78}$$

Here $K(A)$ denotes the convex conical hull of the set A.

In view of the previous representation of the set $\Gamma_1(\sigma)$ we also have (see Dem'yanov et al., 1996):

$$\begin{aligned}\mathcal{N}_P(\sigma) &= \bigcap_{v\in\overline{\partial}F(\sigma)} \bigcup_{w\in\underline{\partial}F(\sigma)} T^0_{v+w}(\sigma) = \\ &= \bigcap_{v\in\overline{\partial}F(\sigma)} T^0_{\underline{\partial}F(\sigma)+w}(\sigma) = \bigcap_{v\in\overline{\partial}F(\sigma)} \{ -T^+_{\underline{\partial}F(\sigma)+w}(\sigma) \}.\end{aligned} \tag{4.79}$$

Here the polar cone to T is denoted by T^0, and the conjugate cone of T is denoted by T^+. The following relations hold true:

$$T^0 = -T^+ = \{g \in \mathbb{R}^n \mid \ < g, x > \leq 1, \ \forall x \in T \}. \tag{4.80}$$

After this short mathematical introduction we define plasticity laws in the form:

$$\epsilon^p \in \mathcal{N}_P(\sigma) \tag{4.81}$$

or equivalently

$$(\epsilon^p, \tau - \sigma) \leq 0, \ \forall \tau - \sigma \in \mathcal{N}^0_P(\sigma) \tag{4.82}$$

or equivalently

$$(\epsilon^p, \tau - \sigma) \leq \mathcal{I}_P(\tau) - \mathcal{I}_P(\sigma), \ \forall \tau \in \Sigma_{ad}, \tag{4.83}$$

where $\mathcal{I}_P(\sigma)$ denotes the indicator function of the set P.

From the relations (3.106), (4.59) and (4.81)–(4.83) we can now formulate the following problems:
Find $\sigma \in P$ such that

$$\max_{w_1^e\in\underline{\partial}\tilde{w}_m(\sigma)} (w_1^e, \tau - \sigma) + \min_{w_2^e\in\overline{\partial}\tilde{w}_m(\sigma)} (w_2^e, \tau - \sigma) - < l, \tau - \sigma > \geq 0,$$

$$\forall \tau \in P, \text{ such that } \tau - \sigma \in \mathcal{N}^0_P(\sigma). \tag{4.84}$$

In view of the relations (4.78), (4.81) and the relations (4.63), (4.64), (4.67) we get from (4.84) the following problem:
Find $\sigma \in \Sigma_{ad}$ such that $\exists w_1^e \in \underline{\partial}\tilde{w}_m(\sigma)$, $\forall w_2^e \in \overline{\partial}\tilde{w}_m(\sigma)$, $\exists w_2 \in \overline{\partial}F(\sigma)$, $\forall w_1 \in \underline{\partial}F(\sigma)$ and the following inequality holds true:

$$(w_1^e, \tau - \sigma) + (w_2^e, \tau - \sigma) + (w_2 + w_1, \tau - \sigma) - < l, \tau - \sigma > \geq 0, \ \forall \tau \in \Sigma_{ad}. \quad (4.85)$$

Remark 1: In the above concise formulation one may perform an iterative convexification of the problem by fixing, in each iteration, one element of $w_2^e \in \overline{\partial}\tilde{w}_m(\sigma)$ and $w_2 \in \overline{\partial}F(\sigma)$. The resulting problems are convex analysis elastoplastic problems of the type discussed in the previous Chapter. Of course this technique can be used if the superpotentials of the functions $\tilde{w}_m(\sigma)$ and $F(\sigma)$ are known. This is the case, for instance, for difference convex sets $\tilde{w}_m(\sigma)$ and $F(\sigma)$ (respectively of a yield function $P(\sigma)$ which is a difference of convex sets, see, among others, Horst et al., 1995, Dem'yanov et al., 1996).

Remark 2: We have to notice here that with respect to the problems discussed in this section, the heuristic nonconvex optimization method presented in the previous Chapters seems to be promising. Indeed, instead of solving the nonconvex optimization problem induced by (4.63), an appropriate sequence of convex problems can be solved, by considering in each step the solution of a problem with a convex yield function of the type of (3.111). Here the convex yield functions that approximate the nonconvex one at each step are defined by applying the rules given by (2.81).

4.2.3 Rate problems

Here we deal with a nonconvex generalization of the generalized standard elastoplasticity problems written in Section 3.2.2 and studied there by means of time discretization schemes and convex analysis techniques.

This generalization is based on the quasidifferentiability theory. Roughly speaking by means of the notion of quasidifferentiability the tangent and the normal cones to a generally nonconvex and nonsmooth set are decomposed in a systematic way as unions and intersections of a finite number of convex cones as explained previously. Note that the more engineering oriented method proposed in Tzaferopoulos and Panagiotopoulos, 1993, Tzaferopoulos and Panagiotopoulos, 1994 and Panagiotopoulos, 1993 Chapt. 11 can also be used.

Let us assume the simple generalized elastoplasticity model of the previous Chapter (relations (3.126) to (3.129)) along with the time-discretization (3.141). If these relations are coupled with the previous nonconvex yield surface model, which has been analyzed by means of the quasidifferentiability approach, we arrive at the following finite step problem:

Find σ_{n+1} such that

$$\partial_{\sigma_{n+1}} f(\sigma_{n+1}) \cap T^+(\sigma_{n+1}) \neq \emptyset, \ \forall v \in \overline{\partial} F(\sigma_{n+1}), \tag{4.86}$$

where $f(\sigma_{n+1})$ has been defined in (3.148) of Section 3.2.2.5.
Relation (4.86) is written equivalently as

$$\partial_{\sigma_{n+1}} f(\sigma_{n+1}) \cap \{-\mathcal{N}_T(\sigma_{n+1})\} \neq \emptyset, \ \forall v \in \overline{\partial} F(\sigma_{n+1}), \tag{4.87}$$

or

$$\partial_{\sigma_{n+1}} f(\sigma_{n+1}) \cap \{-T^0(\sigma_{n+1})\} \neq \emptyset, \ \forall v \in \overline{\partial} F(\sigma_{n+1}). \tag{4.88}$$

In terms of plastic multipliers we first write the flow rule (3.118) as

$$\dot{e}^p = \mathcal{N}_\Omega(\sigma) = \bigcup_{s_j \in \underline{\partial} F(\sigma)} \lambda_j (s_j + v), \ \lambda_j \geq 0, \ \forall v \in \overline{\partial} F(\sigma). \tag{4.89}$$

Note here that, in contrast to what happens in the convex case (see (3.119), (3.124)), the plastic multipliers λ_j depend on the choice of $w_2 \in \overline{\partial}\phi(\sigma)$.

By using the time discretization scheme introduced previously we get the incremental relation (cf. (3.143)):

$$F_0 \sigma_{n+1} - e_{n+1} + e^p_{n+1} + \bigcup_{s_j \in \underline{\partial} F(\sigma_{n+1})} \lambda_j (s_j + v) = 0, \quad \forall v \in \overline{\partial} F(\sigma_{n+1}) \tag{4.90}$$

or the nonconvex minimization problem (cf. (3.144))

$$\min_{\sigma_{n+1} \in P} J_{n+1}(\sigma_{n+1}). \tag{4.91}$$

Note here that the nonconvexity of the optimization problem (4.91) is due to the nonconvexity of the yield surface set P, while the stepwise holonomic potential function $J_{n+1}(\sigma_{n+1})$ is convex and it has been given in (3.145).

In conclusion we have developed here the following extension of the rate problems arising in classical convex elastoplasticity for the cases of nonconvexity and nondifferentiability of a structured type: At each time step find a solution σ_{n+1} of a holonomic problem expressed as the following system of differential inclusions

$$F_0 \sigma_{n+1} - e_{n+1} + e^p_{n+1} + (w + v) = 0, \ \ w \in \underline{\partial} F(\sigma_{n+1}), \ \forall v \in \overline{\partial} F(\sigma_{n+1}). \tag{4.92}$$

If moreover the problem is subdifferentiable, i.e. $\overline{\partial} F(\sigma) = \{0\}$, then from (4.86)–(4.88) one gets similar relations to the ones arising in convex multisurface

plasticity, namely:
Find plastic multipliers λ_i, $i \in \mathcal{I}_0$ such that

$$F_0\sigma_{n+1} - e_{n+1} + e^p_{n+1} + \sum_{i\in\mathcal{I}_0(\sigma_{n+1})} \lambda_i \nabla F_i(\sigma_{n+1}) = 0,\ \lambda_i \geq 0, \tag{4.93}$$

or find the solution of the differential inclusion (multivalued equation)

$$F_0\sigma_{n+1} - e_{n+1} + e^p_{n+1} + w = 0,\ w \in \underline{\partial} F(\sigma_{n+1}). \tag{4.94}$$

Recall that (4.94) is equivalent to a variational inequality, while (4.92) is equivalent to a system of variational inequalities (cf. Panagiotopoulos and Stavroulakis, 1992, Stavroulakis, 1993, Stavroulakis and Panagiotopoulos, 1993, Stavroulakis et al., 1995).

4.2.3.1 Softening effects. Besides the previously described elastoplastic problems with nonconvex yield surfaces, the case of softening behaviour leads also to nonconvex problems. Softening, in contrast to hardening (cf. (3.125)) arises if the initial yield surface does not expand under the effect of loading but, on the contrary, shrinks. Certain assumptions concerning the "hardening" parameters $k(\kappa)$ and α in (3.125) may describe this behaviour. As a consequence, e.g., in the model (3.134), the matrix H of plastic hardening moduli is no more positive semidefinite. Accordingly, the Helmholtz free energy function (e.g., (3.134)) includes a nonconvex term and is therefore a nonconvex function.

The nonconvex optimization approach developed in this book can, in principle, be used for the numerical treatment of these kind of problems as well. Nevertheless, till now, this direction of research has not been followed and it remains open for future efforts.

Among the several attempts to deal with this problem one should mention here the following two techniques.

First, the dynamics of the structural system can be used to "convexify" the problems by means of the inertial (and eventually the viscosity) term. This approach can be found, for instance, in Maier and Perego, 1992, Comi et al., 1992. The underlying idea is that in the time-discretized form of the dynamic problem (i.e., of the previously described static problem enlarged with the inertial and viscous damping terms) a convex contribution arises due to the inclusion of the inertial (and eventually the viscosity) term (cf. (2.88)). Thus, for appropriately chosen small time steps this convex term may cancel the nonconvex contribution of the structural system and may permit the solution of the arising relations by means of convex optimization tools.

The second approach goes through a regularization of the system by means of the inclusion of higher-order terms of the plastic multipliers. This way follows analogous considerations for the treatment of damage mechanics problems and it is connected with the adoption of non-local mechanical theories. For the softening elastoplastic analysis problems reviewed here the reader may consult the resent publications of Comi and Corigliano, 1996, Comi and Perego, 1996.

The fact that sparse nonconvex energy terms do not influence considerably the result of a convex optimization algorithm has been first observed by Panagiotopoulos and Koltsakis, 1987b, Panagiotopoulos, 1993, p. 243, Panagiotopoulos and Koltsakis, 1987a.

4.3 DAMAGE AND FRACTURE MECHANICS

For the modelling of the gradual strength degradation of structures at a macroscopic (phenomenological) level the continuum damage mechanics theory has been developed. This theory can describe certain effects which are not covered by plasticity models (and in fact it can be coupled with plasticity) and, on the other hand, describes the overall (smeared) behaviour in the presence of cracks, if these cracks are not large enough in comparison with the dimensions of the structure so that the fracture mechanics approach is necessary. Note that all the interface models developed in other places of this book are appropriate for the treatment of nonlinear fracture mechanics problems.

Concerning the damage mechanics modelling, roughly speaking, the presence of microcracks and other microdefects in a material element at point x is represented by a scalar damage variable $d \in [0, 1]$. Here $d = 0$ represents the intact material and $d = 1$ represents the fully damaged one. Then the overall strength degradation can be considered by including reduction factors in the elasticity constitutive relations as follows:

$$\bar{\sigma} = \frac{\sigma}{(1-d)}, \quad \bar{\epsilon} = (1-d)\epsilon. \tag{4.95}$$

Here the real stresses and strains $\bar{\sigma}, \bar{\epsilon}$ are referred to a reduced, due to damage, unit volume, area or length, while the nominal quantities σ, ϵ are referred to the initial structure. The latter quantities are used in the continuum mechanics modelling while the damage parameter(s) d serve as internal variables and require special treatment. For instance, a linear elasticity law can be written, in the presence of damage, as follows:

$$\epsilon = \frac{F_0}{(1-d)}\sigma, \quad \sigma = (1-d)K_0\epsilon, \tag{4.96}$$

with the usual notations.

Moreover, if appropriate flow rules for the internal damage parameter d are given, the whole damaging process (possibly coupled with other nonlinear structural effects, e.g., plasticity) can be traced for a given loading history.

Within the theory of damage mechanics several models which follow the previously outlined ideas have been proposed. In particular, while scalar damage parameters cope with isotropic damage, matrix damage parameters D of appropriate order can be used for the description of anisotropic damage effects. On the other hand, within the damage flow (or evolution) rules, combined stress and strain damage criteria can be used, even with unilateral behaviour, as it arises, for instance, in the calculation of concrete structures, composites or masonry structures. The whole damage evolution history can be formulated within a thermodynamical framework with internal (hidden) variables associated with the damage variables and the damage criteria. This way the theoretical and algorithmic treatment follows the remarks given for the internal variable elastoplasticity problem (and permits the coupling of the two models in a straightforward way).

Details of this theory can be found, among others, in Simo and Ju, 1987a, Simo and Ju, 1987b, Lemaitre, 1992, Frémond and Nedjar, 1993, Frémond and Nedjar, 1995, Frémond and Nedjar, 1996. Coupling with other theories is treated in Lemaitre and Chaboche, 1985, Arnold and Saleeb, 1994, large displacement problems in Lubarda and Krajcinovič, 1995 among others. Applications on composite modelling can be found, e.g., in Ladevèze, 1995 and on masonry structures in Gambarotta and Lagomarsino, 1997a, Gambarotta and Lagomarsino, 1997b. Several interface laws for adhesive and frictional interfaces have also been formulated within the general damage mechanics approach (see, among others, Corigliano, 1993, Cannmo et al., 1995).

Here a simple isotropic model damage theory with a scalar damage variable will be formulated and some aspects of this internal variable formulation will be briefly discussed. Reference is done to the material given in the previous Sections of this book concerning the plasticity modelling.

Let us consider a stress-based isotropic continuum damage model. The complementary free energy potential of the body subjected to damage reads (cf. (3.134))

$$\Xi(\sigma, d_\sigma) = \frac{1}{2} d_\sigma \sigma^T F_0 \sigma = \frac{1}{2} d_\sigma \Pi_0^c(\sigma), \tag{4.97}$$

where $d_\sigma = \dfrac{1}{(1-d)}$ and the initial complementary elastic potential is denoted by $\Pi_0^c(\sigma)$.

Following analogous reasoning with the case of the elastoplasticity, we get the governing relations of the problem in the form

$$\epsilon = \frac{\partial \Xi}{\partial \sigma} = d_\sigma \frac{\partial \Pi_0^c(\sigma)}{\partial \sigma}, \tag{4.98}$$

and the nonnegative dissipation inequality

$$\mathcal{D} = \Pi_0^c(\sigma)\dot{d}_\sigma \geq 0. \tag{4.99}$$

Damage initiates and evolves if the damage criterion in stress space holds as equality, where the damage criterion is expressed as:

$$g(\bar{\tau}, r) = \bar{\tau} - r = \sqrt{2\Pi_0^c(\sigma)} - r \leq 0. \tag{4.100}$$

Here r is a damage threshold. Moreover, the complementary energy norm is involved as an equivalent strain measure in (4.100) (the latter is dictated by the fact that this quantity is thermodynamically conjugate to d_σ, as it can be seen by considering the expression $\frac{\partial \Xi}{\partial d_\sigma}$).

Assuming, as in plasticity, that damage evolution follows the principle of maximum dissipation and writing down the optimality conditions for the problem (cf. (4.99), (3.136)–(3.138)):

$$\max_{\{\bar{\tau}: g(\bar{\tau}, r) \leq 0\}} = \bar{\tau}.\dot{d}_\sigma, \tag{4.101}$$

we get the damage evolution law along with the damage loading / unloading conditions:

$$\begin{aligned} \dot{d}_\sigma &= \dot{\mu}\frac{\partial g(\bar{\tau}, r)}{\partial \bar{\tau}}, \\ \dot{\mu} &\geq 0, g(\bar{\tau}, r) \leq 0, \quad \dot{\mu} g(\bar{\tau}, r) = 0. \end{aligned} \tag{4.102}$$

An analogous strain-based theory can be formulated (with a damage criterion written in terms of strains). Moreover, if the damage evolution $\dot{d}_\sigma$ (resp. $\dot{d}$) is known, the current damage level can be found by integration along the previous loading history.

One should observe here that the damage mechanics approach includes stress-strain laws with softening, i.e., nonmonotone ones, but without complete (vertical) falling branches. The inclusion of the latter case within a damage model, which corresponds to the holonomic laws introduced earlier would require the enlargement of the free energy or of the complementary energy function with nonconvex terms, as it has been discussed in other places of this book and will be presented elsewhere.

References

Altenbach, J. and Altenbach, H. (1994). *Eiführung in die Kontinuumsmechanik.* B.G. Teubner Verlag, Stuttgart.

Argyris, J. H. and et. al. (1974). Recent developments in the finite element analysis of PCRV. *Nuclear Engineering*, 28:42–75.

Arnold, S. M. and Saleeb, A. F. (1994). On the thermodynamic framework of generalized coupled thermoelastic-viscoplastic-damage modeling. *International Journal of Plasticity*, 10(3):263–278.

Auchmuty, G. (1989). Duality algorithms for nonconvex variational principles. *Num. Functional Analysis and Optimization*, 10:211–264.

Baniotopoulos, C. C. and Panagiotopoulos, P. D. (1987). A hemivariational approach to the analysis of composite material structures. In Paipetis, S. A. and Papanicolaou, G. C., editors, *Engineering Applications of New Composites.* Omega Publications, London.

Bathe, K. J. (1981). *Finite element procedures in engineering analysis.* Prentice–Hall, New Jersey.

Bažant, Z. P. and Cedolin, L. (1991). *Stability of structures. Elastic, inelastic, fracture and damage theories.* Oxford University Press, New York, Oxford.

Betten, J. (1993). *Kontinuumsmechanik. Elasto-, Plasto- und Kriechmechanik.* Springer Verlag, Berlin - Heidelberg.

Björkman, G. (1991). The solution of large displacement frictionless contact problems using a sequence of linear complementarity problems. *International Journal for Numerical Methods in Engineering*, 31:1553–1566.

Björkman, G., Klarbring, A., and Sjödin, B. (1995). Sequential quadratic programming for non-liner elastic contact problems. *International Journal for Numerical Methods in Engineering*, 38:137–165.

Cannmo, P., Runesson, K., and Ristinmaa, M. (1995). Modelling of plasticity and damage in a polycrystalline microstructure. *International Journal of Plasticity*, 11(8):959–970.

Chen, W. F. (1994). *Constitutive equations for engineering materials. Vol. 1: Elasticity and modeling. Vol. 2: Plasticity and modeling.* Elsevier, Holland.

Chung, T. J. (1996). *Applied continuum mechanics.* Cambridge University Press, Cambridge.

Ciarlet, P. (1983). *Lectures on three-dimensional elasticity.* Springer Verlag, New York.

Clarke, F. H. (1983). *Optimization and nonsmooth analysis.* J. Wiley, New York.

Comi, C. and Corigliano, A. (1996). On uniqueness of the dynamic finite-step problems in gradient-dependent softening plasticity. *International Journal of Solids and Structures*, 33(26):3881–3902.

Comi, C., Corigliano, A., and Maier, G. (1992). Dynamic analysis of elastoplastic softening discretized structues. *ASCE Journal of Engineering Mechanics*, 118(12):2352–2375.

Comi, C. and Perego, U. (1996). A generalized variable formulation for gradient dependent softening plasticity. *International Journal for Numerical Methods in Engineering*, 39:3731–3755.

Cook, R. D. (1978). *Finite element method. Concepts and applications.* J. Wiley, New York.

Corigliano, A. (1993). Formulation, identification and use of interface models in the numerical analysis of composite delamination. *International Journal of Solids and Structures*, 30(20):2779–2811.

Crisfield, M. A. (1991). *Non-linear finite element analysis of solids and structures.* J. Wiley, Chichester.

Curnier, A. (1993). *Méthodes numériques en mécanique des solides.* Presses Polytechniques et Universitaires Romandes, Lausanne. English translation, Kluwer, 1994.

Curnier, A., He, Q. C., and Zysset, P. (1995). Conewise linear elastic materials. *Journal of Elasticity*, 37:1–38.

Dem'yanov, V. F., Stavroulakis, G. E., Polyakova, L. N., and Panagiotopoulos, P. D. (1996). *Quasidifferentiability and nonsmooth modelling in mechanics, engineering and economics.* Kluwer Academic, Dordrecht.

Dem'yanov, V. F. and Vasiliev, L. N. (1985). *Nondifferentiable optimization.* Optimization Software, New York.

Findley, W. N. and Michno, M. J. J. (1987). Concerning cusps and vertices on the yield surface of annealed mild steel. *Zeitschrift für Angewandte Mathematik und Mechanik (ZAMM)*, 67(7):309–312.

Fosdick, R. and Volkmann, E. (1993). Normality and convexity of the yield surface in nonlinear plasticity. *Quarterly of Applied Mathematics*, LI(1):117–127.

Frémond, M. and Nedjar, B. (1993). Endommagement et principe des puissances virtuelles. *C.R. Acad. Sci. Paris Ser. II*, 317:857–864.

Frémond, M. and Nedjar, B. (1995). Damage in concrete: the unilateral phenomenon. *Nuclear Engineering and Design*, 156(1-2):323–336.

Frémond, M. and Nedjar, B. (1996). Damage, gradient of damage and principle of virtual power. *International Journal of Solids and Structures*, 33(8):1083–1103.

Galka, A. and Telega, J. J. (1995a). Duality and the complementary energy principle for a class of nonlinear structures. Part I. Five-parameter shell model. *Archives of Mechanics*, 47(4):677–698.

Galka, A. and Telega, J. J. (1995b). Duality and the complementary energy principle for a class of nonlinear structures. Part II. Anomalous dual variational principles for compressed elastic nonlinear structures. *Archives of Mechanics*, 47(4):699–724.

Gambarotta, L. and Lagomarsino, S. (1997a). Damage models for the seismic response of brick masonry shear walls. Part I: The mortar joint model and its applications. *Earthquake Engineering and Structural Dynamics*, 26:423–439.

Gambarotta, L. and Lagomarsino, S. (1997b). Damage models for the seismic response of brick masonry shear walls. Part II: The continuum model and its applications. *Earthquake Engineering and Structural Dynamics*, 26:441–462.

Gao, D. Y. (1996). Complementary finite-element method for finite deformation nonsmooth mechanics. *Journal of Engineering Mathematics*, 30:339–353.

Gao, Y. (1992). Global extremum criteria for nonlinear elasticity. *Zeitschrift für Angew. Mathematik und Physik (ZAMP)*, 43:742–755.

Gao, Y. and Strang, G. (1989a). Dual extemum principles in finite deformation elastoplastic analysis. *Acta Applicande Mathematica*, 17:257–267.

Gao, Y. and Strang, G. (1989b). Geometric nonlinearity: potential energy, complementary energy and the gap function. *Quarterly of Applied Mathematics*, 47:487–504.

Ghosh, S. (1992). Arbitrary Lagrangian-Eulerian finite element analysis of large deformation in contact bodies. *International Journal of Numerical Methods in Engineering*, 33:1891–1925.

Givoli, D. and Doukhovni, I. (1996). Finite element - quadratic programming approach for contact problems with geometrical nonlinearity. *Computers and Structures*, 61(1):31–41.

Glocker, C. (1995). *Dynamik von Starrkörpersystemen mit Reibung und Stößen.* VDI Fortschritt-Berichte, Düsseldorf. PhD Thesis, Technical University of Munich 1995.

Goeleven, D. (1995). On the hemivariational inequality approach to nonconvex constrained problems in the theory of von Kárman plates. *Zeitschrift für Angewandte Mathematik und Mechanik (ZAMM)*, 75(11):861–866.

Goeleven, D. (1997). A bifurcation theory for nonconvex unilateral laminated plate problem formulated as a hemivariational inequality involving a potential operator. *Zeitschrift für Angewandte Mathematik und Mechanik (ZAMM)*, 77(1):45–51.

Green, A. E. and Naghdi, P. M. (1965). A general theory of an elastic plastic coninuum. *Archive of Rational Mechanics and Analysis*, 18:251–281.

Hill, R. (1979). Theoretical plasticity of textured aggregates. *Mathematical Proceedings of Cambridge Philosophical Society*, 85:179–191.

Hill, R. (1987). Constitutive dual potentials in classical plasticity. *Journal of Mechanics and Physics of Solids*, 3:23–33.

Hill, R. and Rice, J. R. (1973). Elastic potentials and structure of inelastic constitutive laws. *SIAM Journal of Applied Mathematics*, 25:448–461.

Horst, R., Pardalos, P. M., and Thoai, N. Y. (1995). *Introduction to global optimization*. Kluwer Academic, Dordrecht Boston.

Ionescu, J. R. and Paumier, J. C. (1994). On the contact problem with slip rate dependent friction in elastodynamics. *European Journal of Mechanics A / Solids*, 13(4):555–568.

Johnson, A. R. and Quigley, C. J. (1989). Frictionless geometrically non-linear contact using quadratic programming. *International Journal for Numerical Methods in Engineering*, 28:127–144.

Justusson, J. W. and Phillips, A. (1966). Stability and convexity in plasticity. *Acta Mechanica*, 2:251–267.

Khan, A. S. and Juang, S. (1995). *Continuum theory of plasticity*. J. Wiley and Sons, Inc., Chichester.

Kim, S. J. and Oden, J. T. (1984). Generalized potentials in finite elastoplasticity. Part I. *International Journal of Engineering Sciences*, 22:1235–1257.

Kim, S. J. and Oden, J. T. (1985). Generalized potentials in finite elastoplasticity. Part II. *International Journal of Engineering Sciences*, 23:515–530.

Klarbring, A. (1995). Large displacement frictional contact: a continuum framework for finite element discretization. *European Journal of Mechanics A / Solids*, 14:237–253.

Koltsakis, E. K., Mistakidis, E. S., and Tzaferopoulos, M. A. (1995). On the numerical treatment of nonconvex energy problems of mechanics. *Journal of Global Optimization*, 6(4):427–448.

Kuczma, M. S. and Stein, E. (1994). On nonconvex problems in the theory of plasticity. *Archivum Mechanicky*, 46(4):603–627.

Ladevèze, P. (1995). A damage computational approach for composites: basic aspects and micromechanical relations. *Computational Mechanics*, 17:142–150.

Lemaitre, J. (1992). *A course on damage mechanics*. Springer Verlag, Berlin Heidelberg.

Lemaitre, J. and Chaboche, J. L. (1985). *Mécanique des matériaux solides*. Dunod, Paris. English Translation Cambridge Univ. Press 1994.

Lin, F. B. and Bažant, Z. P. (1986). Convexity of smooth yield surface of frictional material. *ASCE Journal of Engineering Mechanics*, 112(11):1259–1262.

Lochegnies, D. and Oudin, J. (1995). External penalized mixed functional algorithms for unilateral contact and friction in a large strain finite element framework. *Engineering Computations*, 12:307–331.

Lubarda, V. A. and Krajcinovič, D. (1995). Some fundamental issues in rate theory of damage-elastoplasticity. *International Journal of Plasticity*, 11(7): 763–797.

Lubliner, L. (1990). *Plasticity theory*. Macmillan Publ., New York, London.

Maier, G. and Perego, U. (1992). Effects of softening in elastic-plastic structural dynamics. *International Journal for Numerical Methods in Engineering*, 34:319–347.

Malvern, L. E. (1969). *Introduction to the mechanics of a continuous medium*. Prentice–Hall Inc., Englewood Cliffs, N.Jersey.

Marsden, J. E. and Hughes, T. J. R. (1983). *Mathematical foundations of elasticity*. Prentice-Hall, N.J.

Miettinen, M. (1993). *Approximation of hemivariational inequalities and optimal control problems*. University of Jyväskylä, Department of Mathematics, Jyväskylä Finnland. PhD Thesis, Report No. 59.

Miettinen, M. (1995). On constrained hemivariational inequalities and their approximation. *Applicable Analysis*, 56:303–326.

Mistakidis, E. S., Baniotopoulos, C. C., and Panagiotopoulos, P. D. (1995). On the numerical treatment of the delamination problem in laminated composites under cleavage loading. *Composite Structures*, 30:153–166.

Mistakidis, E. S. and Panagiotopoulos, P. D. (1993). Numerical treatment of nonmonotone (zig–zag) friction and adhesive contact problems with debonding. Approximation by monotone subproblems. *Computers and Structures*, 47:33–46.

Mistakidis, E. S. and Panagiotopoulos, P. D. (1994). On the approximation of nonmonotone multivalued problems by monotone subproblems. *Computer Methods in Applied Mechanics and Engineering*, 114:55–76.

Mistakidis, E. S. and Panagiotopoulos, P. D. (1997). On the search for substationarity points in the unilateral contact problems with nonmonotone friction. *J. of Math. and Comp. Modelling*, (to appear).

Moreau, J. J. and Panagiotopoulos, P. D., editors (1988). *Nonsmooth mechanics and applications*, volume 302 of *CISM Lect. Notes*. Springer, Wien-New York.

Moreau, J. J., Panagiotopoulos, P. D., and Strang, G., editors (1988). *Topics in nonsmooth mechanics*. Birkhäuser, Basel-Boston.

Motreanu, D. and Panagiotopoulos, P. D. (1993). Hysteresis: the eigenvalue problem for hemivariational inequalities. In Visintin, A., editor, *Moldels of hysteresis*, pages 102–117. Longman Scientific and Technical, New York.

Motreanu, D. and Panagiotopoulos, P. D. (1995). An eigenvalue problem for a hemivariational inequality involving a nonlinear compact operator. *Set Valued Analysis*, 3:157–166.

Nadai, A. (1963). *Theory of flow and fracture of solids. Vol. II.* McGraw Hill, New York.

Naniewicz, Z. (1989). On some nonconvex variational problems related to hemivariational inequalities. *Nonlin. Anal.*, 13:87–100.

Naniewicz, Z. (1994a). Hemivariational inequalities with functions fulfilling directional growth condition. *Applicable Analysis*, 55:259–285.

Naniewicz, Z. (1994b). Hemivariational inequality approach to constrained problems for star-shaped admissible sets. *Journal of Optimization Theory and Applications*, 83(1):97–112.

Naniewicz, Z. (1995). Hemivariational inequalities with functionals which are not locally Lipschitz. *Nonlinear Analysis*, 25(12):1307–1320.

Naniewicz, Z. and Panagiotopoulos, P. D. (1995). *Mathematical theory of hemivariational inequalities and applications.* Marcel Dekker.

Nemat-Nasser, S. (1992). Phenomenological theories of elastoplasticity and localization at high strain rates. *Applied Mechanics Review*, 45(3):S19–S45.

Panagiotopoulos, P. D. (1975). A nonlinear programming approach to the unilateral contact and friction boundary value problem in the theory of elasticity. *Ing. Archiv*, 44:421–432.

Panagiotopoulos, P. D. (1983). Nonconvex energy functions. Hemivariational inequalities and substationary principles. *Acta Mechanica*, 42:160–183.

Panagiotopoulos, P. D. (1985). *Inequality problems in mechanics and applications. Convex and nonconvex energy functions.* Birkhäuser, Basel - Boston - Stuttgart. Russian translation, MIR Publ., Moscow 1988.

Panagiotopoulos, P. D. (1993). *Hemivariational inequalities. Applications in mechanics and engineering.* Springer, Berlin - Heidelberg - New York.

Panagiotopoulos, P. D. and Koltsakis, E. K. (1987a). Hemivariational inequalities in linear and nonlinear elasticity. *Meccanica*, 22:65–75.

Panagiotopoulos, P. D. and Koltsakis, E. K. (1987b). Interlayer slip and delamination effects: a hemivariational inequality approach. *Trans. Canadian Society of Mechanical Engineering*, 11(1):43–52.

Panagiotopoulos, P. D., Panagouli, O. K., and Mistakidis, E. S. (1993). Fractal geometry and fractal material behaviour in solids and structures. *Archive of Applied Mechanics*, 63:1–24.

Panagiotopoulos, P. D. and Stavroulakis, G. E. (1992). New types of variational principles based on the notion of quasidifferentiability. *Acta Mechanica*, 94:171–194.

Pilkey, W. D. and Wunderlich, W. (1994). *Mechanics of structures. Variational and computational methods.* CRC Press, Boca Raton.

Rabinowicz, E. (1959). A study of stick-slip processes. In Daview, R., editor, *Friction and wear*, pages 149–161. Elsevier, London.

Rockafellar, R. T. (1970). *Convex analysis.* Princeton University Press, Princeton.

Rockafellar, R. T. (1979). *La théorie des sous-gradients et ses applications à l' optimization. Fonctions convexes et non-convexes.* Les Presses de l' Université de Montréal, Montréal.

Rohde, A. and Stavroulakis, G. E. (1995). Path following energy optimization in unilateral contact problems. *Journal of Global Optimization*, 6(4):347–365.

Rohde, A. and Stavroulakis, G. E. (1997). Genericity analysis for path-following methods in unilateral contact elastostatics. *Zeitschrift für Angewandte Mathematik und Mechanik (ZAMM)*, 77(6):(to appear).

Rubinov, A. M. and Yagubov, A. A. (1986). The space of star-shaped sets and its applications in nonsmooth optimization. *Mathematical Programming Study*, 29:176–202.

Salençon, J. and Tristan-Lopez, A. (1980). Analyse de la stailité des talus en sols cohérents anisotropes. *C.R. Acad. Sci. Paris*, 290B:493–496.

Simo, J. C. and Ju, J. W. (1987a). Strain- and stress-based continuum damage models. I. Formulation. *International Journal of Solids and Structures*, 23(7):821–840.

Simo, J. C. and Ju, J. W. (1987b). Strain- and stress-based continuum damage models. II. Computational aspects. *International Journal of Solids and Structures*, 23(7):841–869.

Stavroulakis, G. E. (1993). Convex decomposition for nonconvex energy problems in elastostatics and applications. *European Journal of Mechanics A / Solids*, 12(1):1–20.

Stavroulakis, G. E. (1995a). Quasidifferentiability and star–shaped sets. Application in nonconvex, finite dimensional elastoplasticity. *Communications on Applied Nonlinear Analysis*, 2(3):23–46.

Stavroulakis, G. E. (1995b). Variational problems for nonconvex elastoplasticity based on the quasidifferentiability concept. In Theocaris, P. S. and Gdoutos, E. E., editors, *Proc. 4th Greek Nat. Congress on Mechanics*, pages 527–534.

Stavroulakis, G. E., Demyanov, V. F., and Polyakova, L. N. (1995). Quasidifferentiability in mechanics. *Journal of Global Optimization*, 6(4):327–345.

Stavroulakis, G. E. and Mistakidis, E. S. (1995). Numerical treatment of hemivariational inequalities. *Computational Mechanics*, 16:406–416.

Stavroulakis, G. E. and Panagiotopoulos, P. D. (1993). Convex multilevel decomposition algorithms for non-monotone problems. *Int. J. Num. Meth. Engng.*, 36:1945–1966.

Stavroulakis, G. E. and Rohde, A. (1996). Stability of structures with quasidifferentiable energy functions. In Sotiropoulos, D. and Beskos, D., editors, *2nd Greek Conf. on Computational Mechanics*, pages 406–413, Chania.

Stein, E., Wagner, W., and Wriggers, P. (1989). Grundlagen nichtlinearer Berechnungsverfahren in der Strukturmechanik. In Stein, E., editor, *Nichtlineare Berechnungen im Konstruktiven Ingenieurbau*, pages 1–53. Springer, Wien - New York.

Stuart, C. A. and Toland, J. F. (1980). A variational method for boundary value problems with discontinuous nonlinearities. *J. London Math. Soc. (2)*, 21:319–328.

Sun, S. M. and Natori, M. C. (1996). Numerical solution of large deformation problems involving stability and unilateral constraints. *Computers and Structures*, 58(6):1245–1260.

Toland, J. F. (1979). A duality principle for nonconvex optimization and the calculus of variations. *Arch. Rat. Mech. Analysis*, 71:41–61.

Tzaferopoulos, M. A., Mistakidis, E. S., Bisbos, C. D., and Panagiotopoulos, P. D. (1995). Comparison of two methods for the solution of a class of nonconvex energy problems using convex minimization algorithms. *Computers and Structures*, 57(6):959–971.

Tzaferopoulos, M. A. and Panagiotopoulos, P. D. (1993). Delamination of composites as a substationarity problem: Numerical approximation and algorithms. *Computer Methods in Applied Mechanics and Engineering*, 110 (1–2):63–86.

Tzaferopoulos, M. A. and Panagiotopoulos, P. D. (1994). A numerical method for a class of hemivariational inequalities. *Computational Mechanics*, 15:233–248.

Visintin, A. (1994). *Differential models of hysteresis.* Springer Verlag, Berlin Heidelberg.

Wriggers, P., Vu, V. T., and Stein, E. (1990). Finite element formulation of large deformation impact - contact problems. *Computers and Structures*, 37:319–331.

Zeidler, E. (1988). *Nonlinear functional analysis and its applications. IV: Applications to mathematical physics.* Springer Verlag, New York - Heidelberg.

Zhu, C. and Jin, Y. (1994). The solution of frictional contact problems using a finite element - mathematical programming method. *Computers and Structures*, 52:149–155.

Zhu, Y., Dodd, B., Caddel, R. M., and Hosford, W. (1987). Convexity restrictions on non-quadratic anisotropic yield criteria. *International Journal of Mechanical Sciences*, 29(10):733–741.

Zienkiewicz, O. C. and Taylor, R. L. (1991). *The finite element method. Vol. II: Solid and fluid mechanics, dynamics and non-linearity.* McGraw-Hill.

5 OPTIMAL DESIGN PROBLEMS

5.1 MULTILEVEL, ITERATIVE OPTIMAL DESIGN PROBLEMS

5.1.1 Introduction

In this Section related problems which arise in the optimal design of structures are formulated as two level optimization problems and are numerically treated by multilevel iterative techniques. Certain classes of optimal material design problems and topology optimization problems formulated by means of the homogenization approach can be treated with this method. This approach can also be used for the rigourous formulation of optimality criteria methods for optimal design of structures. These methods are popular in engineering applications because, roughly speaking, they decompose the difficult optimal design problem into a number of classical structural analysis problems with appropriate, decentralized (at the finite element level) modification rules.

For more details on this class of problems and in particular on this approach for their solution, the reader is referred to Bendsøe, 1995, Bendsøe and Kikuchi, 1988, Jog et al., 1994, Thierauf, 1996, Allaire et al., 1997 among others. The optimality criteria methods have been described in details in Rozvany, 1989, Rozvany et al., 1995. The link between optimal design of structures and nonsmooth mechanics has been discussed in Stavroulakis and Tzaferopoulos, 1994, Tzaferopoulos and Stavroulakis, 1995. For the problems of static control (or optimal design) of structures including structures governed by variational inequalities see Panagiotopoulos, 1976, Panagiotopoulos, 1977, Panagiotopoulos, 1980, Panagiotopoulos, 1983, Haslinger and Neittaanmäki, 1988, Kočvara and Outrata, 1996, Mäkelä and Neittaanmäki, 1992, Panagiotopoulos and Haslinger, 1992, Stavroulakis, 1995a, Stavroulakis, 1995b. For a review of multilevel approaches to optimal design of structures and applications see, among others, Stavroulakis and Günzel, 1997 (cf. also Panagiotopoulos et al., 1997 for related problems in structural analysis). For an application on optimal design with composites with negative Poisson's ratio see Theocaris et al., 1997b and Theocaris and Stavroulakis, 1997. The link with damage mechanics, which has already been mentioned in a previous Chapter (see also Achtziger and Bendsøe, 1995) is briefly discussed at the end of this Section. Note that this Section is strictly restricted to the discussion of certain models relevant to the topics studied in this book and is by no means a thorough study of optimal structural design problems.

The multilevel approach to structural optimization has recently gained on interest due to the developments in the area of topology optimization based on numerical homogenization techniques. The optimal material design problem is formulated in a general setting, by following recent developments in optimal shape and topology design by means of numerical homogenization, see e.g. Bendsøe, 1995, Bendsøe and Kikuchi, 1988, Jog et al., 1994.

Only simple optimization problems concerning the finding of the stiffest or the more flexible structure are treated here. The stiffest structure is obtained by maximizing the potential energy of the structure at equilibrium or, equivalently, by minimizing the complementary energy at equilibrium. The choice of the material at each point of the structure, or at each finite element in a discretized problem, is governed by the design variables of the optimal design problem. These variables may correspond to the real design variables of, e.g., a composite material whose properties can be taylored at each point of the examined structure, or, may be auxiliary variables which help us to find an optimal material distribution within the given area.

From the formulation of the problem and since an equilibrium configuration makes the potential energy of the structure minimum, one expects that multilevel techniques will be most appropriate. In fact a multilevel splitting of the problem into two subproblems is possible, where the first subproblem is the structural analysis problem itself and the second subproblem tackles with the material adaptation which is performed in a pointwise (or elementwise) sense by means of small scale, decomposed subproblems (adaptation rules).

This approach has its origin in the optimality criteria methods for the solution of optimal layout problems (Prager and Taylor, 1968, Rozvany, 1989, Rozvany et al., 1995) and leads to local adaptation rules and repetitive solution of classical structural analysis problems which are easily implemented in general purpose structural analysis problems with open architecture. These problems belong to the so called class of mathematical programs with equilibrium constraints (Luo et al., 1996), where by the word constraints one understands, in general, a possible set of inequalities and complementarity relations, as the case is in optimum weight problems for unilaterally constrained structures whose equilibrium positions are the solutions of a L.C.P. (cf. e.g. Panagiotopoulos, 1983).

For the material selection problem either classical fibre–reinforced composites and porous, cellular materials with given, analytically obtained effective elastic properties (Theocaris, 1987, Gibson and Ashby, 1982) or look–up tables produced by numerical techniques (Theocaris et al., 1997b, Theocaris et al., 1997a) can be used. For the topology optimization problem a material with elasticity modulus depending on a power of a single design variable can be used (Mlejnek, 1992, Mlejnek and Schirrmacher, 1993, Baier et al., 1994, Bendsøe, 1995).

These problems are briefly treated here only due to their connection with the multilevel optimization techniques discussed in this book. More details can be found in the specialized bibliography. Moreover, one could consider optimal design problems for all nonsmooth material or interface/boundary laws, or in connection with the plasticity and damage models described in other Sections of this work. Most of the latter problems have, to the authors' knowledge, not treated till now.

5.1.2 Formulation of the optimal design problem

A distributed parameter, minimum compliance (or equivalently maximum stiffness) problem is assumed. Let the design vector be denoted by $[\rho(x), \alpha(x)]^T$, where $\rho(x)$ is the density of each point x of the structure and $\alpha(x)$ is the set of design variables that describe the microstructure of the material.

For instance, in a composite material structure the fibre–volume fraction, the elasticity constants of the matrix and the fibre, the mesophase variables etc., may be used as design variables in $\alpha(x)$. Analogously, for a cellular (or foamy) structure $\alpha(x)$ describes the design parameters of the unit cell as it will be discussed in details later on with respect to the concrete applications. Note also that in a discretized structure $[\rho(x), \alpha(x)]^T$ is defined elementwise.

In a displacements based formulation the minimum compliance problem is written in the following form (cf. Bendsøe, 1995, p. 88):

$$\max_{[\rho(x),\alpha(x)]\in\mathcal{A}_{ad},x\in\Omega} \min_{u\in U_{ad}} \left\{\overline{\Pi}(u,\rho,\alpha) = \Pi(e(u),\rho,\alpha) - l(u)\right\}. \tag{5.1}$$

Here $\mathcal{A}_{ad}$ is the admissible set of the design variables which, on the assumption that a structure with maximum mass V is sought, reads:

$$\mathcal{A}_{ad} = \left\{\rho(x), \alpha(x), x \in \Omega \text{ such that } \int_\Omega \rho d\Omega \leq V,\ 0 \leq \rho \leq 1\right\}. \tag{5.2}$$

In the previous problem the inner subproblem is a potential energy optimization problem. The solution of this subproblem, the marginal function, is maximized in the outer subproblem.

The total potential energy $\overline{\Pi}(u,\rho,\alpha)$ is composed of the energy density $\Pi(e(u),\rho,\alpha)$, which, in turn, depends on the chosen design variables ρ, α and the loading potential term $l(u)$. Finally, the admissible displacements set U_{ad} depends on the nature of the analyzed problem and on the boundary conditions.

For the case of linear elastic structures within a small displacements and deformations theory, problem (5.1) reads:

$$\max_{[\rho(x),\alpha(x)]\in\mathcal{A}_{ad}} \min_{u\in U_{ad}} \left\{\frac{1}{2}\int_\Omega E_{ijkl}(x,\rho(x),\alpha(x))\epsilon_{ij}(u)\epsilon_{kl}(u)d\Omega - l(u)\right\}, \tag{5.3}$$

where $E_{ijkl}(x,\rho(x),\alpha(x))$ is the fourth order elastic stiffness tensor.

The stress space problem reads analogously:

$$\min_{[\rho(x),\alpha(x)]\in\mathcal{A}_{ad}} \min_{\sigma\in\Sigma_{ad}} \left\{\Pi^c(\sigma,\rho,\alpha)\right\}. \tag{5.4}$$

For the linear elasticity problem considered here it has the form:

$$\min_{[\rho(x),\alpha(x)]\in\mathcal{A}_{ad}} \min_{\sigma\in\Sigma_{ad}} \left\{\frac{1}{2}\int_\Omega C_{ijkl}(x,\rho(x),\alpha(x))\sigma_{ij}\sigma_{kl}d\Omega\right\}. \tag{5.5}$$

Here Σ_{ad} is the set of admissible stresses which incorporates the stress equilibrium condition: $div\sigma + p = 0$ and the prescribed boundary tractions condition: $\sigma n = t$. Moreover $\Pi^c(\sigma,\rho,\alpha)$ is the complementary energy density and $C_{ijkl}(x,\rho(x),\alpha(x))$ is the fourth order elastic compliance tensor.

In principle, more general cost functions can be assumed in the optimal design problems (5.1) and (5.4). Nevertheless, the marginal function form used here, where the value of a minimization subproblem is maximized in (5.1) or minimized in (5.4) has certain advantages from both the theoretical and the numerical point of view. Without entering into the details here, we mention that continuity and in some cases differentiability or convexity/concavity information for the marginal function (inner subproblem in (5.1) or in (5.4)) as a function of the design variables $(\rho(x), \alpha(x))$ can be proved (see e.g. Stavroulakis and Günzel, 1997).

5.1.3 *Multilevel decomposition and solution algorithm*

For the needs of the applications treated in this book it is sufficient to assume that the density of the structure at each point $x \in \Omega$ (cell or finite element for the numerical application) depends on the local design variables $\alpha(x)$, i.e. $\rho(x) = f(\alpha(x))$, $\forall x \in \Omega$. In order to exploit the pointwise nature of the above relation we decompose the optimal design problem as follows. For problem (5.3):

$$\max_{\rho(x)\in A_1} \max_{\alpha(x)\in A_2(\rho)} \min_{u\in U_{ad}} \left\{ \frac{1}{2}\int_\Omega E_{ijkl}(x,\rho(x),\alpha(x))\epsilon_{ij}(u)\epsilon_{kl}(u)d\Omega - l(u) \right\}. \tag{5.6}$$

For problem (5.5) respectively:

$$\min_{\rho(x)\in A_1} \min_{\alpha(x)\in A_2(\rho)} \min_{\sigma\in\Sigma_{ad}} \left\{ \frac{1}{2}\int_\Omega C_{ijkl}(x,\rho(x),\alpha(x))\sigma_{ij}\sigma_{kl}d\Omega \right\}. \tag{5.7}$$

Here $A_1 = \{\rho(x) \ : \ 0 \le \rho(x) \le 1, \ \int_\Omega \rho(x)d\Omega \le V, \ x \in \Omega\}$ and $A_2 = \{\alpha(x) \ : \ \rho(x) = \text{given} = f(\alpha(x)), \ x \in \Omega\}$. Obviously the following relation holds: $A_{ad} = \{\rho(x), \alpha(x) \ : \ \rho(x) \in A_1, \ \alpha(x) \in A_2(\rho)\}$.

Following Bendsøe, 1995, Jog et al., 1994 we change the order of the second and third operators in problems (5.6) and (5.7) and by observing that the restrictions of set $A_2(\rho)$ above can be assumed to hold elementwise, we get the following problems:

$$\max_{\rho(x)\in A_1} \min_{u\in U_{ad}} \left\{ \int_\Omega \left\{ \max_{\alpha(x) \text{s.t.} \rho(x)=f(\alpha(x))} \frac{1}{2} E_{ijkl}(x,\rho(x),\alpha(x))\epsilon_{ij}(u)\epsilon_{kl}(u)d\Omega \right\} \atop -l(u) \right\} \tag{5.8}$$

and

$$\min_{\rho(x)\in A_1} \min_{\sigma\in\Sigma_{ad}} \left\{ \int_\Omega \left\{ \min_{\alpha(x) \text{s.t.} \rho(x)=f(\alpha(x))} \frac{1}{2} C_{ijkl}(x,\rho(x),\alpha(x))\sigma_{ij}\sigma_{kl}d\Omega \right\} \right\}. \tag{5.9}$$

From the above form of the optimal design problem a hierarchical iterative solution strategy is straightforward (cf. Bendsøe, 1995, p. 83, Jog et al., 1994, see also Al-Fahed and Panagiotopoulos, 1993). The inner maximization (resp. minimization) problem is solved in a decoupled, elementwise form. The second level minimization problem is solved at the structural level by a general purpose finite element code. Both the above mentioned subproblems are solved for a given density distribution $\rho(x)$ which in turn is updated by solving the first level optimization problem in (5.8) (resp. (5.9)) with respect to variable $\rho(x)$.

Let us remark that if the expression of $E_{ijkl}(\alpha(x))$ as a function of $\alpha(x)$ (resp. of $C_{ijkl}(\alpha(x))$) is given analytically then the lower level local optimization problem in (5.8) (resp. (5.9)) can be solved also analytically. This is the case e.g. of optimal topology design problems through homogenization which is studied in details in Bendsøe, 1995, Jog et al., 1994 among others. In this case the second level problem can be formulated as a fictitious non–linear elastic structural analysis problem with a nonlinear and, in general, nondifferentiable strain energy density which is given by the previously mentioned local optimization problem.

In order to avoid restrictive assumptions we assume here that the above mentioned relation is not given explicitly but it is produced from a complete analysis of a unit representative cell or it is interpolated from a set of such solutions which are stored in a database. Then the problems (5.8), (5.9) have to be solved iteratively. Moreover, the second level problems, i.e. the nonlinear and possibly nondifferentiable structural analysis problems are approximated by linear problems which are defined by the current values of E_{ijkl} (resp. C_{ijkl}) as they are given from the solution of the lower level (local) subproblems.

Note that in some applications, like the design of fibre reinforced composites, the mass distribution is approximately constant for the whole structure, irrespectively of the values of the design variables. In this case the previous method is appropriately simplified. Alternatively the "mass" variable may be kept in the formulation to measure the cost associated with the material changes performed in the course of the algorithm. In this case the "mass constraint" should be interpreted as a maximum allowable cost constraint.

One possible solution algorithm for problem (5.8) has the following steps:

Algorithm 5.1 *Multilevel Optimal Design*

- *Step 0*: Initialization
  $l = 0$, choose initial distribution of $\rho^l(x)$, $\alpha^l(x)$.

- *Step 1*: Iteration
  Set iteration counter $l = l + 1$.

- *Step 2*: Structural analysis subproblem
  For given α^l solve the potential energy minimization problem:

$$\begin{aligned}
&\min_{u \in U_{ad}} \left\{ \int_\Omega \pi(\epsilon, \rho^l, \alpha^l) d\Omega - l(u) \right\} && (5.10)\\
= &\min_{u \in U_{ad}} \left\{ \frac{1}{2} \int_\Omega E_{ijkl}(x, \rho^l(x), \alpha^l(x)) \epsilon_{ij} \epsilon_{kl} d\Omega - l(u) \right\},
\end{aligned}$$

  with solution u^{l+1}.

- *Step 3*: Local subproblem
  For each finite element and for given $\epsilon^{l+1}(u^{l+1})$ solve the local problem:

$$\max_{\alpha(x) \in A_2(\rho^l)} \pi(\epsilon^{l+1}, \rho^l, \alpha) = \max_{\alpha(x) \in A_2(\rho^l)} \frac{1}{2} E_{ijkl}(x, \rho^l(x), \alpha(x)) \epsilon_{ij}^{l+1} \epsilon_{kl}^{l+1}. \tag{5.11}$$

  The solution α^{l+1} defines the current design (e.g. microstructure) of the finite element.

- *Step 4*: Density updating by the iterative formula

$$\rho^{l+1} = \max \left\{ \rho_{min}, \min \left\{ \left[\frac{1}{\Lambda_l} \frac{\partial \pi}{\partial \rho} (\epsilon^l, \rho, \alpha^{l+1}) \right]^\eta \rho^l, 1 \right\} \right\}, \tag{5.12}$$

  where η is a nonnegative constant used for numerical efficiency and Λ_l is the Lagrange multiplier which acts as a scaling factor here, and is determined by the maximum mass constraint $\int_\Omega \rho^{l+1} d\Omega = V$.

- *Step 5*: Convergence check
  If no convergence then continue with step 1.

Depending on the specific problem of step 3, it may be performed analytically, in an approximate way or it may be used for an iterative modification of the field α (suboptimal approach).

An analogous algorithm can be followed for the solution of the stress based formulation (5.9).

In the two-dimensional applications which will be presented in Chapter 7, we encountered checkerboard, parasitic patterns in the solution (see relevant discussions in Bendsøe, 1995, Diaz and Sigmund, 1995, Youn and Park, 1997, Křížek and Neitaanmäki, 1990, p. 97). This problem has been suppressed by the simple element-to-nodal-density technique, described e.g. in Youn and Park, 1997.

5.1.4 Applications

5.1.4.1 Fibre reinforced composites. Let us consider a cross-section of a transversally isotropic fibre reinforced composite perpendicular to the direction of the reinforcing fibres. A plane strain elasticity problem is considered, where the in–plane constitutive law reads:

$$\begin{bmatrix} \sigma_x \\ \sigma_y \\ \tau_{xy} \end{bmatrix} = \begin{bmatrix} E(\alpha) & -\frac{E(\alpha)}{\nu(\alpha)} & 0 \\ -\frac{E(\alpha)}{\nu(\alpha)} & E(\alpha) & 0 \\ 0 & 0 & 2(1-\nu(\alpha)) \end{bmatrix} \begin{bmatrix} \epsilon_x \\ \epsilon_y \\ \gamma_{xy} \end{bmatrix}. \tag{5.13}$$

Here $E(\alpha)$ and $\nu(\alpha)$ denote the transversal elasticity modulus and Poisson's ratio of the fibre reinforced composite which depend on the design variables α of the composite.

Let us assume that a fibre reinforced composite with fibres having a circular cross-section is chosen. The design vector α is composed of the fibre volume ratio of the composite $\frac{u_f}{u_c}$ and the fibre to matrix elasticity ratio $\frac{E_f}{E_m}$. An analytical model for the calculation of the transversal elasticity modulus $E(\alpha)$ and the effective transversal Poisson's ratio $\nu(\alpha)$ as functions of the design variables at each point of the composite can be constructed, see, e.g., Theocaris, 1987, Theocaris and Varias, 1986. For a numerical investigation based on homogenization techniques, see, e.g., Theocaris et al., 1997a, Theocaris and Stavroulakis, 1997a.

In this application the mass of the composite is practically constant for all values of the chosen design variables; thus step 4 of the Algorithm (5.1) is not used here. Moreover, the local problem (5.11) is solved numerically by means of a gradient free optimization algorithm. The latter choice is dictated by the complicated form of the formulae which give $E(\alpha)$ and make the analytic calculation of the first derivative $\frac{\partial \pi}{\partial \alpha}$ complicated.

5.1.4.2 Cellular materials. Classical cellular material theory, e.g. Ashby, 1983, provides us with the following relation for the overall elasticity constants of an open–walled cellular solid:

$$E(\alpha) = E_s k_1 \left[\frac{\rho}{\rho_s} \right]^2, \tag{5.14}$$

where $\alpha = \frac{\rho}{\rho_s}$ is the comparative density of the cellular material, ρ is the real density, ρ_s is the density and E_s the elasticity modulus of the matrix material and k_1 is a proportionality factor. Poisson's ratio is usually taken equal to zero. This model has been used for the mechanical behaviour of tissues (see e.g. Niklas, 1992, p. 279). A more complicated lattice continuum model can be extracted from the general anisotropic model of Koiter, 1964 which has been used for bone modelling and remodelling studies (Tanaka et al., 1996).

Moreover, cellular structures with reentrant corner cells which predict even negative effective Poisson's ratio can be used (see a relevant numerical homogenization approach in Theocaris et al., 1997b).

5.1.5 Topology optimization problems

Within classical optimization problems which concern the material distribution within a structure, one usually assumes that the optimal design will be composed of classical composites or adaptive materials with an overall elasticity modulus equal to $E(x, \rho, \alpha)$. In a more general setting, the area which occupies the structure as well as the material distribution within this area are not known. If this is the case, a problem with unknown structural topology arises. Some classes of topology optimization problems can be solved, numerically, by the multilevel iterative optimization approach outlined previously. For this purpose, first, all possible mixtures or microstructures which are controlled by the design variables ρ, α are assumed in the previous optimal design problem.

Since this problem is ill-posed, theoretical investigations are used to restrict the set of admissible microstructures within which the optimal material distribution is sought. In turn, the overall elasticity $E(x, \rho, \alpha)$ is computed by analytic or numerical homogenization techniques for the adopted class of microstructures. The whole optimal design problem is then solved and the resulting material distribution $\rho(x)$ is used for the extraction of topology design information. In fact $0 \leq \rho(x) \leq 1$ for all points $x \in \Omega$ of the initial structure. At the postprocessing phase all points with $\rho(x)$ near the maximum value 1 are interpreted as being areas where material is used in the optimal structure while all other areas are deleted.

The previous approach corresponds, mathematically, to a regularization of the initial ill-posed material design problem (i.e., the discrete value problem $\rho(x) = \{0, 1\}$, where $\rho(x) = 0$ means no material and $\rho(x) = 1$ means that material is needed at $x \in \Omega$), by means of homogenization (i.e., with continuous values $0 \leq \rho(x) \leq 1$). Alternatively, a given microstructure is chosen (not the optimal, it may be unknown) and additional measures are taken in order to get analogous regularization results (e.g., perimeter regularization techniques).

A more detailed discussion on these topics (most of them are currently investigated) lies beyond the scope of this book and can be found, among others, in Bendsøe and Kikuchi, 1988, Baier et al., 1994, Sigmund, 1994, Bendsøe, 1995, Rozvany et al., 1995, Haslinger and Neittaanmäki, 1996. If the optimality conditions of the inner (elementwise) subproblem of step 3 are used to provide local update rules for the involved design variables, the proposed algorithm belongs to the class of the optimality criteria optimal design methods (see, e.g., Rozvany, 1989).

In the rest of this Section some simple mixture laws or heuristic approaches for the construction of $E(x, \rho, \alpha)$ are reviewed from the recent literature. These laws can be used for the solution of some topology optimization problems.

Power Law

An isotropic elastic material with a power elasticity law

$$E(x, \rho) = \alpha^n E_0, \; 0 \leq \rho(x) \leq 1, \; \nu = \text{const.} \tag{5.15}$$

has been proposed (with $n = 4$ in Mlejnek, 1992, Mlejnek and Schirrmacher, 1993 and with $n = 2$ in Yang and Chuang, 1994, Gea, 1996, see also Baier et al., 1994). One may easily verify that the external properties $\rho(x) = 0$, resp. $\rho(x) = 1$, are approximated quite satisfactorily by (5.15).

Microvoid composite laws

Starting from the classical mixture theories for composites with a simplified composite made of spherical microvoids, the following simplified law for the Young's modulus and the shear modulus is proposed in Gea, 1996:

$$E(x, \rho) = E_0 \frac{\rho}{2 - \rho}, \; \mu(x, \rho) = \mu_0 \frac{8\rho}{15 - 7\rho}, \tag{5.16}$$

where $\rho = v_m$ is the volume fraction of the matrix in the composite, E_0, μ_0 are the material constants of the matrix material and a Poisson's ratio equal to $\nu = 0.33$ is assumed.

Analogous formulae can be found in Youn and Park, 1997, which are derived by an artificial (design) material with a cell-like microstructure and use of the Hashin-Shtrikman bounds for the derivation of the effective elastic properties.

5.1.6 *A simple damage mechanics model*

Finally, one may pose the problem of finding the more flexible structure which can be designed to carry a given loading, i.e., a maximum compliance problem of the form (cf. (5.1)):

$$\min_{[\rho(x), \alpha(x)] \in \mathcal{A}_{ad}, x \in \Omega} \; \min_{u \in U_{ad}} \left\{ \overline{\Pi}(u, \rho, \alpha) = \Pi(e(u), \rho, \alpha) - l(u) \right\}. \tag{5.17}$$

Clearly the resource constraint (cf. (5.2)) reads

$$\mathcal{A}_{ad} = \left\{ \rho(x), \alpha(x), x \in \Omega \text{ such that } V_{\min} \leq \int_\Omega \rho d\Omega, \; 0 \leq \rho \leq 1 \right\}. \tag{5.18}$$

In this case the constraint $V_{\min}$ can be interpreted to be a maximum amount of degradation (or the remained intact area) in the structure. Note that the latter lower bound is essential for the calculation of a bounded solution of the optimization problem (5.17).

Following Bendsøe, 1995, p. 224, Achtziger and Bendsøe, 1995, the latter problem is coupled with a material law of the form

$$E(x,\rho) = \rho(x)E_0(x) \tag{5.19}$$

with the constraint $V_{\min} \leq \int_\Omega \rho(x)d\Omega$, $0 \leq \rho(x) \leq 1$. Equivalently one may use the law

$$E(x,\alpha) = (1-\alpha(x))\, E_0(x) \tag{5.20}$$

with the constraint $\int_\Omega \alpha(x)d\Omega \leq \int_\Omega d\Omega - \int_\Omega V_{\min} d\Omega = A$, $0 \leq \alpha(x) \leq 1$. In these relations, as in classical damage theory, $\alpha(x) = 0$ means no damage, $\alpha(x) = 1$ denotes a fully damaged element and A is the maximum amount of damage assumed in a given analysis.

The needed modifications in the previously given Algorithm are straightforward and will not be discussed in details.

Using the above simple model of damage and solving the structure for various levels of A, i.e., $0 \leq A_1 \leq A_2 \leq \cdots \leq A_m$ a fairly complete damage scenario can be calculated. Links with the classical damage theory can be established to provide an energetic mechanical explanation of the previous simple model. Moreover the introduction of more complicated damage models is possible. In the numerical results given in Chapter 7 such a mechanical interpretation has not been attempted, thus we prefer to speak about the design of the most flexible structure calculated from the algorithm for a given design level $V_{\min}$.

References

Achtziger, W. and Bendsøe, M. P. (1995). Design for maximal flexibility as a simple computational model of damage. *Structural Optimization*, 10:258–268.

Al-Fahed, A. and Panagiotopoulos, P. D. (1993). A linear complementarity approach to the articulated multifingered friction gripper. *Journal of Robotics Research*, 10(6):871–887.

Allaire, G., Bonnetier, E., Francfort, G., and Jouve, F. (1997). Shape optimization by the homogenization method. *Numerische Mathematik*, 76(1):27–68.

Ashby, M. F. (1983). The mechanical properties of cellular solids. *Amer. Inst. Mech. Eng. Metal. Trans.*, A 14:1755–1769.

Baier, H., Seeßelberg, C., and Specht, B. (1994). *Optimierung in der Strukturmechanik.* Vieweg Verlag, Braunschweig, Wiesbaden.

Bendsøe, M. P. (1995). *Optimization of structural topology, shape and material.* Springer, Berlin.

Bendsøe, M. P. and Kikuchi, N. (1988). Generating optimal topologies in structural design using a homogenization method. *Computer Methods in Applied Mechanics and Engineering*, 71:197–224.

Diaz, A. and Sigmund, O. (1995). Checkerboard patterns in layout optimization. *Structural Optimization*, 10:40–45.

Gea, H. C. (1996). Topology optimization: a new microstructure based design domain method. *Computers and Structures*, 61:781–788.

Gibson, L. J. and Ashby, M. F. (1982). The mechanics of two–dimensional cellular materials. *Proc. Royal Society London*, A382:43–59.

Haslinger, J. and Neittaanmäki, P. (1988). *Finite element approximation for optimal shape design. Theory and applications.* J. Wiley and Sons, Chichester.

Haslinger, J. and Neittaanmäki, P. (1996). *Finite element approximation for optimal shape, material and topology design.* J. Wiley and Sons, Chichester. (2nd edition).

Jog, C. S., Haber, R. B., and Bendsøe, M. P. (1994). Topology design with optimized, self–adaptive materials. *Computer Methods in Applied Mechanics and Engineering*, 37:1323–1350.

Kočvara, M. and Outrata, J. V. (1996). *A nonsmooth approach to optimization problems with equilibrium constraints.* Universität Erlangen–Nürnberg, Institut für Angewandte Mathematik, Erlangen–Nürnberg Germany. Preprint No. 175.

Koiter, W. (1964). Couple–stresses in the theory of elasticity. *Proc. K. Ned. Akad., Wet.*, B–67:17.

Křížek, M. and Neitaanmäki, P. (1990). *Finite element approximation of variational problems and applications.* Longman Scientific and Technical, Essex U.K.

Luo, Z. Q., Pang, J. S., and Ralph, D. (1996). *Mathematical programs with equilibrium constraints.* Cambridge University Press, Cambridge.

Mäkelä, M. M. and Neittaanmäki, P. (1992). *Nonsmooth optimization: analysis and algorithms with applications to optimal control.* Word Scientific Publ. Co.

Mlejnek, H. P. (1992). Some aspects of the genesis of structures. *Structural Optimization*, 5:64–69.

Mlejnek, H. P. and Schirrmacher, R. (1993). An engineers approach to optimal material distribution and shape finding. *Computer Methods in Applied Mechanics and Engineering*, 106:1–2.

Niklas, K. (1992). *Plant Biomechanics. An engineering approach to plant form and function.* The University of Chicago Press, Chicago and London.

Panagiotopoulos, P. D. (1976). Subdifferentials and optimal control in unilateral elasticity. *Mechanics Research Communication*, 3:91–96.

Panagiotopoulos, P. D. (1977). Optimal control in the unilateral thin plate theory. *Archives of Mechanics*, 29:25–39.

Panagiotopoulos, P. D. (1980). Optimal control in the theory of the unilateral von-Karman plates. In *Proc. IUTAM Conf. 1978 on Variational Methods in Mechanics of Solids*, pages 344–348. Pergamon, Evanston Illinois.

Panagiotopoulos, P. D. (1983). Optimal control and parameter identification of structures with convex and non-convex strain energy density. Applications in elastoplasticity and to contact problems. *Solid Mechanics Archives*, 8:160–183.

Panagiotopoulos, P. D. and Haslinger, J. (1992). Optimal control and identification of structures involving multivalued nonmonotonicities. Existence and approximation results. *European Journal of Mechanics A / Solids*, 11(4):4240–4450.

Panagiotopoulos, P. D., Stavroulakis, G. E., Mistakidis, E. S., and Panagouli, O. K. (1997). Multilevel algorithms for structural analysis problems. In Migdalas, A., Pardalos, P., and Värbrand, P., editors, *Multilevel Optimization: Algorithms and Applications*, Dordrecht Boston. Kluwer Academic. (to appear).

Prager, W. and Taylor, J. E. (1968). Problems of optimal structural design. *Journal of Applied Mechanics*, 35:102–106.

Rozvany, G. I. N. (1989). *Structural design via optimality criteria. The Prager approach to structural optimization.* Kluwer Academic, Dordrecht.

Rozvany, G. I. N., Bendsøe, M. P., and Kirsch, U. (1995). Layout optimization of structures. *Applied Mechanics Review*, 48(2):41–119.

Sigmund, O. (1994). Materials with prescribed constitutive parameters: an inverse homogenization problem. *Int. J. Solids and Structures*, 31(17):2313–2329.

Stavroulakis, G. E. (1995a). Optimal prestress of cracked unilateral structures: finite element analysis of an optimal control problem for variational inequalities. *Computer Methods in Applied Mechanics and Engineering*, 123:231–246.

Stavroulakis, G. E. (1995b). Optimal prestress of structures with frictional unilateral contact interfaces. *Archives of Mechanics (Ing. Archiv)*, 66:71–81.

Stavroulakis, G. E. and Günzel, H. (1997). Optimal structural design in nonsmooth mechanics. In Migdalas, A., Pardalos, P., and Värbrand, P., editors, *Multilevel Optimization: Algorithms and Applications.* Kluwer Academic, Dordrecht.

Stavroulakis, G. E. and Tzaferopoulos, M. A. (1994). A variational inequality approach to optimal plastic design of structures via the Prager-Rozvany theory. *Structural Optimization*, 7(3):160–169.

Tanaka, M., Adachi, T., and Tomita, Y. (1996). Mechanical remodeling of bone structure considering residual stress. *JSME Intern. Journal*, A 39(3):297–305.

Theocaris, P. S. (1987). *The mesophase concept in composites.* Springer Verlag, Berlin Heidelberg.

Theocaris, P. S. and Stavroulakis, G. E. (1997a). The effective elastic moduli of plane-weave woven fabric composites by numerical homogenization. *Journal of Reinforced Plastics and Composites*, (to appear).

Theocaris, P. S. and Stavroulakis, G. E. (1997b). Multilevel iterative optimal design of composite structures, including materials with negative Poisson's ratio. *Structural Optimization*, (to appear).

Theocaris, P. S., Stavroulakis, G. E., and Panagiotopoulos, P. D. (1997a). Calculation of effective transverse elastic moduli of fiber-reinforced composites by numerical homogenization. *Composites Science and Technology*, 57:573–586.

Theocaris, P. S., Stavroulakis, G. E., and Panagiotopoulos, P. D. (1997b). Negative Poisson's ratio in materials with a star–shaped microstructure. A numerical homogenization approach. *Archive of Applied Mechanics*, 67(4):274–286.

Theocaris, P. S. and Varias, A. G. (1986). The influence of the mesophase on the transverse and longitudinal moduli and the major Poisson ratio in fibrous composites. *Colloid and Polymer Science*, 264:1–9.

Thierauf, G. (1995). Optimal topologies of structures. Homogenization, pseudo-elastic approximation and the bubble-method. *Engineering Computation*, 13(1):86–102.

Tzaferopoulos, M. A. and Stavroulakis, G. E. (1995). Optimal structural design via optimality criteria as a nonsmooth mechanics problem. *Computers and Structures*, 55(5):761–772.

Yang, R. J. and Chuang, C. H. (1994). Optimal topology using linear programming. *Computers and Structures*, 52:265–275.

Youn, S. K. and Park, S. H. (1997). A study on the shape extraction process in the structural topology optimization using homogenized material. *Computers and Structures*, 62(3):527–538.

IV Computational Mechanics. Computer Implementation, Applications and Examples

6 COMPUTATIONAL MECHANICS ALGORITHMS

6.1 NUMERICAL OPTIMIZATION AND COMPUTATIONAL MECHANICS

As it has already been discussed in previous Chapters of this book, a large number of the nonlinear structural analysis problems can be written in a form of a potential or complementary energy optimization problem. Moreover, unilateral effects, friction, plasticity and damage effects introduce inequality restrictions in the optimization problem or require the consideration of more complicated potentials and dissipation functions.

Nonsmooth mechanics applications involve nondifferentiable, in the classical sense, functions and problems which have inequalities in their definition. Algorithms for classical smooth computational mechanics problems are based on differentiable mathematical optimization techniques. After discretization of the mechanical problem, an energy optimization problem is solved or, equivalently, the optimality conditions of this problem are solved, i.e., a system of nonlinear equations. The appearance of inequality constraints and the nonsmoothness of the involved functions require appropriate modifications of these techniques. Nonsmooth analysis and optimization tools can be used for this purpose.

All these points have been tackled in Chapter 2 of this book and in the concrete applications of Chapter 3. Nevertheless, it should be mentioned that even for classical, differentiable problems, the characteristics of the mechanical problem in connection with the ones of the modelling techniques (e.g., symmetry and sparsity of the involved matrices) must be taken into account for the efficient treatment of large scale problems (see, among others, Zienkiewicz and Taylor, 1991, Papadrakakis, 1997, Papadrakakis and Bitzarakis, 1996). This need is also reflected on the large number of approaches and algorithms cited in the previous Chapters for each particular class of problems.

Moreover, despite its convenience, convexity is not always guaranteed in computational mechanics applications. The theoretical and algorithmic problems connected with nonconvex or global optimization problems have briefly been mentioned in Chapter 2. The causes of nonconvexity in structural analysis models along with the physical mechanisms which are responsible for this behaviour have been discussed in Chapter 4. As with the previous, considerably simpler, class of convex and differentiable problems, the direct use of black-box mathematical optimization algorithms should be avoided in favour of specialized task-taylored algorithms which profit from the structure of the given mechanical problem.

Various iterative convex decomposition techniques have been proposed in this book for the solution of engineering mechanics problems. They apply to both smooth and nonsmooth problems. Moreover, they are in some extend amenable to parallel or distributed computer implementations. In this context the ideas elaborated in this book recall the current revived interest on operator splitting techniques, domain decomposition methods, augmented Lagrangian techniques and methods for large time increments (Glowinski and LeTallec, 1989, LeTallec, 1990, Ladèveze, 1995) which are used to obtain a structured partitioning of large scale computational mechanics problems. The gain is either an efficient treatment of highly nonlinear problems, or the treatment of problems where a physical decomposition is already given by specialized algorithms for each part of the problem, or even the formulation of algorithms amenable to parallelization.

6.1.1 Smooth problems

A model potential energy optimization problem is considered here which arises in an elastic structure with nonlinear adhesive interfaces. Moreover, let us assume that the governing relations of the elastostatic analysis problem can be derived by appropriate differentiation of a potential energy function. A finite element discretization of the direct stiffness method with nodal displacements as the primary variables of the problem is assumed here. Let $\mathbf{u}$ be the n-

dimensional vector of displacement degrees of freedom and $\mathbf{e}$ the m-vector of element deformations. The discrete potential energy optimization problem in elastostatics reads:

$$\min_{\mathbf{u}\in V_{ad}} \{\overline{\Pi}(\mathbf{u}) = \Pi(\mathbf{e}(\mathbf{u})) + \Phi(\mathbf{u}) + p(\mathbf{u})\}, \tag{6.1}$$

where $\Pi(\mathbf{e})$ is the elastic internal deformation energy, $\Phi(\mathbf{u})$ is the potential that counts for boundary or interface effects and $p(\mathbf{u})$ is the potential that generates the external loading vector. The geometric compatibility transformation is written in the form of a generally nonlinear but differentiable operator $\mathcal{A}(\mathbf{u}) : \mathbb{R}^n \to \mathbb{R}^m$, $\mathbf{e} = \mathcal{A}(\mathbf{u})$. Assume that the set of kinematically admissible displacements is $V_{ad} = \mathbb{R}^n$.

Variational formulations for the elastostatic analysis problem described by (6.1) are produced by expressing the optimality conditions for the potential minimization problem and using appropriate smooth or nonsmooth calculus rules for the expression of the involved (generalized) gradients of the composite function $\overline{\Pi}(\mathbf{u})$ (see also Panagiotopoulos, 1985, Dem'yanov et al., 1996).

We recall that classical nonlinear elastostatic analysis problems are written in the form: Find $\mathbf{u} \in \mathbb{R}^n$ such that

$$\nabla\overline{\Pi}(\mathbf{u}) = \mathbf{0}. \tag{6.2}$$

The variational equality form of the problem reads:
Find $\mathbf{u} \in \mathbb{R}^n$ such that

$$\overline{\Pi}'(\mathbf{u}, \Delta\mathbf{u}) = \nabla\overline{\Pi}(\mathbf{u})^T\Delta\mathbf{u} = \left[\frac{\partial\mathcal{A}(\mathbf{u})}{\partial\mathbf{u}}\right]^T \frac{\partial\Pi(\mathbf{e})}{\partial\mathbf{e}}\Delta\mathbf{u} + \\ + \left[\frac{\partial\Phi(\mathbf{u})}{\partial\mathbf{u}}\right]^T \Delta\mathbf{u} + \left[\frac{\partial p(\mathbf{u})}{\partial\mathbf{u}}\right]^T \Delta\mathbf{u} = 0,\ \forall\Delta\mathbf{u} \in \mathbb{R}^n. \tag{6.3}$$

Finite element methods are based on an appropriate discretization of (6.3) and on the solution of the arising system of nonlinear equations. Moreover, the above interpretation of the virtual work as the directional derivative of the potential energy function makes the link between computational mechanics and numerical optimization straightforward.

6.1.2 *Algorithms vs. numerical optimization*

Nonlinear structural analysis problems are usually solved by incremental iterative techniques (see e.g. Ortega and Rheinboldt, 1970, Stein et al., 1989, Zienkiewicz and Taylor, 1991, Crisfield, 1991, Wriggers and Wagner, 1991, Curnier, 1993). Incrementation refers to the gradual application of the external action (e.g. the loading) in a path following like way. A good initial point for the iterations which follow, hopefully in the neighbourhood of the solution, is provided by the solution at the end of the previous increment. These iterations are usually based on linearization techniques which lead to the solution of a series of linear subproblems. The solution of the initial nonlinear problem is approached along a prescribed incrementation and by performing iterations within each incremental step.

A direct connection between structural analysis methods and minimization techniques exists for potential problems, i.e., for problems where the solution renders minimum a potential energy function (Matthies and Strang, 1979). This assumption will hold throughout this Section in order to facilitate the comparative presentation of structural analysis and mathematical optimization techniques.

6.1.3 Iterative linearization

Let us first consider that all external actions (loading) are applied in one loading step. Moreover, let us consider the unconstrained potential energy minimization problem (6.1) written in compact form:
Find kinematically admissible displacements $u \in \mathbb{R}^n$ such that

$$\overline{\Pi}(\mathbf{u}) = \min_{\mathbf{v} \in \mathbb{R}^n} \overline{\Pi}(\mathbf{v}). \tag{6.4}$$

On the assumption that $\overline{\Pi}(\mathbf{u})$ is differentiable the optimality conditions of (6.4) are written in the (strong) form, i.e. the system of nonlinear equations (6.2). The weak form of the above optimality conditions, which in mechanics is known as the principle of virtual work reads (cf. (6.3)):
Find kinematically admissible displacements $u \in \mathbb{R}^n$ such that

$$\overline{\Pi}(\mathbf{u},\ \boldsymbol{\eta}\ ) = \overline{\Pi}'(\mathbf{u},\ \boldsymbol{\eta}\ ) = \left[\nabla\overline{\Pi}(\mathbf{u})\right]^T \boldsymbol{\eta} = 0,\ \forall \boldsymbol{\eta} \in \mathbb{R}^n . \tag{6.5}$$

Vector $\boldsymbol{\eta}$ in (6.5) is the vector of virtual displacements. Moreover, $\overline{\Pi}'(\mathbf{u},\ \boldsymbol{\eta}\ )$ denotes the directional derivative of the function $\overline{\Pi}$ at the point $\mathbf{u}$ and along the direction $\boldsymbol{\eta}$. Recall that by appropriate finite element discretization of the fields $\mathbf{u}$ and $\boldsymbol{\eta}$, equation (6.5) leads to a system of nonlinear equations with respect to the chosen degrees of freedom.

For the solution of the nonlinear equations (6.5) the general method of iterative linearization can be used (Curnier, 1993, Crisfield, 1991, Wriggers, 1993, Stein et al., 1989). For this purpose a linearization of (6.5) around a given point $\overline{\mathbf{u}}$ is performed and the iterative solution of the corresponding linearized equation is considered (Curnier, 1993, Wriggers, 1993):

$$\overline{\Pi}(\overline{\mathbf{u}}, \boldsymbol{\eta}, \Delta\mathbf{u}) = \overline{\Pi}(\overline{\mathbf{u}}, \boldsymbol{\eta}) + \left[\nabla_u\overline{\Pi}(\overline{\mathbf{u}}, \boldsymbol{\eta}\ )\right]^T \Delta\mathbf{u} = 0. \tag{6.6}$$

By applying the chain rule of differentiation on (6.5) and by using (6.6) we get:

$$\overline{\Pi}(\overline{\mathbf{u}}, \boldsymbol{\eta}, \Delta\mathbf{u}) = \overline{\Pi}(\overline{\mathbf{u}}, \boldsymbol{\eta}) + \boldsymbol{\eta}^T \nabla^2_{uu}\overline{\Pi}(\overline{\mathbf{u}})\Delta\mathbf{u}. \tag{6.7}$$

Equivalently the linearization reads:

$$\nabla\overline{\Pi}(\mathbf{u}) = \nabla\overline{\Pi}(\overline{\mathbf{u}}) + \nabla^2_{uu}\overline{\Pi}(\overline{\mathbf{u}})\Delta\mathbf{u} = \mathbf{0}. \tag{6.8}$$

The Hessian of the potential energy function $\mathbf{H}(\overline{\mathbf{u}}) = \nabla^2_{uu}\overline{\Pi}(\overline{\mathbf{u}})$ is the tangential stiffness matrix in the structural analysis terminology. Solution of (6.8) for $\Delta\mathbf{u}$ requires the inversion of $\mathbf{H}(\overline{\mathbf{u}})$:

$$\Delta\mathbf{u} = -\mathbf{H}(\overline{\mathbf{u}})^{-1}\nabla\overline{\Pi}(\overline{\mathbf{u}}). \tag{6.9}$$

In turn $\overline{\mathbf{u}}$ is updated by the formula

$$\overline{\mathbf{u}} = \mathbf{u} + \Delta\mathbf{u} \tag{6.10}$$

and the procedure continues until $\|\Delta\mathbf{u}\|$ becomes smaller than a specified accuracy.

Variants of the previously outlined scheme which follow the parallel developments of numerical optimization techniques have also been proposed (see, e.g., Chapter 2 and the references there).

Within the general scheme of numerical optimization methods, a method of the Newton family, as the one previously given, can be characterized by the choice of a search direction (at the k–th iteration step):

$$\mathbf{d}^{(k)} = -\mathbf{H}^{(k)-1}\nabla\overline{\Pi}(\mathbf{u}^{(k)}) \tag{6.11}$$

and subsequently the solution of a one–dimensional minimization problem along the direction $\mathbf{d}^{(k)}$:

$$\min_{\alpha\geq 0}\overline{\Pi}(\mathbf{u}^{(k)} + \alpha\mathbf{d}^{(k)}) = \alpha^{(k)}. \tag{6.12}$$

The next iteration starts from the updated new point:

$$\mathbf{u} = \overline{\mathbf{u}} + \alpha^{(k)}\mathbf{d}^{(k)}. \tag{6.13}$$

In (6.11) $\mathbf{H}^{(k)} = \nabla^2_{uu}\overline{\Pi}(\mathbf{u}^{(k)})$ corresponds to the Newton method, $\mathbf{H}^{(k)} = \nabla^2_{uu}\overline{\Pi}(\mathbf{u}^{(0)})$ is used in the initial stress, quasi–Newton method, etc. More details on the techniques used in this area can be found, among others, in Ortega and Rheinboldt, 1970, Fletcher, 1990, Papadrakakis and Pantazopoulos, 1993.

It is instructive to mention here that the quantity:

$$\mathbf{d}^{(k)} = \nabla\overline{\Pi}(\mathbf{u}^{(k)}) \tag{6.14}$$

has the meaning of unbalanced nodal forces for the discretized structure at the iteration step (k) (i.e. the trial deformed configuration of this step is defined by the vector $\mathbf{u}^{(k)}$).

The extensions proposed for nonsmooth problems and for nonconvex problems can easily be followed from the material given in the previous Chapters. Special purpose algorithms which are effective for certain classes of problems are given in the sequel.

6.1.4 Path-following

The local convergence properties of the most of the numerical solution algorithms, the path dependent effects in mechanics which have to be taken into account and the need to follow the structural response along a given loading history, lead to the use of incremental and path–following methods. These techniques are based on theoretical results from parametric optimization (see, among others, Allgower and Georg, 1990, Guddat et al., 1990).

Let us briefly discuss this technique by considering that the loading vector $\mathbf{p}(\mathbf{u})$ in the potential energy miminization problem (6.1) has been parametrized by a scalar parameter $\lambda \in \mathbb{R}^1$. For each loading stage λ the structural response is given as the solution of the energy optimization problem:
Find displacements $\mathbf{u} \in \mathbb{R}^n$ such that

$$\overline{\Pi}(\mathbf{u},\lambda) = \min_{\mathbf{v}\in\mathbb{R}^n} \left\{\overline{\Pi}(\mathbf{v},\lambda) = \Pi(\mathbf{e}(\mathbf{v})) + \Phi(\mathbf{v}) + \lambda\mathbf{p}(\mathbf{v})\right\}. \tag{6.15}$$

An incremental application of $\mathbf{p}(\mathbf{u})$ consists in the consideration of problem (6.15) for all values of λ within a given interval. In a more general setting the load increments $l = 0, 1, \ldots, l_t$ can be assumed by the loading history

$$\lambda_{l+1} = \lambda_l + \Delta\lambda_l. \tag{6.16}$$

For smooth problems a modification of the solution scheme (6.4)–(6.10) which corresponds to the load incrementation scheme of structural analysis is required. Within each increment, i.e., for fixed $\lambda = \lambda_{i+1}$ the potential minimization problem is considered (cf. (6.4)):

$$\overline{\Pi}(\mathbf{u},\lambda_{l+1}) = \min_{\mathbf{v}\in\mathbb{R}^n} \overline{\Pi}(\mathbf{v},\lambda_{l+1}). \tag{6.17}$$

For this problem the optimality conditions read (cf. (6.2)):

$$\nabla\overline{\Pi}(\mathbf{u},\lambda_{l+1}) = \mathbf{0}. \tag{6.18}$$

Within each loading step, iterations are performed for the solution of (6.18). As a starting point $\mathbf{u}_{(\lambda_l)}$ is used. The general scheme (6.9) can be used:

$$\Delta\mathbf{u} = -\mathbf{H}(\overline{\mathbf{u}},\lambda_{l+1})^{-1}\nabla\overline{\Pi}(\overline{\mathbf{u}},\lambda_{l+1}), \tag{6.19}$$

where **H** is the tangential stiffness matrix.

For more details on these algorithms the reader is referred to the specialized literature (for unilateral contact problems see, among others, the recent contributions of Björkman, 1992, Park and Kwack, 1994). Nevertheless, it should be mentioned that especially for nonconvex and nondifferentiable problems there exist many open theoretical and practical questions (see relevant discussions in Dem'yanov et al., 1996, Rohde and Stavroulakis, 1995, Rohde and Stavroulakis, 1997, Stavroulakis and Rohde, 1996).

6.2 SPECIAL PURPOSE ALGORITHMS FOR INTERFACE PROBLEMS

In this Section a family of algorithms which are targeted to the solution of problems involving interfaces, is presented. Problems leading to convex and nonconvex potential and complementary energy optimization problems are considered. For the most of the cases presented here, this approach is the only one available today to solve the corresponding problems.

As it has already been demonstrated in Chapters 3 and 4, the mathematical formulation of mechanical problems containing interfaces, does not lead directly into independent optimization problems. This is a result of the fact that there exists a strong interaction between the forces arising in the normal and tangential to the interface directions, i.e. the tangential forces depend on the value of the normal forces, and the values of the tangential forces may alter the contact area or the values of the contact forces. In the following Sections we use techniques similar to the ones used in the field of multilevel optimization in order to treat numerically the above problems.

The aim of multilevel optimization is to define with respect to an optimization problem, appropriate mutually independent subproblems. Each of these, when solved independently, yields the optimum of the overall problem after an iterative procedure which is called second - level controller. The decomposition into subproblems is achieved by choosing some variables, called coordinating variables, which are freely manipulated by the second - level controller in such a way that the subproblems (first - level of the problem) have solutions which in fact yield the optimum of the initial problem, i.e. before its decomposition into subproblems (Bauman, 1971, Schoeffler, 1971, Wismer, 1971).

In interface problems the ideas from the field of multilevel optimization are applied by splitting the actions in the normal and tangential to the interface directions (see e.g. the pioneering work of Panagiotopoulos, 1975 where such a decomposition was first presented). Therefore, the unknowns of the problem are split into two groups corresponding to each interface direction. New optimization problems are then formulated involving only the unknowns corresponding to the one direction (normal or tangential), while the unknowns that correspond to the other direction are assumed as known. The subproblems are solved iteratively until some predefined convergence criteria are fulfilled. The above technique is analogous to a multilevel optimization algorithm, in the sense that in each level a minimum problem is solved involving only the unknowns of the one group, while the values of the variables of the other group are considered as known. Therefore, the overall minimum is achieved only in the case of convergence.

Notice in this respect, that in general we have the possibility to consider for each subproblem the primal or the dual formulation. For each subproblem, the formulation selected is the one leading to the most efficient numerical determination of the solution.

6.2.1 Convex problems

The general formulation of the interface problems has been given in Section 3.1.2. Following the symbols and the notation introduced therein, we remind that we have to minimize either the potential energy of the structure (cf. (3.34)):

$$\Pi(\mathbf{u}) = \frac{1}{2}\mathbf{u}^T\mathbf{K}\mathbf{u} - \overline{\mathbf{p}}^T\mathbf{u} + \Phi_N(\mathbf{u}, \mathbf{S}_T) + \Phi_T(\mathbf{u}, \mathbf{S}_N), \ \mathbf{u} \in V_{ad} \tag{6.20}$$

or the complementary energy (cf. (3.35)):

$$\Pi^c(\mathbf{s}) = \frac{1}{2}\mathbf{s}^T\mathbf{F}_0\mathbf{s} + \mathbf{e}_0^T\mathbf{s} + \overline{\Phi}_N(\mathbf{s}, \mathbf{S}_T) + \overline{\Phi}_T(\mathbf{s}, \mathbf{S}_N), \ \mathbf{s} \in \Sigma_{ad} \tag{6.21}$$

where $\Phi_N(\mathbf{u}, \mathbf{S}_T)$, $\Phi_T(\mathbf{u}, \mathbf{S}_N)$, $\overline{\Phi}_N(\mathbf{s}, \mathbf{S}_T)$ and $\overline{\Phi}_T(\mathbf{s}, \mathbf{S}_N)$ are appropriately defined convex superpotentials and V_{ad}, Σ_{ad} are appropriate kinematically and statically admissible sets respectively. Here, the superpotentials holding in the normal (resp. the tangential) to the interface direction $\Phi_N(\mathbf{u}, \mathbf{S}_T)$, $\overline{\Phi}_N(\mathbf{s}, \mathbf{S}_T)$ (resp. $\Phi_T(\mathbf{u}, \mathbf{S}_N)$, $\overline{\Phi}_T(\mathbf{s}, \mathbf{S}_N)$) depend on the values of $\mathbf{S}_T$ (resp. $\mathbf{S}_N$).

Following the ideas of Panagiotopoulos, 1975, problem (6.20) is split in the following subproblems

$$\Pi_1(\mathbf{u}) = \frac{1}{2}\mathbf{u}^T\mathbf{K}\mathbf{u} - \overline{\mathbf{p}}_1^T\mathbf{u} + \Phi_N(\mathbf{u}), \ \mathbf{u} \in V_{ad}^1 \tag{6.22}$$

and

$$\Pi_2(\mathbf{u}) = \frac{1}{2}\mathbf{u}^T\mathbf{K}\mathbf{u} - \overline{\mathbf{p}}_2^T\mathbf{u} + \Phi_T(\mathbf{u}), \ \mathbf{u} \in V_{ad}^2 \tag{6.23}$$

by assuming that in the first (resp. the second) the values of $\mathbf{S}_T$ (resp. $\mathbf{S}_N$) are fixed.

In the previous relations, $\overline{\mathbf{p}}_1$ is a new force vector containing the effect of the assumed known tangential forces $\mathbf{S}_T$ on the structure

$$\overline{\mathbf{p}}_1 = \overline{\mathbf{p}} + \mathbf{G}_T\mathbf{S}_T \tag{6.24}$$

and $\overline{\mathbf{p}}_2$ is a force vector containing the effect of the assumed known normal forces $\mathbf{S}_N$

$$\overline{\mathbf{p}}_2 = \overline{\mathbf{p}} + \mathbf{G}_N\mathbf{S}_N. \tag{6.25}$$

Due to the assumptions for $\mathbf{S}_N$, $\mathbf{S}_T$ we used $\Phi_N(\mathbf{u})$ instead of $\Phi_N(\mathbf{u}, \mathbf{S}_T)$ and $\Phi_T(\mathbf{u})$ instead of $\Phi_T(\mathbf{u}, \mathbf{S}_N)$. Moreover, V^1_{ad} and V^2_{ad} are appropriately defined kinematically admissible sets as we will see in the following.

In the same manner, problem (6.21) is split in the following subproblems

$$\Pi_1^c(\mathbf{s}) = \frac{1}{2}\mathbf{s}^T\mathbf{F}_0\mathbf{s} + \mathbf{e}_0^T\mathbf{s} + \overline{\Phi}_N(\mathbf{s}), \;\; \mathbf{s} \in \Sigma^1_{ad} \tag{6.26}$$

and

$$\Pi_2^c(\mathbf{s}) = \frac{1}{2}\mathbf{s}^T\mathbf{F}_0\mathbf{s} + \mathbf{e}_0^T\mathbf{s} + \overline{\Phi}_T(\mathbf{s}), \;\; \mathbf{s} \in \Sigma^2_{ad} \tag{6.27}$$

where Σ^1_{ad} and Σ^2_{ad} are appropriately defined statically admissible sets resulting from the statically admissible set Σ_{ad} with the assumption that the tangential forces $\mathbf{S}_T$ or the normal forces $\mathbf{S}_N$ are applied as external loading respectively.

Based on the above decompositions of the normal and tangential actions, multilevel–type algorithms can be formulated for the solution of the initial problems. As it is obvious from the form of the above subproblems, the solution of (6.22) and (6.26) depends only on the assumed known values of the tangential forces $\mathbf{S}_T$ and the solution of (6.23) and (6.27) depends only on the assumed known values of the normal forces $\mathbf{S}_N$. Thus, it is possible to formulate mixed schemes of multilevel problems by minimizing for example in the first step the potential energy and in the second step the complementary energy or the opposite. This possibility gives a great flexibility to the computational scheme, as it is possible in this way to avoid the solution of difficult numerical problems, as we will see in the following Sections.

6.2.1.1 Unilateral contact with monotone debonding. In this Section we consider a unilateral contact problem with the assumption that in the case of opening of the interface a monotone law relates the debonding force S_D with the relative normal to the interface displacement $[u]_D$. This law gives rise to the convex superpotential $\phi_D([u]_D)$, i.e. the differential inclusion $-S_D \in \partial\phi_D([u]_D)$ holds. These conditions describe for example the behaviour of an adhesive material connecting the two parts of the interface. In this Section we neglect the effect of the tangential action as this is the case in several engineering problems (Davies and Benzeggagh, 1989). Here we have to distinguish two parts of the interface, the contact part Γ_C and the non-contact part Γ_D. On the contact part the contact forces $\mathbf{S}_N$ develop whereas on the non-contact part the debonding forces $\mathbf{S}_D$ develop. Moreover, in the discretized structure we denote with c the nodes belonging in Γ_C and with d the nodes of Γ_D. Thus, the potential energy of the structure takes the form:

$$\Pi(\mathbf{u}) = \frac{1}{2}\mathbf{u}^T\mathbf{K}\mathbf{u} - \overline{\mathbf{p}}^T\mathbf{u} + \Phi_D(\mathbf{u}), \ \mathbf{u} \in V_{ad} \tag{6.28}$$

where Φ_D is the global convex superpotential of the interface, i.e.

$$\Phi_D(\mathbf{u}) = \sum_{i=1}^{d} \phi_D^{(i)}([\mathbf{u}]_D) \tag{6.29}$$

and V_{ad} is the kinematically admissible set, i.e. (see Section 3.1.2 for the notation)

$$V_{ad} = \{\mathbf{u} \in \mathbf{R}^n \,|\mathbf{E}\mathbf{u} = \mathbf{u}_0, \ \mathbf{N}\mathbf{u} - \mathbf{g} \leq \mathbf{0}\}. \tag{6.30}$$

Notice that due to the fact that the contact and non-contact areas (and thus the energy term Φ_D which depends on the summation index d) are not *a priori* known, the direct solution of the minimization problem implied by (6.28) is not possible.

Following a strategy analogous to the one proposed by Panagiotopoulos, 1975, we split the problem in two subproblems. In the first level, we consider the subproblem corresponding to the solution of the unilateral contact problem with known debonding forces $\mathbf{S}_D$, i.e. we consider the minimization of the potential energy

$$\Pi_1(\mathbf{u}) = \frac{1}{2}\mathbf{u}^T\mathbf{K}\mathbf{u} - \overline{\mathbf{p}}_1^T\mathbf{u}, \ \mathbf{u} \in V_{ad} \tag{6.31}$$

where

$$\overline{\mathbf{p}}_1 = \overline{\mathbf{p}} + \mathbf{G}_D\mathbf{S}_D. \tag{6.32}$$

Next we consider the subproblem corresponding to the solution of the debonding problem with known contact forces, i.e. we consider the potential energy minimization subproblem

$$\Pi_2(\mathbf{u}) = \frac{1}{2}\mathbf{u}^T\mathbf{K}\mathbf{u} - \overline{\mathbf{p}}_2^T\mathbf{u} + \Phi_D(\mathbf{u}), \ \mathbf{u} \in V_{ad}^2 \tag{6.33}$$

where

$$V_{ad}^2 = \{\mathbf{u} \in \mathbf{R}^n \,|\mathbf{E}\mathbf{u} = \mathbf{u}_0\} \tag{6.34}$$

i.e. V^2_{ad} is V_{ad} without the unilateral contact conditions.

The interaction between the two subproblems is very strong because different contact and non-contact areas lead to different values of the summation index d in (6.29). Based on the above decomposition, the following algorithm is proposed for the solution of the problem.

Algorithm 6.1 *Unilateral contact and monotone debonding solver*

1. Starting the algorithm we consider that the debonding forces are constant. Usually we assume that $\mathbf{S}_D^{(0)} = \mathbf{0}$. Initialize i to 1.

2. Calculate the structure with the debonding forces $\mathbf{S}_D^{(i-1)}$ assumed as known. The solution is obtained by minimizing the potential energy expression (6.31). The solution procedure yields the contact and non-contact areas and the values of the contact forces $\mathbf{S}_N^{(i)}$.

3. Solve the debonding problem assuming the contact forces $\mathbf{S}_N^{(i)}$ as constant. The solution is obtained by minimizing the potential energy expression (6.33) and the values of $\mathbf{S}_D^{(i)}$ are obtained.

4. If $\dfrac{||\mathbf{S}^{(i)} - \mathbf{S}^{(i-i)}||}{||\mathbf{S}^{(i)}||} \leq \varepsilon$ where $\mathbf{S} = [\mathbf{S}_N \ \mathbf{S}_D]^T$ and ε is an appropriately small number, then convergence has been achieved and terminate the algorithm, else continue for another iteration setting $i = i + 1$ and returning to step 2.

6.2.1.2 Unilateral contact with Coulomb friction. Here we will study the classical unilateral contact problem with Coulomb friction. The theoretical formulation of the problem has been presented in Section 3.1.3.1 (cf. (3.60)). Here we propose an algorithm for the numerical solution based on a multilevel (in the sense explained before) decomposition of the normal and tangential actions. The problem arising in the normal to the interface direction is solved by minimizing the potential energy

$$\Pi_1(\mathbf{u}) = \frac{1}{2}\mathbf{u}^T\mathbf{K}\mathbf{u} - \overline{\mathbf{p}}_1^T\mathbf{u} \ , \ \mathbf{u} \in V_{ad} \tag{6.35}$$

where V_{ad} is the kinematically admissible set of the displacements given by (6.30) and $\overline{\mathbf{p}}_1$ is defined by (6.24). For the problem arising in the tangential to the interface direction, we prefer to minimize the complementary energy of the structure in order to avoid the nondiferentiability conditions introduced by the frictional constraints. Thus, in the second step the complementary energy (cf. (3.61))

$$\Pi_2^c(\mathbf{s}) = \frac{1}{2}\mathbf{s}^T\mathbf{F}_0\mathbf{s} + \mathbf{e}_0^T\mathbf{s} \ , \ \mathbf{s} \in \Sigma^2_{ad} \tag{6.36}$$

is minimized, where Σ^2_{ad} is the statically admissible set, i.e.

$$\Sigma^2_{ad} = \{\mathbf{s} \in \mathbf{R}^n \,|\mathbf{Z}\mathbf{s} = \mathbf{F},\ \mathbf{G}\mathbf{s} = \mathbf{p} + \mathbf{G}_N\mathbf{S}_N,\ |\mathbf{S}_T| \leq \mathbf{T}_0\} \tag{6.37}$$

and $\mathbf{T}_0 = \mu\mathbf{S}_N$, μ is the Coulomb friction coefficient.

From the viewpoint of numerical optimization, both subproblems (6.35) and (6.36) are constrained quadratic minimization problems and are easily solved using a quadratic programming (Q.P.) algorithm. Notice that due to the convexity of the expressions (6.35) and (6.36), the minimum of each of them is uniquely defined. The following algorithm is formulated for the solution of the problem:

Algorithm 6.2 *Unilateral contact and Coulomb friction solver*

1. Starting the algorithm we consider that the tangential forces are constant. Usually we assume that $\mathbf{S}_T^{(0)} = \mathbf{0}$. Initialize i to 1.

2. Calculate the structure with the given tangential forces $\mathbf{S}_T^{(i-1)}$. The solution is obtained by minimizing the potential energy expression (6.35). The solution procedure yields the contact and non-contact areas and the values of the contact forces $\mathbf{S}_N^{(i)}$.

3. Solve the problem arising in the tangential to the interface direction assuming the contact forces $\mathbf{S}_N^{(i)}$ as constant. The solution is obtained by minimizing the complementary energy (6.36) and the values of $\mathbf{S}_T^{(i)}$ are obtained.

4. Convergence check: if $\dfrac{||\mathbf{S}^{(i)} - \mathbf{S}^{(i-i)}||}{||\mathbf{S}^{(i)}||} \leq \varepsilon$ where $\mathbf{S} = [\mathbf{S}_N\ \mathbf{S}_T]^T$ and ε is an appropriately small number, then convergence has been achieved and terminate the algorithm, else continue for another iteration setting $i = i + 1$ and returning to step 2.

Notice that the above algorithm is also capable to treat adhesive contact problems, i.e. unilateral contact problems where in the tangential direction the interface is glued with an adhesive material obeying to a monotone reaction-displacement law similar to the frictional one. This case is similar to the previous one with the only difference that the values of the tangential forces $\mathbf{S}_T$ do not depend on the values of the contact forces $\mathbf{S}_N$ but they are a monotone function of the relative tangential displacements. In this case the values $\mathbf{T}_0$ in (6.37) are obtained directly from this function. Notice also that this fact does not make independent the problems arising in the two directions but a strong coupling of the two actions still exists.

6.2.1.3 Unilateral contact with Coulomb friction and monotone debonding. In this Section unilateral contact problems are treated where it is assumed that on the interface monotone reaction-displacement laws hold in both the normal and tangential directions. More specifically, for the normal direction we assume that a differential inclusion of the form $-S_D \in \partial\phi_D([u]_D)$ holds. For the tangential direction we assume that a diagram similar to the one of Fig. 3.2 holds which can be expressed in the form $-S_T \in \partial\phi_T([u]_T)$. This is the case of e.g. an interface glued with an adhesive material.

This problem is much more complicated than the previous one because one first has to determine the active contact area Γ_C where the contact forces S_N develop and the non-contact area Γ_D where the debonding forces S_D develop due to the presence of the adhesive material. Moreover, for the parts of the interface which come in contact, we assume that a monotone frictional or adhesive law holds in the tangential direction.

As it was demonstrated in Chapter 3 (cf. Section 3.1.2), the solution of this problem can be obtained by minimizing the potential energy of the structure

$$\Pi(\mathbf{u}) = \frac{1}{2}\mathbf{u}^T\mathbf{K}\mathbf{u} - \overline{\mathbf{p}}^T\mathbf{u} + \Phi_D(\mathbf{u}) + \Phi_T(\mathbf{u}), \quad \mathbf{u} \in V_{ad} \tag{6.38}$$

where $\Phi_D(\mathbf{u})$ and $\Phi_T(\mathbf{u})$ are the convex superpotentials of the monotone reaction-displacement laws holding in the normal and tangential directions respectively and V_{ad} is the kinematically admissible set given by (6.30). Denoting with c the nodes of Γ_C and with d the nodes of Γ_D we have that

$$\Phi_D(\mathbf{u}) = \sum_{i=1}^{d} \phi_D^{(i)}([\mathbf{u}]_D) \tag{6.39}$$

and

$$\Phi_T(\mathbf{u}) = \sum_{i=1}^{c} \phi_T^{(i)}([\mathbf{u}]_T). \tag{6.40}$$

The method for the solution of the above problem is similar to the one presented in the previous Section. Here we split further the normal to the interface problem into two subproblems.

Three minimization subproblems are now formulated. The first subproblem corresponds to the solution of the unilateral contact problem under the assumption that the forces $\mathbf{S}_D$ and $\mathbf{S}_T$ are given. The potential energy

$$\Pi_1(\mathbf{u}) = \frac{1}{2}\mathbf{u}^T\mathbf{K}\mathbf{u} - \overline{\mathbf{p}}_1^T\mathbf{u}, \quad \mathbf{u} \in V_{ad} \tag{6.41}$$

is minimized where

$$\overline{\mathbf{p}}_1 = \overline{\mathbf{p}} + \mathbf{G}_D\mathbf{S}_D + \mathbf{G}_T\mathbf{S}_T \tag{6.42}$$

and V_{ad} is given by (6.30).

The second subproblem corresponds to the solution of the debonding problem with the assumption that the forces $\mathbf{S}_N$ and $\mathbf{S}_T$ are given. Here the potential energy

$$\Pi_2(\mathbf{u}) = \frac{1}{2}\mathbf{u}^T\mathbf{K}\mathbf{u} - \overline{\mathbf{p}}_2^T\mathbf{u} + \Phi_D(\mathbf{u}), \ \mathbf{u} \in V_{ad}^2 \tag{6.43}$$

is minimized, where

$$\overline{\mathbf{p}}_2 = \overline{\mathbf{p}} + \mathbf{G}_N\mathbf{S}_N + \mathbf{G}_T\mathbf{S}_T \tag{6.44}$$

and V_{ad}^2 is given by (6.34).

The third subproblem corresponds to the problem arising in the tangential direction assuming that the normal to the interface direction forces $\mathbf{S}_N$ and $\mathbf{S}_D$ are given, i.e. we minimize the complementary energy

$$\Pi_3^c(\mathbf{s}) = \frac{1}{2}\mathbf{s}^T\mathbf{F}_0\mathbf{s} + \mathbf{e}_0^T\mathbf{s}, \ \mathbf{s} \in \Sigma_{ad}^3 \tag{6.45}$$

where

$$\Sigma_{ad}^3 = \{\mathbf{s} \in \mathbb{R}^n \,|\mathbf{G}\mathbf{s} = \mathbf{p} + \mathbf{G}_N\mathbf{S}_N + \mathbf{G}_D\mathbf{S}_D, \ \mathbf{Z}\mathbf{s} = \mathbf{F}, \ |\mathbf{S}_T| \leq \mathbf{T}_0\}. \tag{6.46}$$

In the third subproblem we passed to the dual problem of minimizing the complementary energy of the structure in order to avoid the nondifferentiability conditions introduced by the corresponding reaction–displacement law (cf. Fig. 3.2).

Based on the above subproblems we can propose the following algorithm for the solution of the initial problem.

Algorithm 6.3 *General interface problem solver (monotone interface laws)*

1. In the first step we consider that the tangential forces $\mathbf{S}_T$ and the debonding forces $\mathbf{S}_D$ are constant. Usually it is assumed that $\mathbf{S}_T^{(0)} = \mathbf{0}$ and $\mathbf{S}_D^{(0)} = \mathbf{0}$. Set $i = 0$ and $j = 0$ where index i (resp. index j) is associated to the iterations for the solution of the problem arising in the tangential (resp. normal) to the interface direction.

2. Set $i = i + 1$.

3. Set $j = j + 1$.

4. Calculate the structure with the given tangential and debonding forces $\mathbf{S}_T^{(i-1)}$ and $\mathbf{S}_D^{(j-1)}$. The solution is obtained by minimizing the potential energy

expression (6.41). As a result we get the contact and noncontact areas and the values of the contact forces $\mathbf{S}_N^{(j)}$.

5. In this step the potential energy expression (6.43) is minimized assuming given the contact and the tangential forces $\mathbf{S}_N^{(j)}$ and $\mathbf{S}_T^{(i-1)}$. As a result the values of $\mathbf{S}_D^{(j)}$ are obtained.

6. Instead of proceeding to the minimization of the complementary energy expression (6.45), we check for the convergence of the results obtained for the normal direction, i.e. we test the expression $\dfrac{\|\mathbf{S}^{(j)} - \mathbf{S}^{(j-i)}\|}{\|\mathbf{S}^{(j)}\|} \leq \varepsilon$ where $\mathbf{S} = [\mathbf{S}_N \ \mathbf{S}_D]^T$. The reason for this preliminary check is that the numerical experimentation has shown a high sensitivity of the whole algorithm in the case of unstable contact regions. If the check is fulfilled we proceed to the next step, if not we return to step 3.

7. In this step the problem arising in the tangential to the interface direction is solved by minimizing the complementary energy expression (6.45), assuming that the normal to the interface forces $\mathbf{S}_N^{(j)}$ and $\mathbf{S}_D^{(j)}$ are given. As a result the values of $\mathbf{S}_T^{(i)}$ are obtained.

8. In this step we check the global convergence: If $\dfrac{\|\mathbf{S}^{(i)} - \mathbf{S}^{(i-i)}\|}{\|\mathbf{S}^{(i)}\|} \leq \varepsilon$ where $\mathbf{S} = [\mathbf{S}_N \ \mathbf{S}_D \ \mathbf{S}_T]^T$ and ε the desired accuracy, then convergence has been achieved and the algorithm is terminated, else it is continued for another iteration going to step 2.

6.2.2 Nonconvex problems

In this Section, the algorithms developed for the case of convex problems are extended in order to solve the corresponding nonconvex problems. The general formulation is analogous to the one given in the previous Section. Here the potential energy of the structure has the form (cf. (4.38))

$$\Pi(\mathbf{u}) = \frac{1}{2}\mathbf{u}^T\mathbf{K}\mathbf{u} - \overline{\mathbf{p}}^T\mathbf{u} + \widetilde{\Phi}_N(\mathbf{u}, \mathbf{S}_T) + \widetilde{\Phi}_T(\mathbf{u}, \mathbf{S}_N), \ \mathbf{u} \in V_{ad} \tag{6.47}$$

and the complementary energy the form

$$\Pi^c(\mathbf{s}) = \frac{1}{2}\mathbf{s}^T\mathbf{F}_0\mathbf{s} + \mathbf{e}_0^T\mathbf{s} + \overline{\overline{\Phi}}_N(\mathbf{s}, \mathbf{S}_T) + \overline{\overline{\Phi}}_T(\mathbf{s}, \mathbf{S}_N), \ \mathbf{s} \in \Sigma_{ad} \tag{6.48}$$

where $\overline{\Phi}_N(\mathbf{u}, \mathbf{S}_T)$, $\overline{\Phi}_T(\mathbf{u}, \mathbf{S}_N)$, $\overline{\overline{\Phi}}_N(\mathbf{s}, \mathbf{S}_T)$ and $\overline{\overline{\Phi}}_T(\mathbf{s}, \mathbf{S}_N)$ are appropriately defined nonconvex superpotentials. Moreover, V_{ad}, Σ_{ad} are the same kinematically and statically admissible sets as in the previous Section. Following the approach of the previous Section, problem (6.47) is split in the subproblems

$$\Pi_1(\mathbf{u}) = \frac{1}{2}\mathbf{u}^T\mathbf{K}\mathbf{u} - \overline{\mathbf{p}}_1^T\mathbf{u} + \widetilde{\Phi}_N(\mathbf{u}), \ \mathbf{u} \in V_{ad}^1 \tag{6.49}$$

and

$$\Pi_2(\mathbf{u}) = \frac{1}{2}\mathbf{u}^T\mathbf{K}\mathbf{u} - \overline{\mathbf{p}}_2^T\mathbf{u} + \widetilde{\Phi}_T(\mathbf{u}), \ \mathbf{u} \in V_{ad}^2 \tag{6.50}$$

by fixing in the first (resp. the second) the values of $\mathbf{S}_T$ (resp. $\mathbf{S}_N$).

Here, $\overline{\mathbf{p}}_1$ and $\overline{\mathbf{p}}_2$ are defined by (6.24) and (6.25) and V_{ad}^1 , V_{ad}^2 are appropriately defined kinematically admissible sets.

Similarly, problem (6.48) is split in the following subproblems

$$\Pi_1^c(\mathbf{s}) = \frac{1}{2}\mathbf{s}^T\mathbf{F}_0\mathbf{s} + \mathbf{e}_0^T\mathbf{s} + \overline{\overline{\Phi}}_N(\mathbf{s}), \ \mathbf{s} \in \Sigma_{ad}^1 \tag{6.51}$$

and

$$\Pi_2^c(\mathbf{s}) = \frac{1}{2}\mathbf{s}^T\mathbf{F}_0\mathbf{s} + \mathbf{e}_0^T\mathbf{s} + \overline{\overline{\Phi}}_T(\mathbf{s}), \ \mathbf{s} \in \Sigma_{ad}^2 \tag{6.52}$$

where Σ_{ad}^1 and Σ_{ad}^2 are appropriately defined statically admissible sets as in the previous Section.

Based on the above decompositions of the normal and tangential actions, multilevel algorithms can be formulated for the solution of the initial problems which are similar to the algorithms proposed for the solution of the convex problems.

6.2.2.1 Unilateral contact with nonmonotone debonding. In this Section we consider a unilateral contact problem with a nonmonotone law describing the debonding process in the normal to the interface direction, i.e. we assume that the conditions

$$-S_D \in \bar{\partial}\widetilde{\phi}_D([u]_D) \text{ or } [u]_D \in \bar{\partial}\overline{\overline{\phi}}_D(-S_D)$$

hold on each interface point, where $\widetilde{\phi}_D$, $\overline{\overline{\phi}}_D$ are appropriately defined nonconvex superpotentials yielding the adhesive law.

The theoretical formulation of the above problem has been given in detail in Chapter 4. As in the case of the corresponding convex problem, we distinguish the contact interface part Γ_C consisting of c nodes in the discretized structure and the non-contact part Γ_D consisting of d nodes. Thus, the potential energy of the structure takes the form

$$\Pi(\mathbf{u}) = \frac{1}{2}\mathbf{u}^T\mathbf{K}\mathbf{u} - \overline{\mathbf{p}}^T\mathbf{u} + \widetilde{\Phi}_D(\mathbf{u}), \ \mathbf{u} \in V_{ad} \tag{6.53}$$

and the complementary energy the form

$$\Pi^c(\mathbf{s}) = \frac{1}{2}\mathbf{s}^T\mathbf{F}_0\mathbf{s} + \mathbf{e}_0^T\mathbf{s} + \overline{\overline{\Phi}}_D(\mathbf{s}), \ \mathbf{s} \in \Sigma_{ad}. \tag{6.54}$$

In the previous relations $\widetilde{\Phi}_D$, $\overline{\overline{\Phi}}_D$ are the global interface nonconvex superpotentials of the adhesive law, i.e.

$$\widetilde{\Phi}_D(\mathbf{u}) = \sum_{i=1}^{d} \widetilde{\phi}_D^{(i)}([\mathbf{u}]_D) \tag{6.55}$$

$$\overline{\overline{\Phi}}_D(\mathbf{s}) = \sum_{i=1}^{d} \overline{\overline{\phi}}_D^{(i)}(-\mathbf{S}_D) \tag{6.56}$$

V_{ad} is the kinematically admissible set given by (6.30) and Σ_{ad} is the statically admissible set given by

$$\Sigma_{ad} = \{\mathbf{s} \in \mathbf{R}^n \,|\mathbf{Z}\mathbf{s} = \mathbf{F}, \ \mathbf{G}\mathbf{s} = \mathbf{p}\}. \tag{6.57}$$

In order to solve the problem we consider first the subproblem corresponding to the solution of the unilateral contact problem with known debonding forces $\mathbf{S}_D$. The potential energy expression for this problem reads:

$$\Pi_1(\mathbf{u}) = \frac{1}{2}\mathbf{u}^T\mathbf{K}\mathbf{u} - \overline{\mathbf{p}}_1^T\mathbf{u}, \ \mathbf{u} \in V_{ad} \tag{6.58}$$

where

$$\overline{\mathbf{p}}_1 = \overline{\mathbf{p}} + \mathbf{G}_D\mathbf{S}_D. \tag{6.59}$$

For the subproblem corresponding to the solution of the debonding problem with known contact forces we can consider either the potential energy substationarity problem

$$\Pi_2(\mathbf{u}) = \frac{1}{2}\mathbf{u}^T\mathbf{K}\mathbf{u} - \overline{\mathbf{p}}_2^T\mathbf{u} + \widetilde{\Phi}_D(\mathbf{u}), \ \mathbf{u} \in V_{ad}^2 \tag{6.60}$$

where

$$\overline{\mathbf{p}}_2 = \overline{\mathbf{p}} + \mathbf{G}_N\mathbf{S}_N \tag{6.61}$$

and

$$V_{ad}^2 = \{\mathbf{u} \in \mathbf{R}^n \,|\mathbf{E}\mathbf{u} = \mathbf{u}_0\} \tag{6.62}$$

or the complementary energy substationarity problem

$$\Pi_2^c(\mathbf{s}) = \frac{1}{2}\mathbf{s}^T\mathbf{F}_0\mathbf{s} + \mathbf{e}_0^T\mathbf{s} + \overline{\overline{\Phi}}_D(\mathbf{s}), \ \mathbf{s} \in \Sigma_{ad}^2 \tag{6.63}$$

where

$$\Sigma_{ad}^2 = \{\mathbf{s} \in \mathbf{R}^n \,|\mathbf{G}\mathbf{s} = \mathbf{p} + \mathbf{G}_N\mathbf{S}_N, \ \mathbf{Z}\mathbf{s} = \mathbf{F}\}. \tag{6.64}$$

Both problems (6.60) and (6.63) are nonconvex nonsmooth substationarity problems and for their numerical solution we can apply the algorithms presented in Chapter 2, i.e. either the d.c. decomposition approach or the heuristic nonconvex optimization approach. The application of both approaches to the examined problem was presented in detail in Chapter 4. Notice here that both approaches are based on the determination of an initial starting point which leads to a certain solution of the substationarity problem. In order to follow a certain loading path, an incremental technique is applied, assuming that the final value of the loading is attained through a certain number of load increments N. Then, the starting point for each substationarity subproblem is obtained from the solution of the previous increment.

As in the case of the respective convex problem, the interaction between the two subproblems is very strong. Based on the above decomposition, the following multilevel algorithm is proposed for the solution of problem.

Algorithm 6.4 *Unilateral contact and nonmonotone debonding solver*

1. Decide for the number N of load increments that will be used. This selection is made in such a way that in the first increment the nonlinearities caused by the nonmonotone debonding diagram are not present. This fact makes necessary some preliminary solutions of the problem in order to select an appropriate value for the load increment $\delta\mathbf{p} = \mathbf{p}/N$.

2. For $k = 1, N$

   (a) Set $\mathbf{p} = k\delta\mathbf{p}$.

   (b) Consider that the debonding forces are constant and more specifically that $\mathbf{S}_D^{(0)} = \mathbf{0}$. Set $i = 1$.

   (c) Calculate the structure with the debonding forces $\mathbf{S}_D^{(i-1)}$ applied as external loading. The solution is obtained through the minimization of the potential energy of the structure (6.31). This procedure yields the contact and non-contact areas and the values of the contact forces $\mathbf{S}_N^{(i)}$.

(d) Solve the debonding problem assuming the contact forces $\mathbf{S}_N^{(i)}$ as constant. The solution is obtained by solving numerically the substationarity problems (6.60) or (6.63). For this reason, one of the Algorithms (2.4), (2.5) or (2.6) can be used. As results the values of $\mathbf{S}_T^{(i)}$ are obtained.

(e) If $\dfrac{\|\mathbf{S}^{(i)} - \mathbf{S}^{(i-i)}\|}{\|\mathbf{S}^{(i)}\|} \leq \varepsilon$ where $\mathbf{S} = [\mathbf{S}_N \ \mathbf{S}_D]^T$ and ε is an appropriately small number, then convergence has been achieved and terminate the algorithm, else continue for another iteration by setting $i = i + 1$ and going to step 2c.

6.2.2.2 Unilateral contact with nonmonotone friction.

Here a unilateral contact problem with nonmonotone friction is studied, i.e. the conditions

$$-S_T \in \bar{\partial}\widetilde{\phi}_T([u]_T) \quad \text{or} \quad [u]_T \in \bar{\partial}\widetilde{\widetilde{\phi}}_T(-S_T)$$

are assumed to hold in the tangential to the interface direction, where $\widetilde{\phi}_T$, $\widetilde{\widetilde{\phi}}_T$ are appropriately defined nonconvex superpotential producing the nonmonotone frictional law.

The theoretical formulation of the problem has been presented in Chapter 4. Here we propose an algorithm for the numerical solution based on a multilevel decomposition of the normal and tangential actions. The problem arising in the normal to the interface direction is the same as in the case of the Coulomb friction diagram (see (6.35)).

In the second step the substationarity problems (6.50) or (6.52) are solved with

$$V_{ad}^2 = \{\mathbf{u} \in \mathbb{R}^n \,|\, \mathbf{E}\mathbf{u} = \mathbf{u}_0\} \tag{6.65}$$

and

$$\Sigma_{ad}^2 = \{\mathbf{s} \in \mathbb{R}^n \,|\, \mathbf{G}\mathbf{s} = \mathbf{p} + \mathbf{G}_N\mathbf{S}_N, \mathbf{Z}\mathbf{s} = \mathbf{F}\}. \tag{6.66}$$

For the numerical treatment of these substationarity problems either the d.c. or the heuristic nonconvex optimization approaches can be applied. As in the previous Section, an incremental technique is applied, assuming that the final value of the loading is attained through a certain number of load increments N. The solution procedure can be summarized in the following algorithm:

Algorithm 6.5 *Unilateral contact and nonmonotone friction solver*

1. Decide for the number N of load increments that will be used. Then, $\delta\mathbf{p} = \mathbf{p}/N$.

2. For $k = 1, N$

   (a) Set $\mathbf{p} = k\delta\mathbf{p}$.

   (b) Consider that the tangential forces are constant. Usually we assume that $\mathbf{S}_T^{(0)} = \mathbf{0}$. Set $i = 1$ where index i denotes the number of normal and tangential subproblems solved.

   (c) Calculate the structure with the given tangential forces $\mathbf{S}_T^{(i-1)}$. The solution is obtained by minimizing the potential energy of the structure (6.35). The solution procedure yields the contact and non-contact areas and the values of the contact forces $\mathbf{S}_N^{(i)}$.

   (d) Solve the problem arising in the tangential to the interface direction assuming that the contact forces $\mathbf{S}_N^{(i)}$ are constant. The solution is obtained by solving the substationarity problem expressed either in terms of the potential or the complementary energy. This step is solved by applying one of the Algorithms (2.4), (2.5) or (2.6). For the application of these algorithms we use as a starting point the solution obtained at the previous step. The solution of this problem gives the values of $\mathbf{S}_T^{(i)}$.

   (e) Convergence check: if $\dfrac{||\mathbf{S}^{(i)} - \mathbf{S}^{(i-i)}||}{||\mathbf{S}^{(i)}||} \leq \varepsilon$ where $\mathbf{S} = [\mathbf{S}_N \ \mathbf{S}_T]^T$ and ε is an appropriately small number, then convergence has been achieved and terminate the algorithm, else continue for another iteration by setting $i = i + 1$ and going to step 2c.

6.2.2.3 Unilateral contact with nonmonotone friction and nonmonotone debonding. In this Section a problem will be treated where it is assumed that on the interface nonmonotone reaction-displacement laws hold both for the normal and tangential directions. For the normal to the interface direction the conditions

$$-S_D \in \bar{\partial}\tilde{\phi}_D([u]_D) \text{ or } [u]_D \in \bar{\partial}\bar{\bar{\phi}}_D(-S_D)$$

are assumed to hold. For the tangential direction the conditions

$$-S_T \in \bar{\partial}\tilde{\phi}_T([u]_T) \text{ or } [u]_T \in \bar{\partial}\bar{\bar{\phi}}_T(-S_T)$$

hold. As in the previous sections, $\tilde{\phi}_D$, $\bar{\bar{\phi}}_D$, $\tilde{\phi}_T$, $\bar{\bar{\phi}}_T$ are nonconvex superpotentials producing the respective nonmonotone laws.

These conditions describe a realistic interface problem where, e.g., the two interface parts are glued together with an adhesive material. Thus, after the determination of the active contact area Γ_C where the contact forces S_N develop, and the non-contact area Γ_D where the debonding forces S_D develop we have to determine the exact distribution of the tangential forces.

This general case has been treated theoretically in Chapter 4. The solution of this problem is obtained through the substationarity problem of the potential (resp. the complementary) energy of the structure. As in the case of the respective convex problem, the direct solution of the aforementioned substationarity problems is not possible.

Three subproblems are formulated:

- The first subproblem is a unilateral contact problem with the assumption of given debonding and tangential forces (cf. (6.41), (6.42)).

- The second subproblem corresponds to the solution of the debonding problem with the assumption that the forces $\mathbf{S}_N$ and $\mathbf{S}_T$ are given. The substationarity problem in terms of the displacements reads:

$$\Pi_2(\mathbf{u}) = \frac{1}{2}\mathbf{u}^T\mathbf{K}\mathbf{u} - \overline{\mathbf{p}}_2^T\mathbf{u} + \widetilde{\Phi}_D(\mathbf{u}), \ \ \mathbf{u} \in V_{ad}^2 \tag{6.67}$$

where

$$\overline{\mathbf{p}}_2 = \overline{\mathbf{p}} + \mathbf{G}_N\mathbf{S}_N + \mathbf{G}_T\mathbf{S}_T, \tag{6.68}$$

and

$$V_{ad}^2 = \{\mathbf{u} \in \mathbf{R}^n \,|\mathbf{E}\mathbf{u} = \mathbf{u}_0\}. \tag{6.69}$$

In terms of the stresses the substationarity problem reads:

$$\Pi_2^c(\mathbf{s}) = \frac{1}{2}\mathbf{s}^T\mathbf{F}_0\mathbf{s} + \mathbf{e}_0^T\mathbf{s} + \overline{\Phi}_D(\mathbf{s}), \ \ \mathbf{s} \in \Sigma_{ad}^2 \tag{6.70}$$

where

$$\Sigma_{ad}^2 = \{\mathbf{s} \in \mathbf{R}^n \,|\mathbf{G}\mathbf{s} = \mathbf{p} + \mathbf{G}_N\mathbf{S}_N + \mathbf{G}_T\mathbf{S}_T, \ \mathbf{Z}\mathbf{s} = \mathbf{F}\}. \tag{6.71}$$

- The third subproblem corresponds to the problem arising in the tangential direction assuming that the normal to the interface forces $\mathbf{S}_N$ and $\mathbf{S}_D$ are given. In terms of the displacements, the potential energy

$$\Pi_3(\mathbf{u}) = \frac{1}{2}\mathbf{u}^T\mathbf{K}\mathbf{u} - \overline{\mathbf{p}}_3^T\mathbf{u} + \widetilde{\Phi}_T(\mathbf{u}), \ \ \mathbf{u} \in V_{ad}^3 \tag{6.72}$$

has to be substationary, where

$$\overline{\mathbf{p}}_3 = \overline{\mathbf{p}} + \mathbf{G}_N\mathbf{S}_N + \mathbf{G}_D\mathbf{S}_D \tag{6.73}$$

and V^3_{ad} is the same as V^2_{ad}. In terms of the stresses, the complementary energy

$$\Pi^c_3(\mathbf{s}) = \frac{1}{2}\mathbf{s}^T\mathbf{F}_0\mathbf{s} + \mathbf{e}_0^T\mathbf{s} + \overline{\overline{\Phi}}_T(\mathbf{s}), \ \mathbf{s} \in \Sigma^3_{ad} \tag{6.74}$$

has to be substationary, where

$$\Sigma^3_{ad} = \{\mathbf{s} \in \mathbf{R}^n \,|\mathbf{G}\mathbf{s} = \mathbf{p} + \mathbf{G}_N\mathbf{S}_N + \mathbf{G}_D\mathbf{S}_D, \ \mathbf{Z}\mathbf{s} = \mathbf{F}\}. \tag{6.75}$$

The previous subproblems are organized in the following solution scheme:

Algorithm 6.6 *General interface problem solver (nonmonotone interface laws)*

1. Decide for the number N of load increments that will be used. Then, $\boldsymbol{\delta}\mathbf{p} = \mathbf{p}/N$.

2. For $k = 1, N$

   (a) Set $\mathbf{p} = k\boldsymbol{\delta}\mathbf{p}$.

   (b) In the first step we consider that the tangential forces $\mathbf{S}_T$ and the debonding forces $\mathbf{S}_D$ are constant. Usually it is assumed that $\mathbf{S}_T^{(0)} = \mathbf{0}$ and $\mathbf{S}_D^{(0)} = \mathbf{0}$. Set $i = 0$ and $j = 0$ where index i (resp. index j) is associated to the iterations for the solution of the problem arising in the tangential (resp. normal) to the interface direction.

   (c) Set $i = i + 1$.

   (d) Set $j = j + 1$.

   (e) Calculate the structure with the given tangential and debonding forces $\mathbf{S}_T^{(i-1)}$ and $\mathbf{S}_D^{(j-1)}$. The solution is obtained by minimizing the potential energy expression (6.41). As a result we get the contact and non-contact areas and the values of the contact forces $\mathbf{S}_N^{(j)}$.

   (f) This step corresponds to the substationarity problem of the potential or the complementary energy expressions (6.67) or (6.70) with the given contact and the tangential forces $\mathbf{S}_N^{(j)}$ and $\mathbf{S}_T^{(i-1)}$ and is solved by applying one of the Algorithms (2.4), (2.5) or (2.6). For the application of these algorithms we use as a starting point the solution obtained at the previous step. As a result the values of $\mathbf{S}_D^{(j)}$ are obtained.

   (g) Check for the convergence of the results obtained for the normal direction, i.e. we test the expression $\dfrac{||\mathbf{S}^{(j)} - \mathbf{S}^{(j-i)}||}{||\mathbf{S}^{(j)}||} \leq \varepsilon$ where $\mathbf{S} = [\mathbf{S}_N \ \mathbf{S}_D]^T$. If the check is fulfilled proceed to the next step, if not return to step 2d.

(h) Here the problem arising in the tangential to the interface direction is considered. The solution of the substationarity problems (6.72) or (6.74) is obtained through Algorithms (2.4), (2.5) or (2.6). Again as a starting point the solution obtained from the previous step is used. As a result the values of $\mathbf{S}_T^{(i)}$ are obtained.

(i) In this step we check the global convergence: If $\dfrac{||\mathbf{S}^{(i)} - \mathbf{S}^{(i-i)}||}{||\mathbf{S}^{(i)}||} \leq \varepsilon$ where $\mathbf{S} = [\mathbf{S}_N \ \mathbf{S}_D \ \mathbf{S}_T]^T$ and ε the desired accuracy, then convergence has been achieved and the algorithm is terminated, else it is continued for another iteration going to step 2c.

6.2.3 Fractals and interface problems

The revolution in geometry, which has already created the notion of the fractional dimension and has formed fractal geometry, has substantially influenced many sciences such as biology, physics, geology and material science. In structural analysis and in applied mechanics fractal geometries appear very often (Mandelbrot, 1972, Falconer, 1985, Barnsley, 1988 Feder, 1988). We can mention here the crack interfaces in natural bodies (Takayasu, 1990, Scholz and Mandelbrot, 1989), the interfaces of fractured bones, metals and cracks, the geometry of metallic surfaces after sandblasting (Le Mehauté, 1990), the crashed interfaces in composite and granular materials (Bunde and Halvin, 1991) which are of fractal geometry. Moreover, real stress-strain laws in alloys and in composite materials (Bell, 1973, Jones, 1975) may have a zig-zag form which is of fractal geometry. Also, fractal type friction laws have been registered lately by various experiments (Feder and Feder, 1991).

Several mathematical approaches to the theory of fractals have been developed (Mandelbrot, 1972, Barnsley, 1988, Falconer, 1985). The approach proposed by Barnsley, 1988 is adopted here because it seems to be more appropriate for the needs of engineering sciences. According to this approach, it is assumed that all the fractals considered are derived by an appropriately defined *iterated function system* (I.F.S.) or that they are the *fractal interpolation functions* (F.I.) of a given set of data, as it is explained in the Appendix of this Chapter. In both cases the fractal is the fixed point of a given set of transformations. This fixed point procedure is the basis for the numerical methods presented in this Section.

In the following Sections the influence of a boundary of fractal geometry, as well as the influence of the fractality of a stress-strain or reaction-displacement law on the displacement and stress fields of a deformable body are investigated. The methods presented here are of heuristic nature and their convergence, in the strict mathematical sense, has not yet been proved. However, they are engineering oriented and they represent realistic solution methods for the respective problems.

6.2.3.1 Fractal interfaces. First, our effort is concentrated on the study of the influence of the irregular fracture of structures on the arising stress and strain fields.

As it has been already mentioned, the crack interfaces in natural bodies are of fractal geometry. We give as an example the crack interfaces in structures of brittle behaviour (MacLeod and el Magd, 1980), e.g. in masonry structures which may have either a zig-zag shape, following the joints of the stones (or the blocks) or pass through the blocks. The first kind of cracks arises in lightly compressed structures with relatively weak mortars. On the contrary the second kind of cracks appears in structures constructed with mortars of better quality (strong mortars) and especially in highly compressed regions of the structure. Moreover, various researchers attest that the real geometry of cracks and interfaces in fracture mechanics, is of fractal type.

The central problem of this Section is the analysis of a structure Ω having certain interfaces or boundaries of fractal type denoted by Q. On the interfaces monotone or nonmonotone laws are assumed to relate the normal and tangential to the interface forces with the respective displacements. The solution of the problem results from the description of the fractal as the I.F.S. or the F.I. of a given set of data (see the Appendix of this Chapter). According to this procedure, the fractal interface or boundary, results as the fixed point of a given set of transformations. In this case the fractal interface Q is the *attractor* of a given transformation denoted by W i.e.

$$Q = W(Q) \text{ and } Q^{(j)} = W(Q^{(j-1)}), \text{ where } Q^{(j)} \to Q, \text{ as } j \to \infty. \tag{6.76}$$

Thus, for each approximation $Q^{(j)}$ of the fractal interface Q we have to solve a structure $\Omega^{(j)}$. Here $Q^{(j)}$ is an interface set with classical geometry. The solutions $u^{(j)}$ and $\sigma^{(j)}$ for the latter problem are obtained using the algorithms presented in the previous Sections. As $j \to \infty$ the displacement and stress fields tend to the solution of the fractal interface problem.

Therefore, the following algorithm is formulated for the solution of this problem.

Algorithm 6.7 *Fractal interface solver*

1. Starting the algorithm, set $j = 1$

2. Use the transformation $Q^{(j)} = W(Q^{(j-1)})$ to approximate the problem involving the fractal curves Q with a problem involving the classical curves $Q^{(j)}$.

3. Solve the arising problem. For the numerical treatment of this step, one of the Algorithms (6.1)–(6.6) can be applied according to the specific interface conditions.

4. If the displacement and stress fields of the current approximation j of the fractal interface Q are identical (within a prescribed accuracy) to the respective fields of the previous approximation $j-1$, stop the algorithm, else set $j = j + 1$ and goto step 2.

6.2.3.2 The case of fractal friction laws. As it has already been mentioned, fractal geometry in structural analysis has to do not only with the geometry of the bodies but also with their mechanical behaviour. An interesting category of problems in this area are the stick-slip phenomena in frictional contact of solids with rough surfaces. In this case, fractal type friction laws have been registered by various experiments (Feder and Feder, 1991). It is important to notice here that the softening interface behaviour observed in these problems is caused by the partial cracking and crushing of the asperities of the interface. In this way, a realistic description of the interface behaviour is achieved.

Here, our attention is focused on the fractal nature of friction laws in contact mechanics. The basic idea for the mathematical description of these laws is the use of the fractal interpolation functions which approximate the fractal geometry of the friction law by means of classical C^0-elements (curves of nonmonotone type generally), in the sense of a limit process. This approximation of the fractal friction laws leads to variational formulations in inequality form, or equivalently to inequality constrained, nonconvex, generally nondifferentiable optimization problems for the potential and the complementary energy of the structure. For the numerical treatment of these problems the heuristic nonconvex optimization approach is preferred due to the complexity of the arising nonmonotone laws (the d.c. decomposition is not always possible in this case).

In the general case where the friction law G is of fractal geometry the various approximations $G^{(k)}$ of the fractal law, where

$$G^{(k)} = T(G^{(k-1)}) \tag{6.77}$$

must be considered and equivalently a series of problems involving laws with classical geometry should be solved. At the limit, the solution of the fractal law problem is obtained.

Algorithm 6.8 *Treatment of fractal friction laws in contact problems*

1. Starting the algorithm, set $k = 1$

2. Use the transformation $G^{(k)} = T(G^{(k-1)})$ to approximate the fractal friction law G with the classical (i.e. non-fractal) law $G^{(k)}$.

3. Solve the arising problem. For this reason Algorithm (6.5) is used.

4. If the displacement and stress fields of the current approximation k of the fractal law G are identical to the respective fields of the previous approximation $k-1$, stop the algorithm, else set $k = k+1$ and goto step 2.

Notice that Algorithms (6.7) and (6.8) can be combined in order to treat the much more complicated problem of a structure having a fractal interface on which fractal friction laws are assumed to hold. In this case the solutions $u^{(j,k)}$ and $\sigma^{(j,k)}$ (where index j denotes the approximations of the fractal interface and index k denotes the approximations of the fractal friction law) are obtained applying Algorithm (6.5). As $k \to \infty$, the displacements and stress fields $u^{(j,k)}$, $\sigma^{(j,k)}$ tend to the solution of the structure $\Omega^{(j)}$ with the fractal friction conditions. Moreover, as $j \to \infty$ the displacement and stress fields tend to the solution of the fractal interface problem with fractal friction conditions.

For more details about the mathematical background of the above iterative procedures the reader is referred to Panagiotopoulos, 1992a, Panagiotopoulos, 1992b, Panagiotopoulos and Panagouli, 1992, Mistakidis et al., 1992, Panagouli et al., 1992, Panagiotopoulos et al., 1993, Panagiotopoulos et al., 1994c, Panagiotopoulos et al., 1994a, Panagiotopoulos et al., 1994b, Panagouli et al., 1995, Panagiotopoulos and Panagouli, 1996, Mistakidis, 1997a, Panagouli, 1997, Panagiotopoulos and Panagouli, 1993.

6.3 ALGORITHMS FOR STRUCTURAL ANALYSIS PROBLEMS

6.3.1 *Convex problems: solution of the classical plasticity problem*

In this Section the structural analysis plasticity problem is briefly discussed. Here we follow the holonomic plasticity approach presented in Maier, 1970, Maier, 1971, Panagiotopoulos et al., 1984. Let us assume a structure within the geometrical linearity framework, for the material of which an elastic-plastic relation is assumed to hold as e.g. the one depicted in Fig. 3.4. The equations of holonomic plasticity read:

$$\mathbf{e} = \mathbf{F}_0 \ \mathbf{s}, \tag{6.78}$$

$$\mathbf{e} = \mathbf{e}_0 + \mathbf{e}_E + \mathbf{e}_P \tag{6.79}$$

$$\mathbf{e}_P = \mathbf{N}\boldsymbol{\lambda} \tag{6.80}$$

$$\boldsymbol{\phi} = \mathbf{N}^T\mathbf{s} - \mathbf{H}\boldsymbol{\lambda} - \mathbf{k} \tag{6.81}$$

$$\boldsymbol{\lambda} \geq 0 \,,\ \boldsymbol{\phi} \leq 0 \,,\ \boldsymbol{\phi}^T\boldsymbol{\lambda} = 0 \tag{6.82}$$

where $\mathbf{F}_0$ is the natural flexibility matrix of the structure, $\mathbf{e}$ the strain vector consisting of three parts, the initial strain $\mathbf{e}_0$, the elastic strain $\mathbf{e}_E$ and the plastic strain $\mathbf{e}_P$, $\boldsymbol{\lambda}$ are the plastic multipliers vector, $\boldsymbol{\phi}$ the yield functions, $\mathbf{H}$ the horkhardening matrix, $\mathbf{N}$ is the matrix of the gradients of the yield functions with respect to the stresses and $\mathbf{k}$ is a vector of positive constants. Moreover, the classical boundary conditions (4.6), (4.7) are assumed to hold.

The solution of this problem can be obtained by minimizing the potential or the complementary energy of the discretized structure. In terms of the displacements we have:
Find kinematically admissible displacements $\mathbf{u} \in V_{ad}$ such that

$$\begin{aligned}\Pi(\mathbf{u}) = \min\Big\{\Pi(\mathbf{v}) \;&=\; \frac{1}{2}\mathbf{v}^T\mathbf{K}\mathbf{v} + \frac{1}{2}\boldsymbol{\lambda}^T\mathbf{H}\boldsymbol{\lambda} - \mathbf{v}^T\mathbf{G}\mathbf{K}_0\mathbf{N}\boldsymbol{\lambda} + \frac{1}{2}\boldsymbol{\lambda}^T\mathbf{N}^T\mathbf{K}_0\mathbf{N}\boldsymbol{\lambda} + \\ &+\; \mathbf{e}_0^T\mathbf{K}_0(\mathbf{N}\boldsymbol{\lambda} - \mathbf{G}^T\mathbf{v}) - \mathbf{p}^T\mathbf{v} + \mathbf{k}\boldsymbol{\lambda} \mid \boldsymbol{\lambda} \geq 0\Big\}\end{aligned} \tag{6.83}$$

where $\mathbf{K}$ is the stiffness matrix of the discrete model and $\mathbf{K}_0$ is the inverse of $\mathbf{F}_0$.

In terms of the stresses we have:
Find $\mathbf{s} \in \Sigma_{ad}$ such that

$$\Pi^c(\mathbf{s}) = \min\left\{\Pi^c(\mathbf{t}) = \frac{1}{2}\mathbf{t}^T\mathbf{F}_0\mathbf{t} + \frac{1}{2}\boldsymbol{\lambda}^T\mathbf{H}\boldsymbol{\lambda} + \mathbf{t}^T\mathbf{e}_0\right\} \tag{6.84}$$

where

$$\Sigma_{ad} = \left\{\mathbf{s} \in \mathbf{R}^n \,|\mathbf{N}^T\mathbf{s} - \mathbf{H}\boldsymbol{\lambda} \leq \mathbf{k},\ \mathbf{G}\mathbf{s} = \mathbf{p},\ \mathbf{Z}\mathbf{s} = \mathbf{F}\right\}. \tag{6.85}$$

Both problems can be numerically treated by applying a quadratic programming algorithm.

6.3.2 Nonconvex problems: structures with elements involving softening moment-rotation relationships

Nonmonotone stress-strain or reaction displacement laws appear very often in structural analysis problems. A very common case is that of a structure with connections obeying to a nonlinear moment-rotation law resulting from the pure or incomplete "cooperation" between the various structural components.

Without loosing generality let us assume a structure Ω consisting of bar elements. For the material of m of these elements we assume that the classic elastoplastic law of Fig. 6.1a holds. This law yields the convex superpotential $q(.)$ of Fig. 6.1b.

Moreover, let us assume that the structure contains l elements exhibiting a softening behaviour, i.e. a nonmonotone law $M = g(\varphi)$ (Fig. 6.1c) relates the moment M with the rotational capacity φ. In this case the respective nonconvex superpotential $\tilde{w}(\varphi)$ is one-dimensional and $\tilde{w}(\xi)$ is the area between the horizontal axis and the graph until the point ξ of the horizontal axis, i.e. $\tilde{w}(\xi) = \int_0^\xi g(\varphi)d\varphi$ (see Fig. 6.1d).

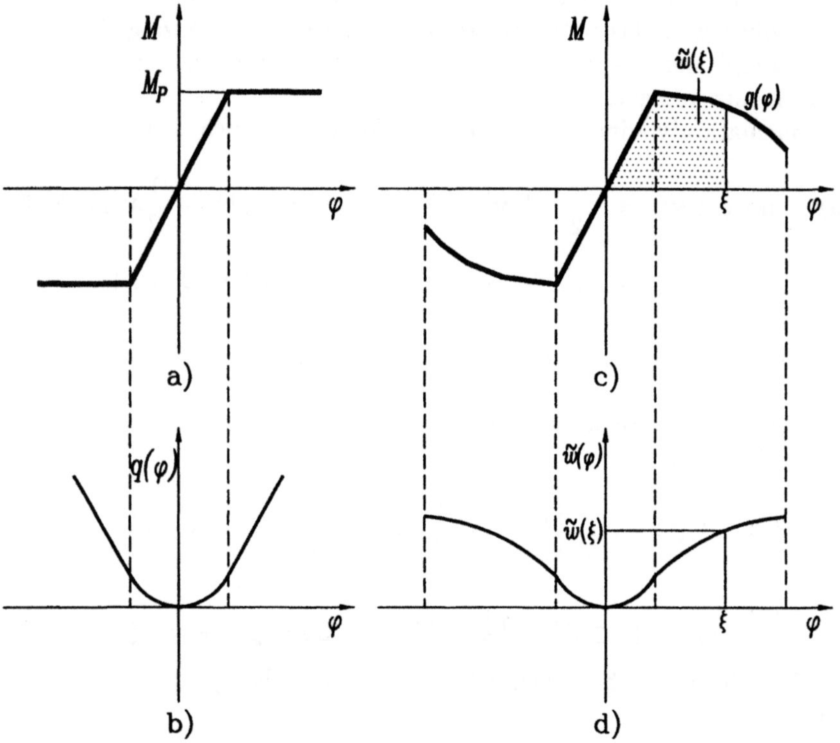

Figure 6.1. Monotone and nonmonotone moment-rotation relationships and the corresponding superpotentials

We define now the set S_1 consisting of the m elements of the first category and the set S_2 consisting of the l elements of the second category. For this problem, the principle of virtual work is written in the form:

$$\mathbf{s}^T(\mathbf{e}(\mathbf{u}^*) - \mathbf{e}(\mathbf{u})) = \mathbf{p}^T(\mathbf{u}^* - \mathbf{u}) \ \forall \ \mathbf{u}^* \in V_{ad} \tag{6.86}$$

where **s**, **e**, **p**, **u** are the stress, strain, loading and displacement vectors respectively and V_{ad} is the kinematically admissible set. Suppose now that **s** (resp. **e**) consists of three parts $\mathbf{s}_1$, $\mathbf{s}_2$ and $\mathbf{s}_3$ (resp. $\mathbf{e}_1$, $\mathbf{e}_2$ and $\mathbf{e}_3$) where $\mathbf{s}_1$ (resp $\mathbf{e}_1$) corresponds to the moments (resp. the rotations) of the elements of set $\mathcal{S}_1$, $\mathbf{s}_2$ (resp $\mathbf{e}_2$) corresponds to the moments (resp. the rotations) of the elements of set $\mathcal{S}_2$ and $\mathbf{s}_3$ (resp $\mathbf{e}_3$) corresponds to the rest stresses (resp. strains) of both sets.

Equation (6.86) is now written in the form

$$\mathbf{s}_1^T(\mathbf{e}_1(\mathbf{u}^*) - \mathbf{e}_1(\mathbf{u})) + \mathbf{s}_2^T(\mathbf{e}_2(\mathbf{u}^*) - \mathbf{e}_2(\mathbf{u})) + \mathbf{s}_3^T(\mathbf{e}_3(\mathbf{u}^*) - \mathbf{e}_3(\mathbf{u})) = \mathbf{p}^T(\mathbf{u}^* - \mathbf{u}) \quad \forall \ \mathbf{u}^* \in V_{ad}. \tag{6.87}$$

For the m elements of $\mathcal{S}_1$ we can write $s_{1k} \in \partial q(e_{1k})$, $k = 1, \ldots, m$ which is equivalent to the variational inequality

$$q(e_{1k}(u^*)) - q(e_{1k}(u)) \geq s_{1k}^T(e_{1k}(u^*) - e_{1k}(u)) \ k = 1, \ldots, m \ \forall \ u^* \in \mathbb{R}. \tag{6.88}$$

Similarly, for the l elements of $\mathcal{S}_2$ we can write $s_{2j} \in \bar{\partial}\widetilde{w}(e_{2j})$, $j = 1, \ldots, l$ which is equivalent to the hemivariational inequality

$$\widetilde{w}^0(e_{2j}(u^*) - e_{2j}(u)) \geq s_{2j}^T(e_{2j}(u^*) - e_{2j}(u)) \ j = 1, \ldots, l \ \forall \ u^* \in \mathbb{R} \tag{6.89}$$

where $\widetilde{w}^0$ is the directional derivative of the nonconvex superpotential $\widetilde{w}$.

Using the inequalities (6.88), (6.89), equation (6.87) is transformed in the form

$$\sum_{k=1}^{m}[q(e_{1k}(\mathbf{u}^*)) - q(e_{1k}(\mathbf{u}))] + \sum_{j=1}^{l}\widetilde{w}^0(e_{2j}(\mathbf{u}^*) - e_{2j}(\mathbf{u})) + \mathbf{s}_3^T(\mathbf{e}_3(\mathbf{u}^*) - \mathbf{e}_3(\mathbf{u})) \geq \mathbf{p}^T(\mathbf{u}^* - \mathbf{u}) \ \forall \ \mathbf{u}^* \in V_{ad}, \tag{6.90}$$

which is a variational-hemivariational inequality. The corresponding substationarity problem reads:

Find $\mathbf{u} \in V_{ad}$ such that the potential energy

$$\Pi_{nc}(\mathbf{u}) = \frac{1}{2}\mathbf{u}^T\mathbf{K}\mathbf{u} + \sum_{k=1}^{m} q(e_{1k}(\mathbf{u})) + \sum_{j=1}^{l}\widetilde{w}(e_{2j}(\mathbf{u})) - \mathbf{p}^T\mathbf{u} \tag{6.91}$$

is substationary, where $\mathbf{K} = \mathbf{G}_3\mathbf{K}_0\mathbf{G}_3^T$, $\mathbf{K}_0$ is the inverse of the natural flexibility matrix $\mathbf{F}_0$ and $\mathbf{G}_3$ is the equilibrium matrix that corresponds to $\mathbf{s}_3$.

The solution of the above problem can be found by applying either the d.c. decomposition approach or the heuristic nonconvex optimization approach presented in Chapter 2. Here, due to the nature of the nonmonotone moment-rotation law, it is preferable to follow the second one. Using approximating monotone laws having the same shape with the classical plasticity law of Fig. 6.1a, the problem arising in each step of the iterative procedure (see Algorithm (2.6)) is a classical plasticity problem which can be treated numerically by minimizing the potential or complementary energy expressions (6.83) or (6.84).

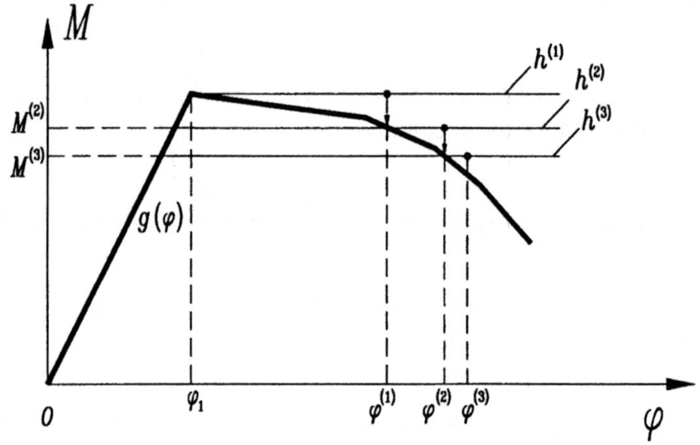

Figure 6.2. Graphical explanation of the proposed iterative scheme

Therefore, the following algorithm is formulated for the treatment of structures with elements exhibiting softening behaviour:

Algorithm 6.9 *Treatment of structures having elements with softening moment-rotation relationship*

1. Set $i = 1$, $\varphi^{(1)} = 0$.

2. Assume that the simplified law $h^{(i)}$ (Fig. 6.2) holds between the moment and the rotation on any element exhibiting softening behaviour. Solve the arising plasticity problem using any algorithm suitable for the treatment of plasticity problems.

3. With the obtained rotation values $\varphi^{(i)}$ determine the new elastic - plastic laws approximating the initial softening law.

4. Check if the quantity $\frac{\|\varphi^{(i)} - \varphi^{(i-1)}\|}{\|\varphi^{(i)}\|}$ is less than a predefined accuracy. If the convergence criterion is fulfilled, then terminate the algorithm, if not, set $i = i + 1$ and go to step 2.

In the following we shall briefly demonstrate the graphical explanation of the previous algorithm by means of Fig. 6.2. First we assume that instead of g, the fictitious plasticity law $h^{(1)}$ holds. Then the structure has a unique solution due to the convexity of all the underlying potentials. Suppose that the solution of this problem gives as a result certain $\varphi^{(1)}$. Now we select $h^{(2)}$ such as $h^{(2)}(\varphi^{(1)}) = g(\varphi^{(1)})$ and we solve the new plasticity problem. The solution of this problem is $\varphi^{(2)}$, which, in turn, gives rise to a new plasticity law $h^{(3)}$ and so on, until after n steps $\frac{\|\varphi^{(n)} - \varphi^{(n-1)}\|}{\|\varphi^{(n)}\||}$ becomes small enough. In the previous example we used $\varphi^{(0)} = \varphi_1$. As it was previously mentioned, a different starting point $\varphi^{(0)}$ may lead to a different solution $\varphi^{(n)}$ due to the nonconvexity of the initial problem.

Analogous algorithms can be formulated for the case where other quantities of stress and the respective strain are connected with a nonmonotone law, as e.g. the axial stress with the axial elongation, the shear stress with the shear deformation, etc. (cf. in this respect Mistakidis and Panagiotopoulos, 1994, Mistakidis, 1997b, Mistakidis et al., 1994).

Appendix: The essentials of fractal geometry

According to the definition of Mandelbrot, 1972 and Wallin, 1989 a set $F \subset \mathbb{R}^n$ is called a "fractal set" if its "Hausdorff" dimension is non-integer or if it is an integer strictly larger than its topological dimension. For a rigourous mathematical definition of the Hausdorff dimension the reader is referred to Barnsley, 1988 and Falconer, 1985.

Here, it is assumed that all the fractals considered, are obtained by an appropriately defined iterated function system (I.F.S.) or that they are the graph of a fractal interpolation function (F.I.) interpolating a given set of data. Thus, in both cases, the fractal results as the fixed point of a given set of functions. This fixed point approximation, which is equivalent to the approximation of the fractal by classical curves (i.e. curves with integer Hausdorff dimensions), is combined with the numerical techniques of structural analysis.

6-A.1 ITERATED FUNCTION SYSTEMS

Let $\{X, d\}$ be a complete metric space with the metric d. We denote by $H(X)$ the space of the compact subsets of X. If $d(A, B)$ is the distance between the sets $A \subset X$ and $B \subset X$ defined by the formula (cf. Barnsley, 1988)

$$d(A, B) = \max_{x \in A} \min_{y \in B} d(x, y), \tag{6-A.1}$$

then the space $H(X)$ endowed with the Hausdorff metric

$$h(A, B) = \max\{d(A, B), d(B, A)\} \quad \forall A, B \in H(X) \tag{6-A.2}$$

is a complete metric space and is called space of deterministic fractals. Here $h(A, B)$ denotes the Hausdorff distance between the sets A, B in $H(X)$. An iterated function system (I.F.S.) on X consists of n contractive mappings $w_i : X \to X$ with contractivity factors $0 \leq s_i < 1$, $i = 1, 2, \ldots, n$, i.e.

$$d(w_i(x), w_i(y)) \leq s_i d(x, y) \quad \forall x, y \in X, \ 0 \leq s_i < 1. \tag{6-A.3}$$

It can be easily shown that, if a new set-valued function $W_i : H(X) \to H(X)$ is defined by setting

$$W_i(B) = \{w_i(x); x \in B\} \quad \forall B \in H(X) \tag{6-A.4}$$

and

$$W(B) = W_1(B) \cup W_2(B) \cup \ldots \cup W_n(B) \quad \forall B \in H(X), \tag{6-A.5}$$

then this set-valued function is a contraction mapping on $H(X)$ with contractivity factor $s = \max\{s_1, s_2, \ldots, s_n\}$. The unique "fixed point" of W is the set

$A \subset H(X)$ such that

$$A = W(A) = \bigcup_{i=1}^{n} W_i(A) \tag{6-A.6}$$

and is given by the relation

$$A = \lim_{m \to \infty} W^{(m)}(B) \quad \forall B \in H(X), \tag{6-A.7}$$

where

$$W^{(0)}(X) = X,\ W^{(m)}(X) = W(W^{(m-1)}(X)),\quad m = 1, 2, \ldots \tag{6-A.8}$$

are the forward iterates of W. The set A is called the "attractor" of the I.F.S. $\{X; w_i,\ i = 1, 2, \ldots, n\}$ and is the deterministic fractal corresponding to the considered I.F.S.

6-A.2 APPROXIMATION OF FRACTALS BY C^0 CURVES

In the sequel we will define the fractal geometry by means of the notion of the fractal interpolation. When we investigate an experimental curve given by a finite number of points $\{(x_i, y_i),\ i = 0, 1, \ldots, N\}$ we have to choose in advance the type of mathematical model that can be associated with this curve. Thus, we pass from a discrete set of data to a continuous model. Such a passage can be done in our case by using a fractal interpolation function $f : [x_0, x_N] \to \mathbf{R}$, i.e. a function f such that $f(x_i) = y_i,\ i = 0, 1, \ldots, N$. The I.F.S. $\{\mathbf{R}^2\, |w_i,\ i = 1, 2, \ldots, N\}$ defined by the transformations

$$(x, y) \to w_i(x, y) = \begin{pmatrix} a_i & 0 \\ c_i & d_i \end{pmatrix} \begin{pmatrix} x \\ y \end{pmatrix} + \begin{pmatrix} b_i \\ f_i \end{pmatrix} \quad i = 1, \ldots, N, \tag{6-A.9}$$

determines this function. Here the factors d_i are the hidden variables of the problem and they have to satisfy $0 \leq d_i < 1$. The remaining coefficients are given by

$$a_i = \frac{(x_i - x_{i-1})}{(x_N - x_0)},\quad b_i = \frac{(x_N x_{i-1} - x_0 x_i)}{(x_N - x_0)} \tag{6-A.10}$$

$$c_i = \frac{(y_i - y_{i-1})}{(x_N - x_0)} - d_i \frac{(y_N - y_0)}{(x_N - x_0)} \tag{6-A.11}$$

$$f_i = \frac{(x_N y_{i-1} - x_0 y_i)}{(x_N - x_0)} - d_i \frac{(x_N y_0 - y_N x_0)}{(x_N - x_0)}. \tag{6-A.12}$$

It has been proved (Barnsley, 1988) that if F is the attractor of the I.F.S. defined by (6-A.10) – (6-A.12) then F is the graph of a continuous function

$y\colon [x_0, x_N] \to \mathbb{R}$ interpolating the data $\{x_i, y_i\}, i = 0, 1, \ldots, N$. If C^0 is the set of all continuous functions $y\colon [x_0, x_N] \to \mathbb{R}$ then the sequence of functions $\tilde{y}_{m+1}(x) = (T\tilde{y}_m)(x)$, where the operator $T\colon C^0 \to C^0$ is defined by

$$T(\tilde{y}(a_i x + b_i)) = c_i x + d_i \tilde{y}(x) + f_i \quad i = 1, 2, \ldots, N \tag{6-A.13}$$

converges to the attractor F as $m \to \infty$.

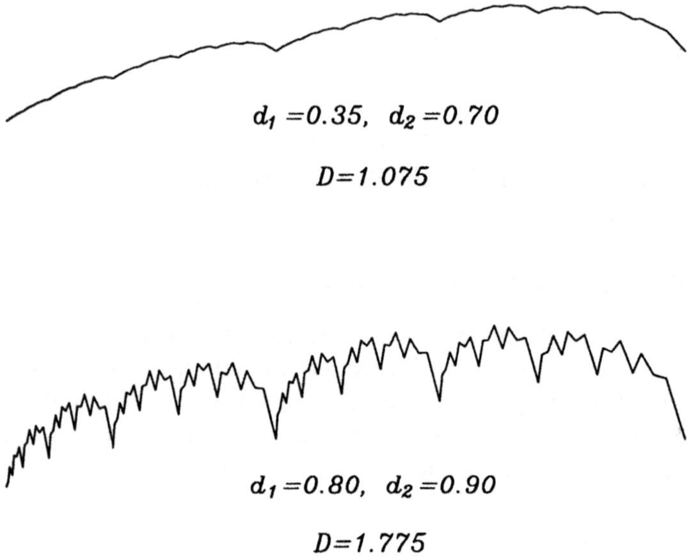

Figure 6-A.1. C^0 - interpolating functions with fractal dimensions $D_1 = 1.075$ and $D_2 = 1.775$

It is important to mention here that the Hausdorff dimension has a restricted application to the study of fractal curves resulting in sciences such as physics, biology or engineering. For that reason in these cases we use the fractal dimension D of the fractal interpolation function. For the definition of this dimension D let us assume that $\epsilon > 0$ and $N(\epsilon)$ denotes the number of square boxes of side length ϵ which intersect the graph of the fractal interpolation function. Then the fractal dimension D is given by the relation $N(\epsilon) \approx \textit{constant}\ \epsilon^{-D}$ as $\epsilon \to 0$. It has been shown that, if $\sum_{i=1}^{N} |d_i| > 1$ and the interpolation points do not lie on a single straight line, then the fractal dimension of F is the unique solution D of

$$\sum_{i=1}^{N} |d_i| a_i^{D-1} = 1. \tag{6-A.14}$$

The proper choice of the hidden variables d_i may make D very close to 1, (line-like fractal), or very close to 2, (surface-like fractal) as shown in Fig. 6-A.1, where the two interpolation functions interpolate the same set of data $\{(0.0, 0.0), (2.0, 0.5), (5.0, 2.0)\}$. In the first case the free parameters d_i have been taken $d_1 = 0.35$ and $d_2 = 0.70$ while in the second case, the corresponding values are 0.80 and 0.90 respectively.

6-A.3 APPROXIMATION OF FRACTALS BY C^1 CURVES

Here we introduce classes of iterated function systems whose attractors are C^1 interpolating functions (cf. Massopust, 1993). These new classes of fractal curves provide a new way of smoothly interpolating and approximating highly complex images. We assume, without loss of generality, that the attractor of the iterated function system is contained in $\mathbb{R}^2$, where a set of data $\{(x_i, y_i) : i = 0, 1, \ldots, N,\ 0 = x_0 < x_1 < \ldots < x_N, y_i \in \mathbb{R}$ with $y_0 = 0\}$ is given. The affine transformations we introduce here are:

$$u_i(x) = a_i x + b_i \tag{6-A.15}$$

where a_i, b_i are given by equation (6-A.10) by setting $x_0 = 0$. Now let $K_i(\xi, n)$ be a symmetric bilinear form on $\mathbb{R}^2$. For $\xi = (x, y) \in \mathbb{R}^2$ let $v_i(\xi) = K_i(\xi, \xi) + d_i$, $d_i \in \mathbb{R}$ and $i = 1, 2, \ldots, N$. Then $v_i : \mathbb{R}^2 \to \mathbb{R}$ and it can be written as

$$v_i(x, y) = c_i x^2 + 2s_i xy + t_i y^2 + d_i. \tag{6-A.16}$$

In order for $w_i(x, y) = (u_i(x), v_i(x, y)), i = 0, 1, \ldots, N$, to be contractive on $\mathbb{R}^2$ so as to have a unique attractor G, it is required that $v_i(., y)$ be Lipschitz for all $y \in \mathbb{R}$ and $v_i(x, .)$ contractive for all $x \in [x_0, x_N]$. Then

$$\begin{aligned} &|v_i(x, y) - v_i(x', y)| = \\ &|c_i(x + x') + 2s_i y||x - x'| \leq |2c_i x_N + 2s_i y_{\max}||x - x'| < l|x - x'| \end{aligned} \tag{6-A.17}$$

for all $x, x' \in [x_0, x_N]$ and $y, y' \in \mathbb{R}$ and $l > \max_i |2c_i x_N + 2s_i y_{\max}|$. Also

$$\begin{aligned} |v_i(x, y) - v_i(x, y')| &= |2s_i x + t_i(y + y')||y - y'| \leq \\ &\quad |2s_i x_N + 2t_i y_{\max}||y - y'| \leq r|y - y'| \end{aligned} \tag{6-A.18}$$

for all $x \in [0, x_N]$ and $y, y' \in \mathbb{R}$, whenever

$$\max_i |s_i x_N + t_i y_{\max}| \leq r < \frac{1}{2}. \tag{6-A.19}$$

In the sequence we denote by $\mathcal{C}^1([x_0, x_N])$ the complete metric space consisting of all $f \in C^1(\mathbb{R}, \mathbb{R})$ such that $f(x_i) = y_i$, $i = 0, 1, \ldots, N$, $f'(x_0) = \alpha$ and

$f'(x_N) = \beta$, for some given $\alpha, \beta \in \mathbb{R}$. We define the operator $T : \hat{C}^1([x_0, x_N]) \to \mathbb{R}$ by

$$(Tf)(x) = v_i(u_i^{-1}(x), f(u_i^{-1}(x))), \quad x \in u_i[x_0, x_N], \quad i = 1, \ldots, N. \qquad (6\text{-A.20})$$

Then

$$(Tf)'(x) = \frac{2}{a_i}(c_i u_i^{-1}(x) \; + \; s_i f(u_i^{-1}(x)) + s_i u_i^{-1}(x) f'(u_i^{-1}(x)) + \\ + \; t_i f(u_i^{-1}(x)) f'(u_i^{-1}(x)), \qquad (6\text{-A.21})$$

where $x \in [x_{i-1}, x_i]$. If we require that

$$v_i(x_0, y_0) = y_{i-1}, \quad v_i(x_N, y_N) = y_i \qquad (6\text{-A.22})$$

and

$$(Tf)'(x_0) = f'(x_0), \quad (Tf)'(x_N) = f'(x_N), \quad \lim_{x \to x_i^-}(Tf)'(x) = \lim_{x \to x_i^+}(Tf)'(x), \qquad (6\text{-A.23})$$

$i = 1, 2, \ldots, N-1$, then these equations imply that $Tf \in C^1(\mathbb{R}, \mathbb{R})$. Furthermore, since u_i and $v_i(x, .)$ are contractive, T is contractive on $\hat{C}^1([x_0, x_N])$ with contractivity factor $2\max_i\{|s_i x_N| + |t_i y_{\max}|\}$. Thus, by the contraction mapping theorem, T has a unique fixed point $f : [x_0, x_N] \to \mathbb{R}$ in $\hat{C}^1([x_0, x_N])$. Moreover, $f(x_i) = y_i$ for all $i = 0, 1, \ldots, N$. We refer to f as a C^1- interpolating fractal function. From the above equations we obtain that:

$$d_i = y_{i-1} \qquad (6\text{-A.24})$$

$$f'(x_0 = 0) = 0, f'(x_N) = \beta \qquad (6\text{-A.25})$$

$$s_N = \frac{\frac{b_N}{2}\beta - c_N x_N - t_N y_N \beta}{x_N \beta + y_N} \qquad (6\text{-A.26})$$

$$c_i = \frac{(y_i - y_{i-1})(x_N \beta + y_N)}{x_N^2 (x_N \beta - y_N)} + t_i \frac{y_N^2}{x_N^2}, \quad i = 1, 2, \ldots, N-1 \qquad (6\text{-A.27})$$

$$s_i = -\frac{c_i x_N + t_i y_N \beta}{x_N \beta + y_N} \qquad (6\text{-A.28})$$

$$c_N = \frac{(y_N - y_{N-1})(x_N \beta + y_N)}{x_N^2 (x_N \beta - y_N)} - \frac{b_N y_N \beta}{x_N (x_N \beta - y_N)} + t_N \frac{y_N^2}{x_N^2} \qquad (6\text{-A.29})$$

where t_i, β are the free parameters of the problem.

In the Fig. 6-A.2 two C^1 interpolating functions $f_1(x), f_2(x)$ which interpolate the same set of data {(0.0,0.0), (0.3,0.2), (0.8,0.3), (1.3,0.45), (2.0,0.5),

(2.7,0.65), (3.2,0.7), (3.9,0.75), (5.0,0.5)} are presented. In the first case the free parameters t_i, β have been taken $t_1 = t_2 = t_3 = 0.08$, $t_4 = t_5 = 0.10$, $t_6 = t_7 = 0.15$, $t_8 = -0.10$ and $\beta = -1.0$, whereas in the second case their values are $t_1 = 0.20$, $t_2 = t_3 = 0.30$, $t_4 = t_5 = 0.20$, $t_6 = t_7 = 0.25$, $t_8 = -0.15$ and $\beta = -1.0$ respectively.

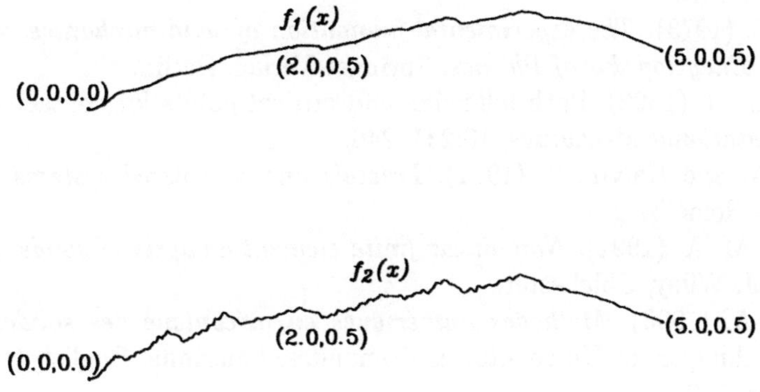

Figure 6-A.2. C^1 - interpolating functions

References

Allgower, E. L. and Georg, K. (1990). *Numerical continuation methods: an introduction*. Springer Verlag, Berlin-Heidelberg.

Barnsley, M. (1988). *Fractals everywhere*. Academic Press, Boston-New York.

Bauman, E. J. (1971). Trajectory decomposition. In Leondes, C. T., editor, *Optimization methods for large scale systems with applications*. McGraw-Hill, N. York.

Bell, J. F. (1973). *The experimental foundation of solid mechanics*, volume VI a/1 of *Encyclopedia of Physics*. Springer Verlag, Berlin.

Björkman, G. (1992). Path following and critical points for contact problems. *Computational Mechanics*, 10:231–246.

Bunde, A. and Halvin, S. (1991). *Fractals and disordered systems*. Springer, Berlin–Heidelberg.

Crisfield, M. A. (1991). *Non–linear finite element analysis of solids and structures*. J. Wiley, Chichester.

Curnier, A. (1993). *Méthodes numériques en mécanique des solides*. Presses Polytechniques et Universitaires Romandes, Lausanne. English translation, Kluwer, 1994.

Davies, P. and Benzeggagh, M. L. (1989). Interlaminal mode-I fracture testing. In Friedrich, K., editor, *Application of fracture mechanics to composite materials*. Elsevier.

Dem'yanov, V. F., Stavroulakis, G. E., Polyakova, L. N., and Panagiotopoulos, P. D. (1996). *Quasidifferentiability and nonsmooth modelling in mechanics, engineering and economics*. Kluwer Academic, Dordrecht.

Falconer, K. J. (1985). *The geometry of fractal sets*. Cambridge Univ. Press, Cambridge.

Feder, H. J. S. and Feder, J. (1991). Self-organized criticality in a stick-slip process. *Phys. Rev. Lett.*, 66:2669–2672.

Feder, J. (1988). *Fractals*. Plenum Press, New York.

Fletcher, R. (1990). *Practical methods of optimization*. J. Wiley, Chichester.

Glowinski, R. and LeTallec, P. (1989). *Augmented Lagrangian and operator–splitting methods for nonlinear mechanics*. SIAM, Philadelphia.

Guddat, J., Vasquez, F. G., and Jongen, J. T. (1990). *Parametric optimization: singularities, path following and jumps*. J. Wiley,, Chichester.

Jones, R. (1975). *Mechanics of composite materials*. MacGraw Hill, N. York.

Ladèveze, P. (1995). *Mecanique des structures nonlineaires*. Hérmes, Paris.

Le Mehauté, A. (1990). *Les Geométries fractales*. Hermes, Paris.

LeTallec, P. (1990). *Numerical analysis of viscoelastic problems*. Masson, Springer, Paris, Berlin.

MacLeod, J. A. and el Magd, S. A. A. (1980). The behaviour of brick walls under conditions of settlement. *The Structural Engineer*, 58A(9):279.

Maier, G. (1970). A matrix structural theory of piecewise linear elastoplasticity with interacting yield planes. *Meccanica*, March:54–66.

Maier, G. (1971). Incremental plastic analysis in the presence of large displacements and physical instabilizing effects. *Int. J. Solids Structures*, 7:345–372.

Mandelbrot, B. (1972). *The Fractal geometry of nature.* W.H. Freeman and Co., New York.

Massopust, P. R. (1993). Smooth interpolating curves and surfaces generated by iterated function systems. *Zeitschrift für Analysis und ihre Anwendungen*, 12:201–210.

Matthies, H. and Strang, G. (1979). The solution of nonlinear finite element equations. *Intern. J. of Numerical Methods in Engineering*, 14:1613–1626.

Mistakidis, E. S. (1997a). Fractal geometry in structural analysis problems: A variational formulation for fractured bodies with nonmonotone interface conditions. *Chaos, Solitons and Fractals*, 8:269–285.

Mistakidis, E. S. (1997b). On the solution of structures involving elements with nonconvex energy potentials. *Structural Optimization*, 13:182–190.

Mistakidis, E. S. and Panagiotopoulos, P. D. (1994). On the approximation of nonmonotone multivalued problems by monotone subproblems. *Computer Methods in Applied Mechanics and Engineering*, 114:55–76.

Mistakidis, E. S., Panagiotopoulos, P. D., and Panagouli, O. K. (1992). Fractal surfaces and interfaces in structures. Methods and algorithms. *Chaos, Solitons and Fractals*, 2:551–574.

Mistakidis, E. S., Thomopoulos, K., Avdelas, A., and Panagiotopoulos, P. D. (1994). Shear connectors in composite beams: A new accurate algorithm. *Thin-Walled Structures*, 18:191–207.

Ortega, J. M. and Rheinboldt, W. C. (1970). *Iterative solution of nonlinear equations in several variables.* Academic Press, New York-London.

Panagiotopoulos, P. D. (1975). A nonlinear programming approach to the unilateral contact and friction boundary value problem in the theory of elasticity. *Ing. Archiv*, 44:421–432.

Panagiotopoulos, P. D. (1985). *Inequality problems in mechanics and applications. Convex and nonconvex energy functions.* Birkhäuser, Basel - Boston - Stuttgart. Russian translation, MIR Publ., Moscow 1988.

Panagiotopoulos, P. D. (1992a). Fractal geometry in solids and structures. *Int. J. Solids and Structures*, 29(17):2159–2175.

Panagiotopoulos, P. D. (1992b). Fractals and fractal approximation in structural mechanics. *Meccanica*, 27:25–33.

Panagiotopoulos, P. D., Baniotopoulos, C. C., and Avdelas, A. V. (1984). Certain propositions on the activation of yield modes in elastiplasticity and their applications to deterministic and stochastic problems. *Zeitschrift für Angewandte Mathematik und Mechanik (ZAMM)*, 64:491–501.

Panagiotopoulos, P. D. and Panagouli, O. K. (1992). Fractal interfaces in structures: Methods of calculation. *Computers and Structures*, 45(2):369–380.

Panagiotopoulos, P. D. and Panagouli, O. K. (1993). The BIEM for fractal boundaries and interfaces. Applications to unilateral problems in geomechanics. In Manolis, G. D. and Davies, T. G., editors, *Boundary Element Techniques in Geomechanics*, pages 477–496. Computational Mechanics.

Panagiotopoulos, P. D. and Panagouli, O. K. (1996). The B.E.M. in plates with boundaries of fractal geometry. *Engineering Analysis with Boundary Elements*, 17:153–160.

Panagiotopoulos, P. D., Panagouli, O. K., and Koltsakis, E. K. (1994a). The B.E.M. in plane elastic bodies with cracks and/or boundaries of fractal geometry. *Computational Mechanics*, 15:350–363.

Panagiotopoulos, P. D., Panagouli, O. K., and Mistakidis, E. S. (1993). Fractal geometry and fractal material behaviour in solids and structures. *Archive of Applied Mechanics*, 63:1–24.

Panagiotopoulos, P. D., Panagouli, O. K., and Mistakidis, E. S. (1994b). Fractal geometry in structures. Numerical methods for convex energy problems. *Int. J. Solids Structures*, 31(16):2211–2228.

Panagiotopoulos, P. D., Panagouli, O. K., and Mistakidis, E. S. (1994c). On the consideration of the geometric and physical fractality in solid mechanics. I: Theoretical results. *Zeitschrift für Angewandte Mathematik und Mechanik (ZAMM)*, 74(3):167–176.

Panagouli, O. K. (1997). On the fractal nature of problems in mechanics. *Chaos, Solitons and Fractals*, 8(2):287–301.

Panagouli, O. K., Panagiotopoulos, P. D., and Mistakidis, E. S. (1992). On the numerical solution of structures with fractal geometry. The F.E. approach. *Meccanica*, 27:263–274.

Panagouli, O. K., Panagiotopoulos, P. D., and Mistakidis, E. S. (1995). Friction laws of fractal type and the corresponding contact problems. *Chaos Solitons and Fractals*, 5:2109–2119.

Papadrakakis, M., editor (1997). *Parallel solution methods in computational mechanics*. Wiley, Chichester.

Papadrakakis, M. and Bitzarakis, S. (1996). Domain decomposition PCG methods for serial and parallel processing. *Advances in Engineering Software*, 25:291–308.

Papadrakakis, M. and Pantazopoulos, G. (1993). A survey of quasi-Newton methods with reduced storage. *International journal for numerical methods in engineering*, 36(9):1573–1596.

Park, J. K. and Kwack, B. M. (1994). Three-dimensional frictional contact analysis using the homotopy method. *ASME Journal of Applied Mechanics*, 61:703–709.

Rohde, A. and Stavroulakis, G. E. (1995). Path following energy optimization in unilateral contact problems. *Journal of Global Optimization*, 6(4):347–365.

Rohde, A. and Stavroulakis, G. E. (1997). Genericity analysis for path-following methods in unilateral contact elastostatics. *Zeitschrift für Angewandte Mathematik und Mechanik (ZAMM)*, 77(6):(to appear).

Schoeffler, J. D. (1971). Static multilevel systems. In Leondes, C. T., editor, *Optimization methods for large scale systems with applications*. McGraw-Hill, N. York.

Scholz, C. H. and Mandelbrot, B. (1989). *Fractals in geophysics*. Birkhäuser, Boston–Basel.

Stavroulakis, G. E. and Rohde, A. (1996). Stability of structures with quasidifferentiable energy functions. In Sotiropoulos, D. and Beskos, D., editors, *2nd Greek Conf. on Computational Mechanics*, pages 406–413, Chania.

Stein, E., Wagner, W., and Wriggers, P. (1989). Grundlagen nichtlinearer Berechnungsverfahren in der Strukturmechanik. In Stein, E., editor, *Nichtlineare Berechnungen im Konstruktiven Ingenieurbau*, pages 1–53. Springer, Wien - New York.

Takayasu, H. (1990). *Fractals in physical sciences*. Manchester Univ. Press, Manchester.

Wallin, H. (1989). The trace to the boundary of Sobolev spaces on a snowflake. *Rep. Dep. of Math. Univ. of Umea.*

Wismer, D. A., editor (1971). *Optimization methods for large scale systems with applications*. McGraw-Hill, N. York.

Wriggers, P. (1993). Continuum mechanics, nonlinear finite element techniques and computational stability. In Stein, E., editor, *Progress in computational analysis of inelastic structures*, volume 321 of *CISM Lect. Notes*. Springer, Wien - New York.

Wriggers, P. and Wagner, W. (1991). *Nonlinear computational mechanics: state of the art*. Springer, Berlin.

Zienkiewicz, O. C. and Taylor, R. L. (1991). *The finite element method. Vol. II: Solid and fluid mechanics, dynamics and non-linearity*. McGraw-Hill.

7 APPLICATIONS

7.1 CHARACTERISTIC ONE- AND TWO- DIMENSIONAL EXAMPLES

The aim of this Section is to investigate the properties and the behaviour of the algorithms presented in the previous Chapters. For this reason, several one- and two- dimensional examples are considered involving very few unknowns only. Although these examples are really very simple and can be solved without a computer, they give a good idea of the results that can be expected in the real engineering examples presented later in this Chapter. Indeed, the results of the algorithms can be followed up and certain conclusions are derived.

7.1.1 One-dimensional springs assembly

In this example, the various properties of the proposed algorithms are investigated by means of the simple structural system of Fig. 7.1a. The structure is composed by a linear spring element E_1 (which simulates for example the linear elastic part of a fictitious structure) and a nonlinear spring element E_2 (which simulates all the elements with a nonlinear behaviour). In a more general setting, simple models like the one adopted here, may result from static condensation of larger structures. The structure is loaded at point G with the load P and u denotes the displacement of G.

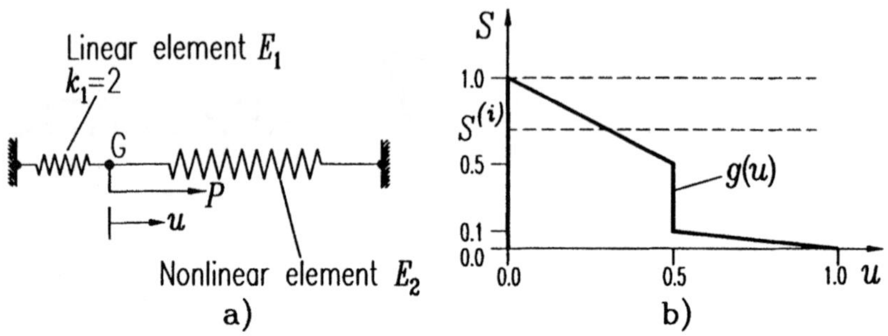

Figure 7.1. Simple springs assembly

The spring constant for E_1 is $k_1 = 2$. For E_2 we assume that the nonmonotone force-displacement diagram of Fig. 7.1b holds. The loading of the structure is $P = 1.0$. The nonconvex energy function $\Pi_{nc}(u)$ is a sum of the quadratic term $\Pi_q(u) = \frac{1}{2}uk_1u - Pu$ and of the nonconvex superpotential $\widetilde{w}(u)$ of the nonmonotone law of Fig. 7.1b, i.e.

$$\Pi_{nc}(u) = \Pi_q(u) + \widetilde{w}(u) = \frac{1}{2}uk_1u - Pu + \widetilde{w}(u) = u^2 - Pu + \widetilde{w}(u). \tag{7.1}$$

It is easily verified that the nonmonotone law $g(u)$ of Fig. 7.1b has the form:

$$g(u) = \begin{cases} [0,1] & \text{if } \ u = 0 \\ 1 - u & \text{if } \ 0 < u < 0.5 \\ [0.1, 0.5] & \text{if } \ u = 0.5 \\ 0.2(1-u) & \text{if } \ 0.5 < u \leq 1.0 \\ 0 & \text{if } \ 1.0 < u \end{cases} \tag{7.2}$$

and the respective nonconvex superpotential $\widetilde{w}(u)$ the form:

$$\widetilde{w}(u) = \begin{cases} 0 & \text{if } u = 0 \\ u - 0.5u^2 & \text{if } 0 < u < 0.5 \\ 0.375 & \text{if } u = 0.5 \\ 0.2u - 0.1u^2 + 0.300 & \text{if } 0.5 < u \leq 1.0 \\ 0.40 & \text{if } 1.0 < u. \end{cases} \tag{7.3}$$

The substationarity points of the nonconvex function Π_{nc} can be found by taking the derivatives with respect to u in the respective intervals. Thus it can be easily verified that Π_{nc} has two minima ($u_1 = 0.300, u_2 = 0.611$). Fig. 7.2

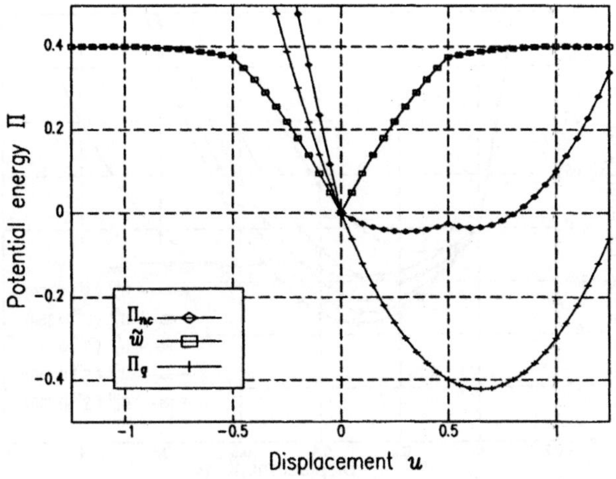

Figure 7.2. The potential energy functions of the structure

depicts the above mentioned functions $\Pi_{nc}(u)$, $\Pi_q(u)$ and $\tilde{w}(u)$ in the interval [-1.25,+1.25] and verifies the results of the previous analysis.

Let us now see how the Algorithm (2.6) presented in Chapter 2 can find the solutions of the problem. In this simple example we select approximating laws similar to the ones depicted with the dashed line in Fig. 7.1b. The superpotential of these laws are written in the form $p^{(i)} = |S^{(i)}||u|$. Therefore, instead of finding the minima of the nonconvex function Π_{nc}, we have to find the minimum of the convex function $\Pi_c^{(i)} = u^2 - Pu + |S^{(i)}||u|$. First we have to define a starting point for the iterations. Let's start with $u = 0$ that corresponds to the maximum force that element 2 can undertake ($S^{(1)} = 1.000$). Applying the algorithm we find the following solutions:

$$\Pi_c^{(1)} : u^{(1)} = 0.15 \rightarrow S^{(2)} = g(0.15) = 0.85$$
$$\Pi_c^{(2)} : u^{(2)} = 0.222 \rightarrow S^{(3)} = g(0.222) = 0.775$$
$$\Pi_c^{(3)} : u^{(3)} = 0.2625 \rightarrow S^{(4)} = g(0.2625) = 0.738$$
$$\vdots$$
$$\Pi_c^{(10)} : u^{(10)} = 0.300 \rightarrow S^{(11)} = g(0.300) = 0.700$$
$$\Pi_c^{(11)} : u^{(11)} = 0.300.$$

Thus we have convergence after 11 steps with accuracy $\varepsilon = 10^{-3}$. Fig. 7.3a depicts the various functions $\Pi_c^{(i)}$. As it is easily verified, at the last iteration the initial nonconvex function and the approximating convex function are minimized for the same u.

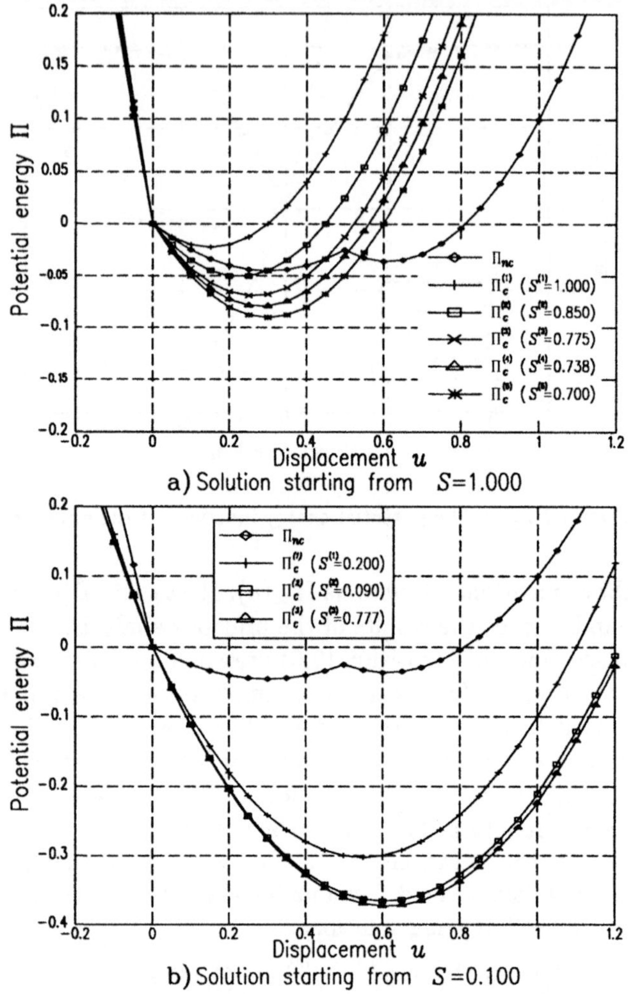

Figure 7.3. Approximation of the two minima of the nonconvex function

Let us try now with another starting point, for example let us start the iterations with $u = 0.501$ which corresponds to $S^{(1)} = 0.1$. Then, the application of the iterative scheme of Algorithm (2.6) yields the following solutions (Fig. 7.3b):

$$\begin{aligned}
\Pi_c^{(1)} &: u^{(1)} = 0.600 \rightarrow S^{(2)} = g(0.600) = 0.080 \\
\Pi_c^{(2)} &: u^{(2)} = 0.610 \rightarrow S^{(3)} = g(0.610) = 0.078 \\
\Pi_c^{(3)} &: u^{(3)} = 0.611 \rightarrow S^{(4)} = g(0.611) = 0.078 \\
\Pi_c^{(4)} &: u^{(4)} = 0.611.
\end{aligned}$$

Thus, different starting points of the iterative scheme, lead to different solutions of the problem. Of course, in this simple example it is easy, after a few tries, to find all the minima of the nonconvex, nonsmooth function. But, if the problem involves even a few unknowns, the determination of the correct starting point in order to find all the minima of the function is a very difficult and demanding task. From the engineering point of view, the problem is such complicated only if we assume that the total load is applied on the structure in one step. Instead, if we assume that the final value of the load is attained by following a prescribed loading history, the situation is much simpler: only solutions in the neighbourhood of the examined loading path are of interest in this case.

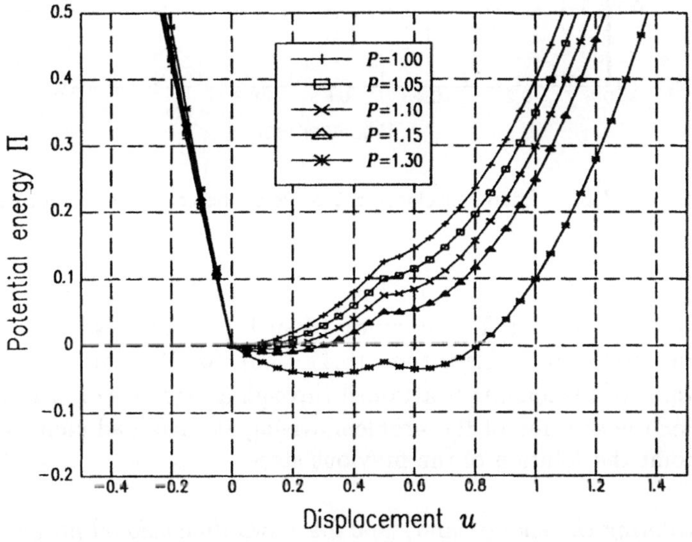

Figure 7.4. The nonconvex potential energy of the structure for various load levels

The things are more clear in Fig. 7.4 which depicts the function Π_{nc} for several values of the external loading. We notice that although the function is always nonconvex, it does not have multiple solutions for all the loadings. For example, for $P = 1.0$ the problem has a unique solution $u = 0$. The same happens for $P = 1.05$ where the unique solution is $u = 0.05$. Multiple solutions exist for $1.1 \leq P \leq 1.5$, as it is easily verified in Fig. 7.5 which gives the load - displacement curve for the examined problem. For $P = 1.1$, the function has as solutions $u = 0.1$ and $u = 0.5$. Although both solutions are mathematically acceptable, the most likely to be realized is the first one, because it results as a "continuation" of the previous solutions.

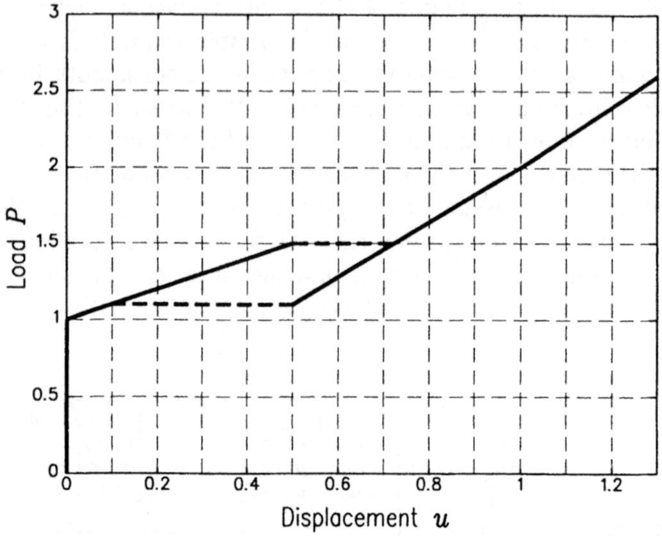

Figure 7.5. Load - displacement curve of the considered structure

Thus, a good practice for the study of such problems (which justifies the incremental procedure of Algorithms (6.4), (6.5), (6.6)) is the assumption that the final value of the loading is attained through a sequence of load increments (steps). For the solution of the problem arising at each load step, we use as a starting point the solution of the previous step.

7.1.2 Exploring the energy landscape for a two-dimensional problem

Here we will consider the simple truss assembly of Fig. 7.6. The structure consists of the linear elastic elements 1 and 3 and of the nonlinear elements 2 and 4 (for which it can be assumed in general that describe nonlinear interface conditions or material laws).

Elements 1 and 3 have a modulus of elasticity $E = 4$. For elements 2 and 4 we assume that the stress-strain diagram of Fig. 7.7a holds. We notice that for $\varepsilon \leq 0.1$ these elements behave linearly with E=10. All the elements have unit length L and cross-section area A, all in compatible units. The structure is loaded with the force P ($P_x = 1$ and $P_y = 1$) at G. For the given loading we want to determine the equilibrium positions of the system.

With the given data we formulate the potential energy of the structure as a function of the displacements u_x, u_y of point G. As in the previous example, the total potential energy Π_{nc} is written as the sum of the potential energy of

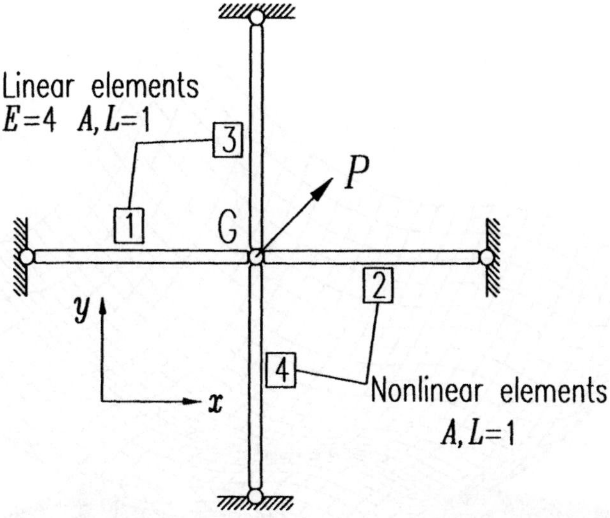

Figure 7.6. Simple truss assembly

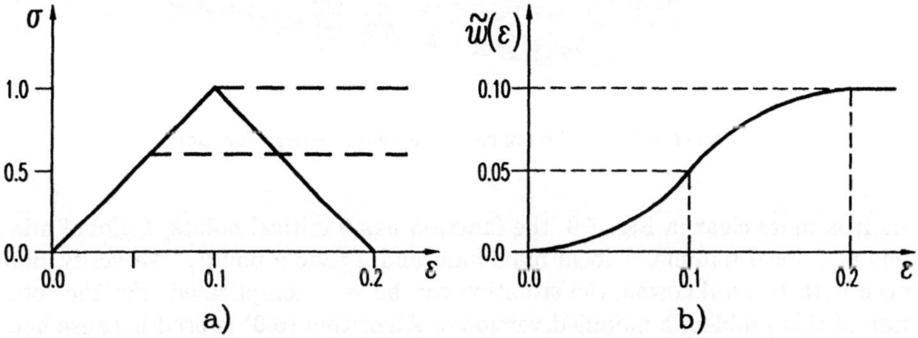

Figure 7.7. The nonmonotone stress-strain law of the nonlinear elements and the respective nonconvex superpotential

the linear elements and of the superpotentials of the nonlinear elements, i.e.

$$\Pi_{nc} = \frac{1}{2}\mathbf{u}^T\mathbf{K}\mathbf{u} + \widetilde{w}(\frac{u_x}{L}) + \widetilde{w}(\frac{u_y}{L}) - \mathbf{u}^T\mathbf{P} \tag{7.4}$$

where $\mathbf{u} = [u_x \quad u_y]^T$, $\mathbf{P} = [P_x \quad P_y]^T$ and $\mathbf{K}$ is the stiffness matrix of the structure which has the very simple form $\mathbf{K} = \begin{bmatrix} 4 & 0 \\ 0 & 4 \end{bmatrix}$.

Moreover, the nonconvex superpotential $\widetilde{w}(.)$ has the form depicted in Fig. 7.7b. The nonconvex function Π_{nc}, for $u_x, u_y \in [0, 0.5]$ is depicted in Fig. 7.8.

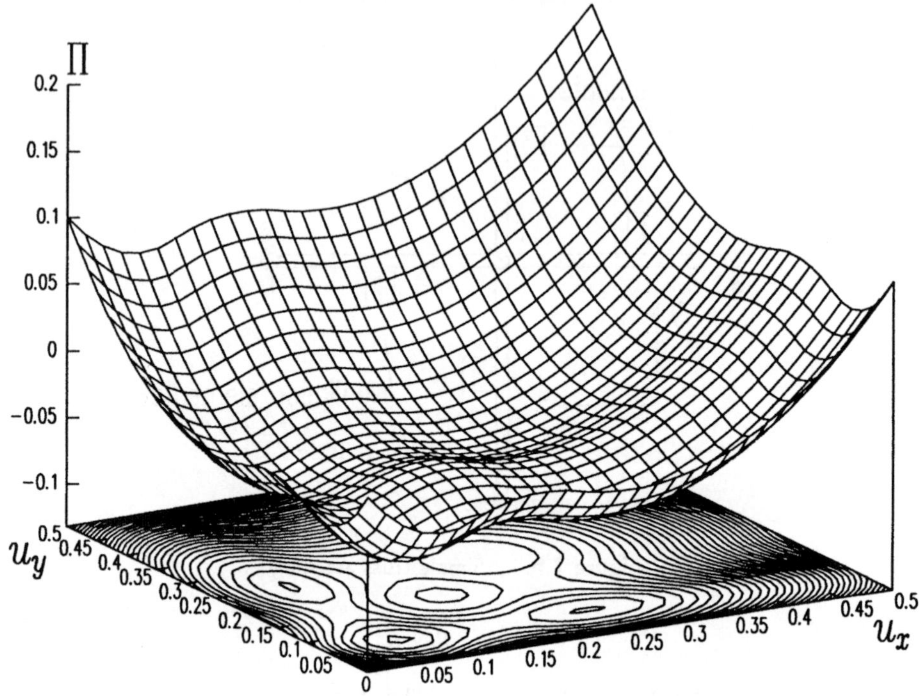

Figure 7.8. The nonconvex potential energy function

As it is more clear in Fig. 7.9, the function has 9 critical points, 1 global minimum, 3 local minima, 1 local maximum and 4 saddle points. We verify that even with two unknowns, the situation can be very complicated. For the solution of this problem a modified version of Algorithm (6.9) is used because here the longitudinal stress σ and the respective strain ε are connected with the nonmonotone diagram. Now we will see how this algorithm can approximate the above critical points. In this problem, in order to approximate the nonmonotone law we select monotone laws of the type depicted with the dashed line in Fig. 7.7a (elastic-perfectly plastic type laws). This selection results to fewer steps of the iterative procedure, as it was verified by the performed numerical experiments.

Starting the algorithm from point $(u_x, u_y) = (0, 0)$ we obtain as a solution the global minimum $(0.0714, 0.0714)$. Defining other starting points, the algorithm converges to other solutions. Thus, if we start from point $(0.18, 0.00)$ we obtain as a solution the point $(0.25, 0.0714)$, if we start with point $(0.18, 0.18)$ we obtain the solution $(0.25, 0.25)$ and if we start with point $(0.00, 0.18)$ we obtain the solution $(0.0714, 0.25)$.

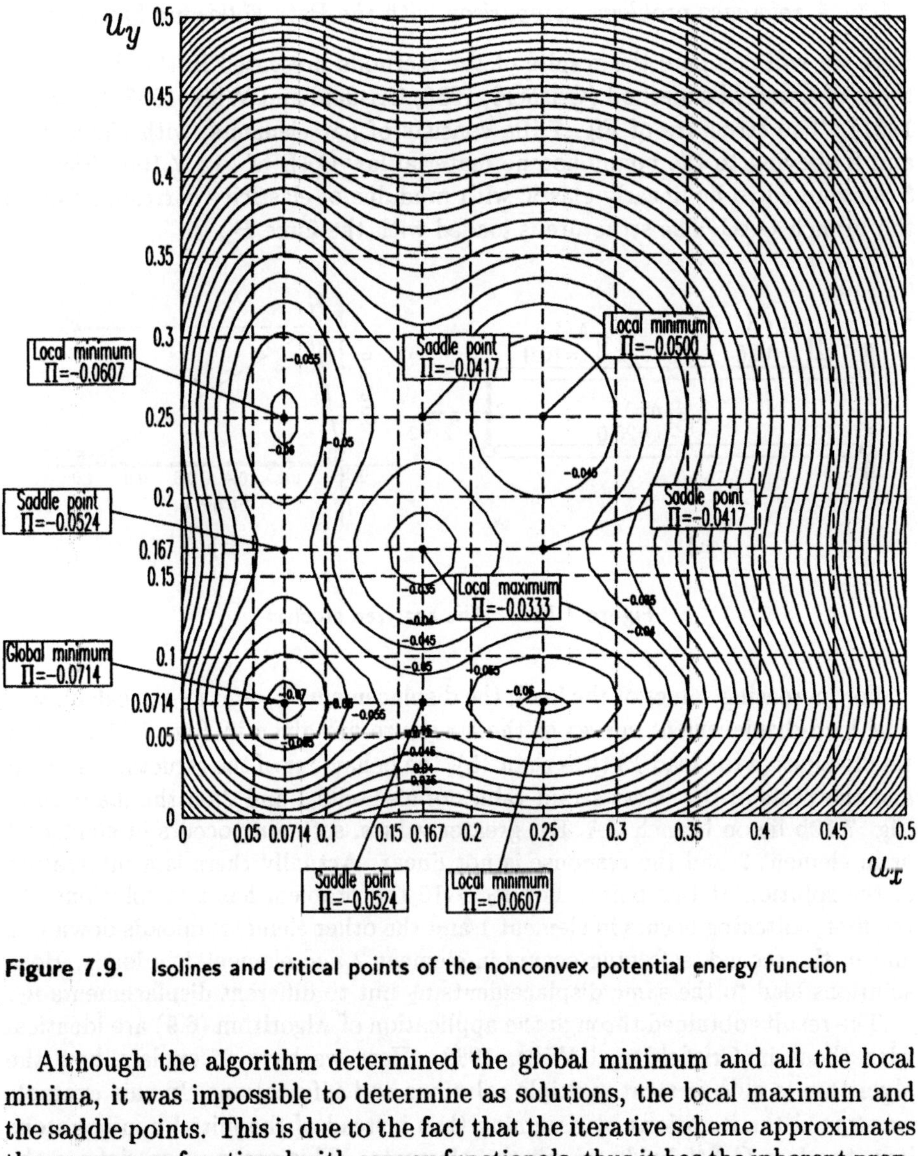

Figure 7.9. Isolines and critical points of the nonconvex potential energy function

Although the algorithm determined the global minimum and all the local minima, it was impossible to determine as solutions, the local maximum and the saddle points. This is due to the fact that the iterative scheme approximates the nonconvex functional with convex functionals, thus it has the inherent property to approximate only substationarity points that give adequate convexity information in their neighborhood. Under these circumstances, it is impossible to obtain as a solution local maxima and saddle points, unless the starting point coincides with one of them. From the engineering point of view this result seems acceptable as these substationarity points are not stable solutions of the problem.

7.1.3 A reference problem: comparison with the Path Following Methods

Here we will compare the results of the methods presented in this book with those of Crisfield's method (Crisfield, 1986, Crisfield and Wills, 1988, Crisfield, 1991). The structure of Fig. 7.10a consists of bar elements, with elements 1 and 2 obeying to the nonmonotone material law OAB of Fig. 7.10b. Element 3 is assumed to be linearly elastic with modulus of elasticity corresponding to the branch OAC. The structure is loaded with the force P_2.

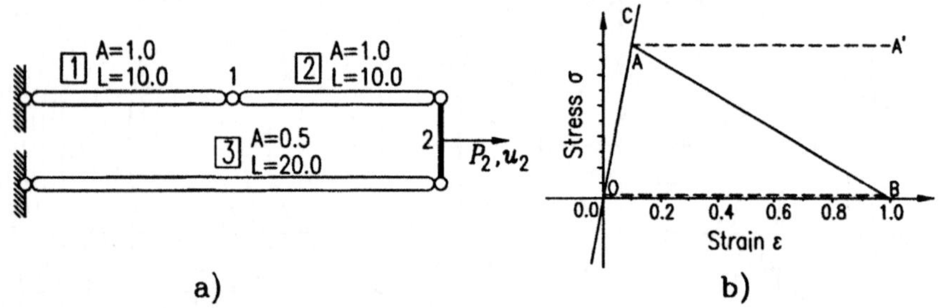

Figure 7.10. The analyzed problem

For increasing values of the load, the displacements u_1 and u_2 are calculated. The load-displacement curves of the structure are given in Figs 7.11a,b. For small values of the load P_2 (less than 15), the behaviour of the structure is linear and the plots of the stress-strain values of elements 1 and 2 on the diagram of Fig. 7.10b lie on branch OA. For greater values, softening occurs in element 1 or in element 2 and the response is not linear. Actually there is a bifurcation of the solution at this point. For $P_2 > 15$ the problem has two solutions. In the first, softening occurs in element 1 and the other element unloads down OA and in the second, softening occurs in element 2 and element 1 unloads. Both solutions lead to the same displacements u_2 but to different displacements u_1.

The results obtained through the application of Algorithm (6.9) are identical with those in Crisfield and Wills, 1988. Here we have to explain how the algorithm is able to treat multiple solutions and bifurcations. In this example we started the iterations by assuming that instead of the initial nonmonotone law, the law OAA' holds, in all the elements. This assumption defines the starting point of the iterations. If we select a different starting point it is possible to obtain a different solution. For example, starting the iterations by assuming that the law OAA' (Fig. 7.10b) holds for both elements, we obtain the first solution while, starting the iterations assuming that the law OAA' holds for element 1 and the law OB (lower dashed line in Fig. 7.10b) holds for elements 2, we obtain the second solution.

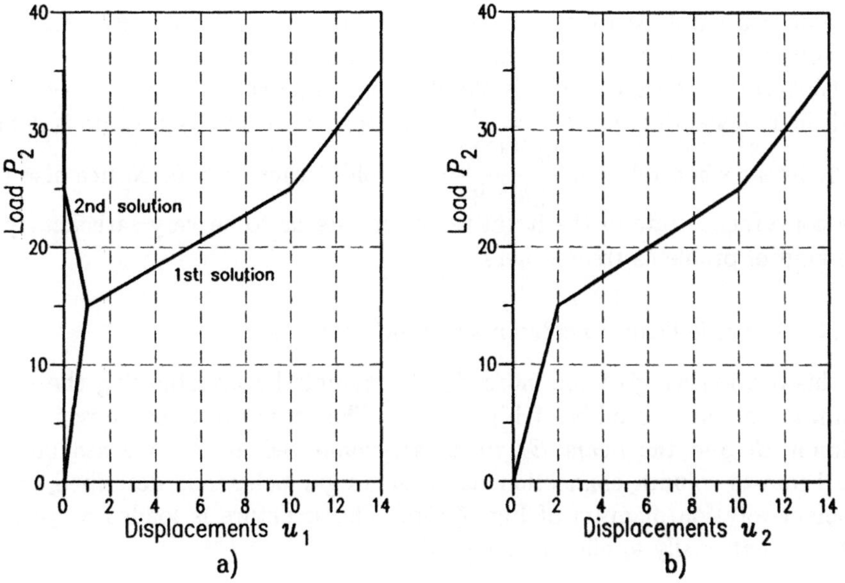

Figure 7.11. Load–displacement curves

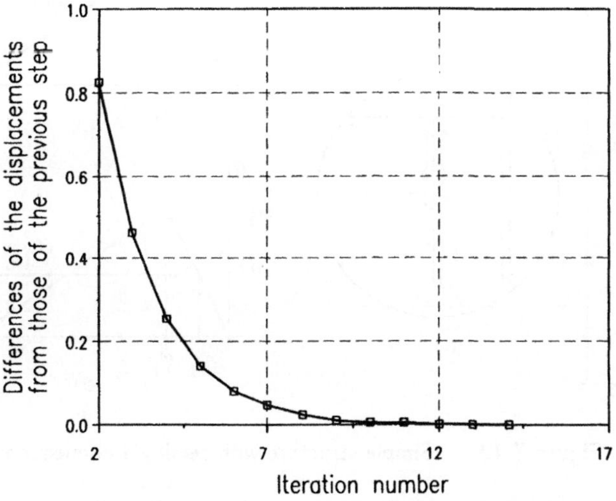

Figure 7.12. Convergence rate of the algorithm

It is interesting to calculate the convergence rate for this simple example. For a given load level ($P_2 = 20$), the differences of the displacements u_2 between two consecutive steps are calculated. The results are depicted in Fig. 7.12. The convergence rate calculated from the numerical data found to be linear.

We recall here that if $h^{(k)}$ is the difference between the solution of step k and the real solution, i.e. $h^{(k)} = u^{(k)} - u^*$, then the convergence rate is defined to be linear when relation $\frac{||h^{(k+1)}||}{||h^{(k)}||} \leq \alpha$ holds, where $\alpha > 0$. Notice also that the convergence rate of the iterative scheme seems to be very satisfactory for the kind of problems treated here.

7.1.4 A simple beam-to-column semirigid connection

In this Section we give one more simple example by investigating the simple beam-to-column assembly of Fig. 7.13a. The structure is composed by the column E_1 and the beam E_2 which are connected at B. We assume that the beam-to-column connection has a semirigid behaviour described by the moment-rotation diagram of Fig. 7.13b. The structure is loaded at point B with the externally applied moment M_e.

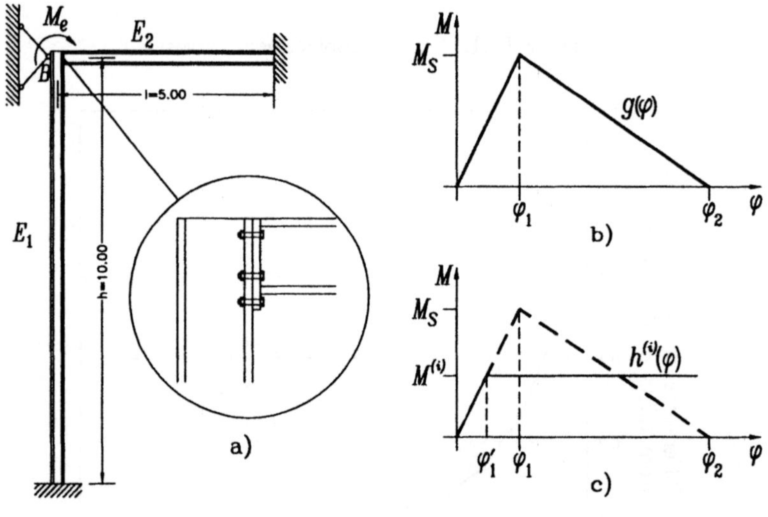

Figure 7.13. Simple structure with semirigid connection

Again, it is easy to formulate the nonconvex energy function Π_{nc} which is the sum of the quadratic term $\Pi_q(\varphi) = \frac{1}{2}\varphi\frac{4EJ_1}{h}\varphi - M_e\varphi$ and of the nonconvex superpotential $\widetilde{w}(\varphi)$ of the nonmonotone law of Fig. 7.13b, i.e.

$$\Pi_{nc}(\varphi) = \Pi_q(\varphi) + \widetilde{w}(\varphi) = \frac{1}{2}\frac{4EJ_1}{h}\varphi^2 - M_e\varphi + \widetilde{w}(\varphi). \tag{7.5}$$

It is easily verified that the nonmonotone law $g(\varphi)$ of Fig. 7.13b has the form:

$$g(\varphi) = \begin{cases} \varphi \frac{M_S}{\varphi_1} & \text{if } 0 < \varphi \leq \varphi_1 \\ M_S(1 - \frac{\varphi - \varphi_1}{\varphi_2 - \varphi_1}) & \text{if } \varphi_1 < \varphi \leq \varphi_2 \\ 0 & \text{if } \varphi_2 < \varphi \end{cases} \tag{7.6}$$

and the respective nonconvex superpotential $\widetilde{w}(\varphi)$ the form:

$$\widetilde{w}(\varphi) = \begin{cases} \frac{1}{2} \frac{M_S}{\varphi_1} \varphi^2 & \text{if } 0 < \varphi \leq \varphi_1 \\ \frac{1}{2} M_S(2\varphi - \varphi_1 - \frac{(\varphi - \varphi_1)^2}{\varphi_2 - \varphi_1}) & \text{if } \varphi_1 < \varphi \leq \varphi_2 \,. \\ \frac{1}{2} M_S \varphi_2 & \text{if } \varphi_2 < \varphi \end{cases} \tag{7.7}$$

We assume that $EJ_1 = 4$, $M_S = 2.0$, $\varphi_1 = 0.0625$. For the value of φ_2 we consider two cases: $\varphi_2 = 0.125$ and $\varphi_2 = 0.1875$. Figures 7.14a,b give the nonconvex energy function $\widetilde{w}(\varphi)$ and the potential energy functions for various values of the externally applied moment M_e, for the two values of φ_2 respectively. It can be easily verified that for $\varphi_2 = 0.125$ and for $2.4 \leq M_e \leq 3.0$, the potential energy function has 3 critical points, 2 local minima which are stable equilibrium points and one local maximum which is an unstable solution of the problem. For $\varphi_2 = 0.1875$, the problem has always a single solution except of the case where $M_e = 3.0$. In this case, the potential energy function has the same value in the interval [0.0625,0.1875] which means that every point in this interval is a stable solution of the problem.

Let us now see how the Algorithm (6.9) of Chapter 6 can find the solutions of the problem. Instead of finding the minima of the nonconvex function Π_{nc}, we have to find the minimum of the convex function $\Pi_c^{(i)}(\varphi) = \frac{1}{2}\varphi \frac{4E_1J_1}{h}\varphi - M_e\varphi + h^{(i)}(\varphi)$ (see Fig. 7.13c) where

$$h^{(i)}(\varphi) = \begin{cases} \frac{1}{2} \frac{M_S}{\varphi_1} \varphi^2 & \text{if } 0 < \varphi \leq \varphi_1' \\ \frac{1}{2} \frac{M_S}{\varphi_1} \varphi_1'^2 + (\varphi - \varphi_1') M^{(i)} & \text{if } \varphi_1' < \varphi \end{cases} \,. \tag{7.8}$$

We study in more detail the case where $\varphi_2 = 0.125$, $M_e = 2.4$. First, we have to define a starting point for the iterations. Starting with $\varphi^{(1)} = \varphi_1$ we obtain as a solution the value $\varphi = 0.500$ while starting with $\varphi^{(1)} = \varphi_2$ we find $\varphi = 1.500$. From the mathematical point of view both solutions are acceptable but the equilibrium configuration that will be finally realized depends on the loading path. We notice also that although the function is always nonconvex, it does not have multiple solutions for all the loadings. For example, for $M_e = 1.0$ the problem has a unique solution $\varphi = 0.2083$. Multiple solutions exist for $2.4 \leq M_e \leq 3.0$ as it is easily verified from Fig. 7.14a. Again, different starting points of the iterative scheme, lead to different solutions of the problem.

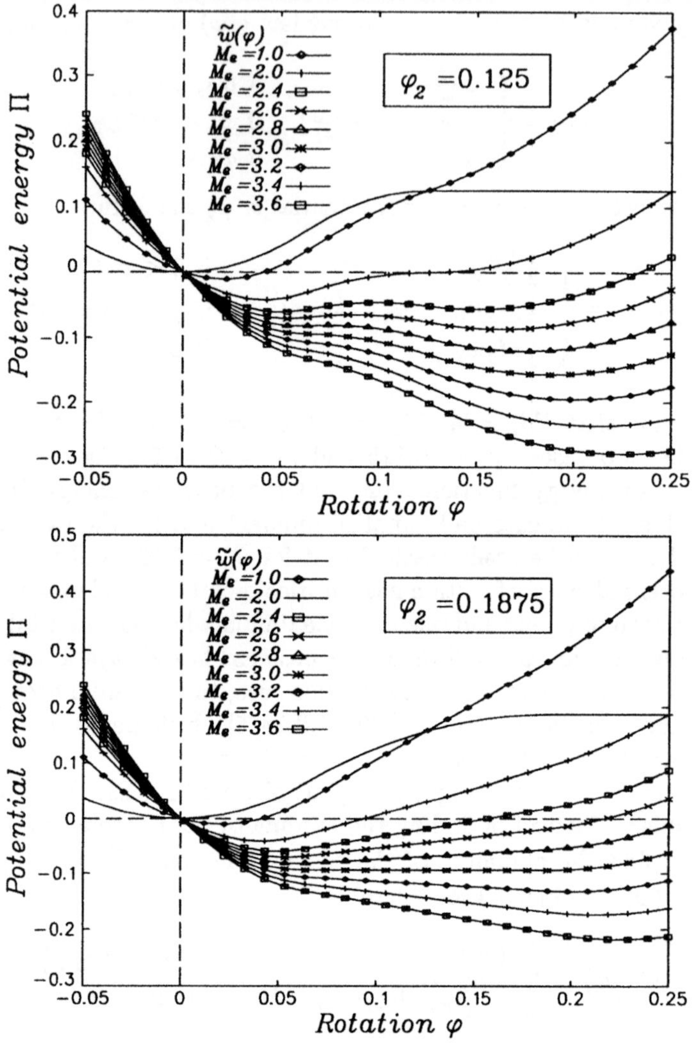

Figure 7.14. The potential energy functions of the structure

7.2 STRUCTURES WITH FRICTIONAL CONTACT AND ADHESIVE CONTACT INTERFACE CONDITIONS

7.2.1 Unilateral contact with the presence of nonmonotone friction

In this example we will show the application of the Algorithm (6.5) of Chapter 6 to the solution of a problem involving unilateral contact with nonmonotone friction interface conditions. The structure is depicted in Fig. 7.15a and consists of a deformable body which is pressed against a rigid support. The normalized diagram of Fig. 7.15b is assumed to hold on the parts of the interface which come in contact in the tangential direction. This law can be put in the form

$$S_T = g([u]_T, \mu, S_N) \tag{7.9}$$

where g is a nonmonotone function, $[u]_T$ are the relative tangential displacements of the two parts of the interface, S_N and S_T are the normal and tangential to the interface forces respectively and μ is the friction coefficient. Note that this diagram may be different from point to point, i.e. we can write (7.9) more generally as

$$S_T = g([u]_T, \mu, S_N, x) \tag{7.10}$$

where x is the position on the interface. At every point the friction force is calculated by multiplying the vertical coordinate of the diagram by $\mu|S_N|$ where $\mu = \mu(x)$ and $|S_N| = |S_N(x)|$ are the friction coefficient and the contact force respectively, at the considered point. Obviously, these conditions hold at the parts of the interface which come in contact. Thus, the interface conditions read:

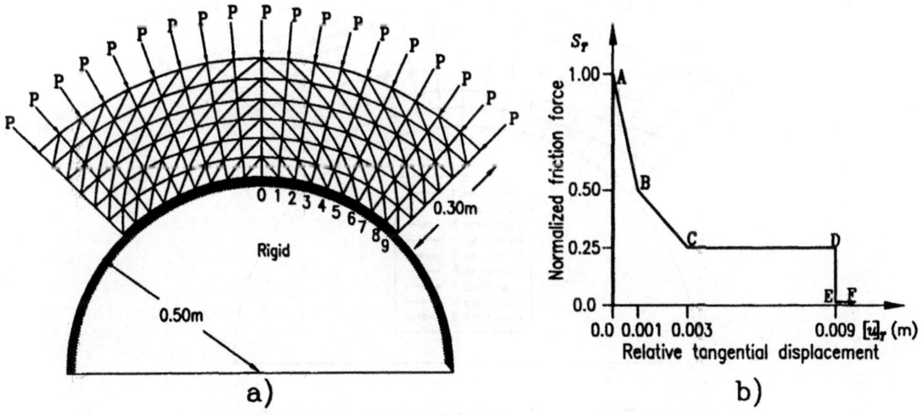

Figure 7.15. F.E. discretization of the considered problem and the adopted nonmonotone friction law

$$\text{If } [u]_N > 0 \text{ then } S_N = 0 \text{ and } S_T = 0 \tag{7.11}$$

$$\text{If } [u]_N = 0 \text{ then } S_N < 0 \text{ and } S_T = g([u]_T, \mu, S_N, x), \tag{7.12}$$

where $[u]_N$ are the relative normal displacements of the interface.

Algorithm (6.5) is now applied to the analysis of the structure which is discretized through constant stress triangular elements. The material of the structure is assumed linear elastic with modulus of elasticity $E = 2.1 \times 10^6$ kN/m^2, Poisson's ratio $\nu = 0.30$ and thickness $t = 0.01$m. Moreover, the friction coefficient μ equals to 0.3. The loading P increases in order to obtain a parametric solution. The nine load cases of Table 7.1 are analyzed.

Table 7.1. The considered load cases

Load case	*P (kN)*	*Load case*	*P (kN)*	*Load case*	*P (kN)*
1	20.00	4	80.00	7	110.00
2	40.00	5	90.00	8	130.00
3	60.00	6	100.00	9	150.00

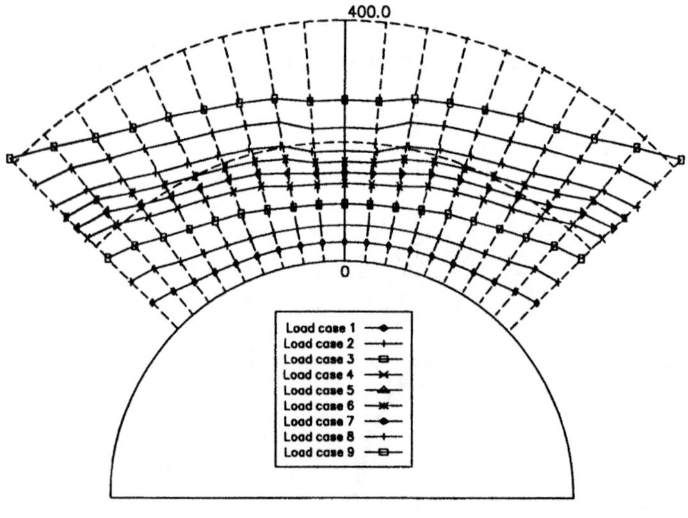

Figure 7.16. Distribution of the contact forces

The distribution of the contact forces along the interface is given in Fig. 7.16. It is noticed that the influence of the nonmonotone diagram to the results is not significant. But the influence of the nonmonotone diagram to the distribution of the frictional forces (Fig. 7.17) is very strong. For the first two load cases, where the loads are small, the friction forces lie on the vertical branch OA of the nonmonotone diagram of Fig. 7.15b. This is the "sticking" friction region,

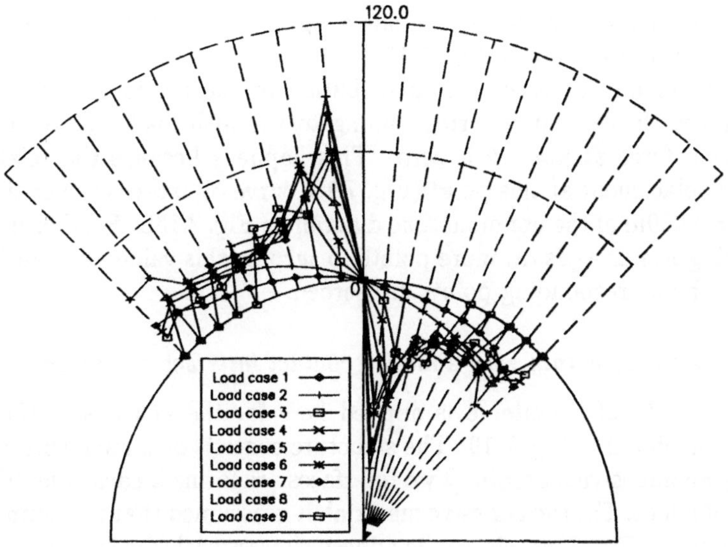

Figure 7.17. Distribution of the tangential forces

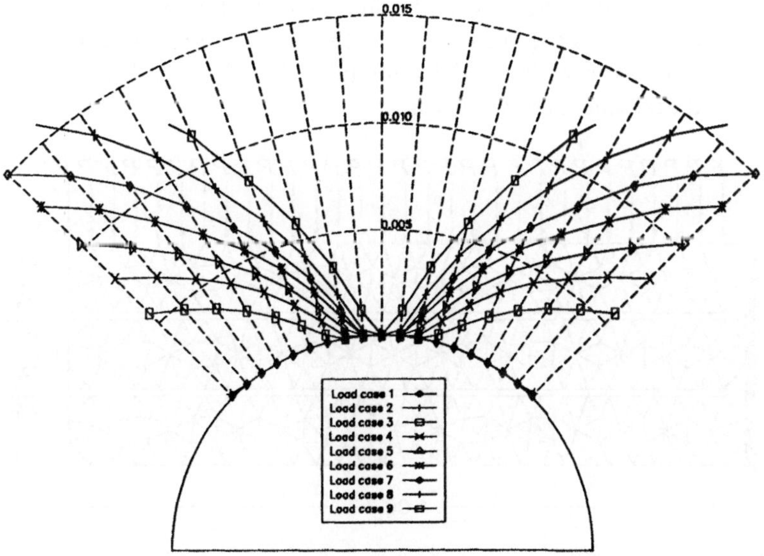

Figure 7.18. Relative tangential displacements along the interface

i.e. the region where the respective tangential displacements are equal to zero (see Fig. 7.18, load cases 1,2). For the rest of the load cases, we note that the distribution of the frictional forces has a zig-zag form which is a result of the softening form of the law of Fig. 7.15b.

As the loading increases, the distribution of the frictional forces changes significantly. In the curve corresponding to the fifth load case we note that the friction force at point 9 is zero. This happens because the relative tangential displacement at this point (Fig. 7.18, curve 5) exceeds the critical value $[u]_T^{max} = 0.009\text{m}$ of the nonmonotone diagram of Fig. 7.15b. Further increase of the loading has as a result, more points to surpass this value and the frictional forces at the corresponding points are zeroed.

7.2.2 Layered structure with adhesive contact interface conditions

Algorithm (6.5) of Chapter 6 is applied here for the analysis of the layered structure depicted in Fig. 7.19. The structure consists of three layers which are glued by an adhesive material. Two interfaces are formed, each consisting of 30 couples of nodes. For the adhesive material, it is assumed that the nonmonotone law depicted in Fig. 7.20 relates the adhesive stress with the relative tangential displacement. In order to find the adhesive force applied on each node of the interface, we have to multiply the stress of the adhesive with the thickness of the structure and with the length of influence of each node.

The structure is loaded with forces in the vertical direction. The layers are assumed to be elastic with elasticity modulus $E = 2.1 \times 10^7 \text{kN/m}^2$, Poisson's ratio $\nu = 0.3$ and thickness $t = 0.1\text{m}$.

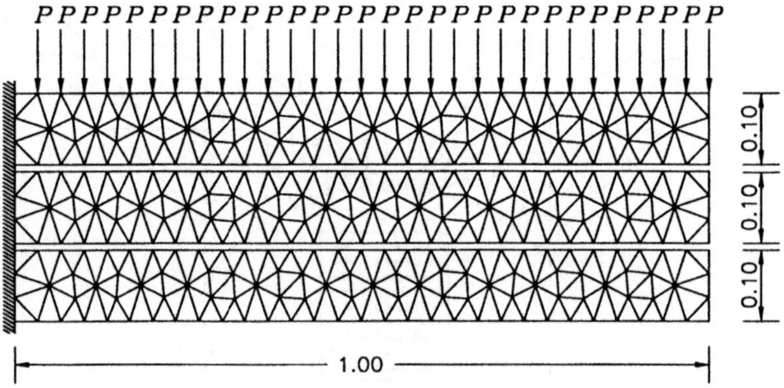

Figure 7.19. Finite element discretization of the structure

This problem is similar to the one presented in the previous example with the difference that the value of the tangential force does not depend on the value of the developed contact force. However, there is an interaction between the

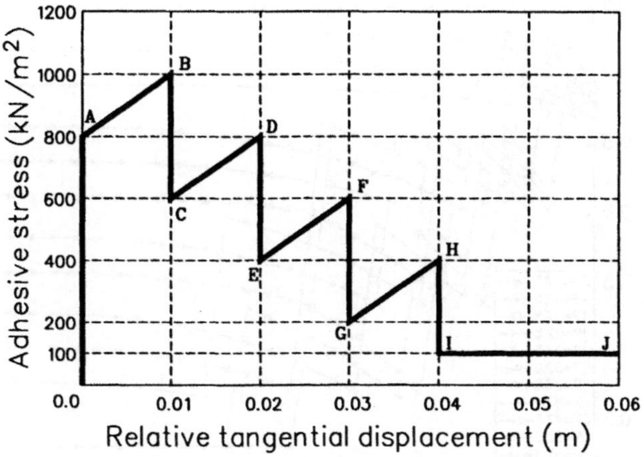

Figure 7.20. **The adopted tangential adhesive law**

normal and the tangential to the interface directions, because the values of the tangential forces can influence the values of the normal forces and vise-versa.

The influence of the nonmonotone diagram of Fig. 7.20 on the distribution of the tangential forces along the upper interface given in Fig. 7.21 is very strong. When the external loading is low, the behaviour of the joint is linear (load cases 1,2) and all the results lie on branch AB of Fig. 7.20. As the loading increases, the equilibrium points on the diagram, move to the branches at the right. More specifically, for the load cases 3,4,5 and for the outer region of the interface, the equilibrium points lie on branch CD, whereas for the inner parts of the interface the equilibrium points remain on branch AB. Increasing the loading, other branches of the nonmonotone diagram are also realized and finally branch IJ is realized at the outer part of the interface (load case 11). Further increase of the loading (load case 12), has as a result the increase of the number of the interface nodes that fall on this branch.

It is easy to interpret the points of abrupt changes of the curves in Fig. 7.21 as the points where the strength of the joint passes from one branch of the zig-zag diagram to another, thus exhibiting the progressive failure as the tangential displacement increases (cf. also Fig. 7.22 that gives the values of the relative tangential displacements along the upper interface). Similar results were obtained for the lower interface but for the sake of brevity are not presented here.

The convergence of the algorithm was very fast. All the previous problems have been solved in about 7-14 cycles of the normal-tangential subproblems procedure using a second order norm of the normal and tangential forces

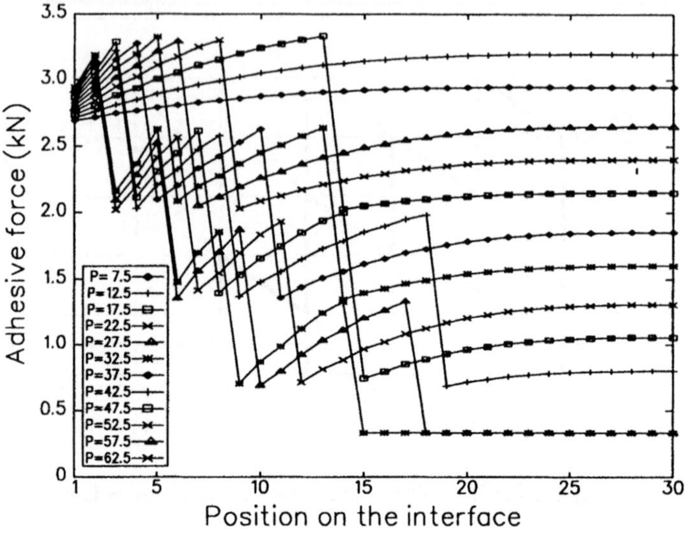

Figure 7.21. Distribution of the tangential forces along the upper interface

as a stopping criterion: the algorithm terminates when $\frac{||S_T^{(i)} - S_T^{(i-1)}||}{||S_T^{(i)}||}$ becomes smaller than 10^{-4} where S_T denotes the vector of the tangential interface forces.

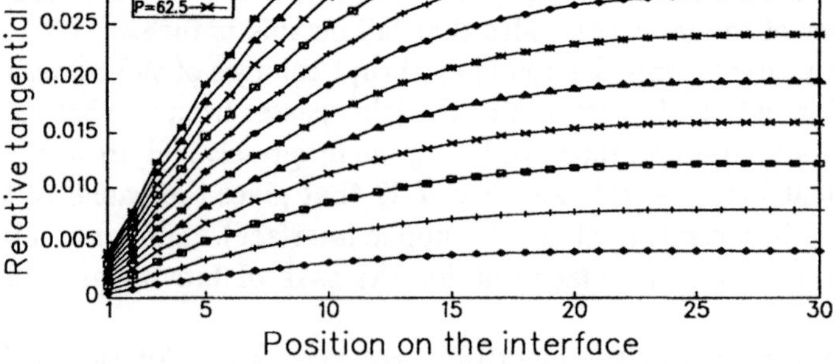

Figure 7.22. Relative tangential displacements along the upper interface

In this example, it is interesting to examine the effect of the finite element discretization on the results. For this reason, four different finite element discretizations of the above problem were considered which are presented in Fig. 7.23. Table 7.2 summarizes the characteristics of each F.E. mesh.

Table 7.2. Characteristics of the F.E. discretizations

Version	Total nodes	Elements	Interface nodes
Model A	231	294	25
Model B	327	450	30
Model C	820	1160	40
Model D	885	1434	50

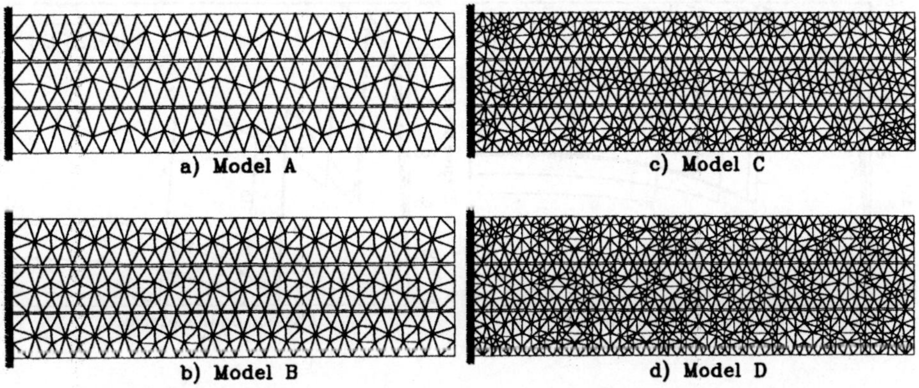

Figure 7.23. Four different F.E. discretizations of the considered problem

The structures are loaded as previously. In this case only the values of the adhesive stresses can be compared and not the forces applied on each node, because the distance between the interface nodes is not the same in all the structures.

The obtained distribution of the adhesive stresses along the upper interface for each structure and for certain load cases are presented in Fig. 7.24a,b. Fig. 7.24a depicts the distribution for the case $P = 10$kN whereas Fig. 7.24b gives this distribution for the case $P = 32.5$kN. It is observed that the results differ between the various models. But, as the F.E. density increases we have convergence of the resulting distributions as it is obvious from the corresponding figures.

Notice that as the F.E. density increases, the adhesive stresses increase as well, and the interface positions where we have the passing from the one branch to the other, move to the left. This happens because, according to the elementary theory of finite elements, for finite element approximations based on a potential energy minimization pronciple which corresponds to the displacements method of classical structural analysis, as the number of elements increases, the structure becomes more flexible. Accordingly, roughly speaking, the relative tangential displacements and the stresses of the interface increase as well.

Thus, we conclude that the calculations seem to be mesh independent, in the sense that no additional assumptions are required, beyond the usual ones holding within the finite element theory. The previous conclusion holds because the nonmonotone diagrams are introduced "accurately", i.e. we consider nonconvex energy functions. The same happens with the theories of damage which describe the softening, i.e. the nonconvexity of the energy function by means of hidden variables.

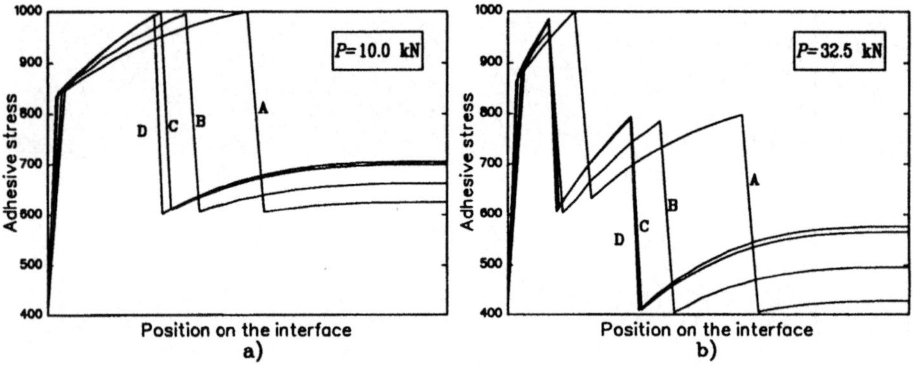

Figure 7.24. Distributions of the tangential adhesive stresses along the upper interface

7.3 DEBONDING OF LAYERS CONNECTED WITH ADHESIVES

Here we will study the problem of laminate structures which are glued with some adhesive material obeying to a nonmonotone law. Such stress-strain or reaction-displacement laws have been obtained as consequence of the use of advanced technology testing machines. We mention here, among others, the force-displacement diagrams registered in Green and Bowyer, 1981 during slow pull-out tests of fibre reinforced specimens of laminated products and the diagrams obtained by Roman et al., 1981 during several types of loading tests of specimens of glass fibre reinforced epoxy laminates (see also Williams and Rhodes, 1982, Schwartz, 1984 and Moyson and Gemert, 1985). In the following, the structural analysis problem for adhesively connected layers is treated by the method presented in Section 6.2.2.1.

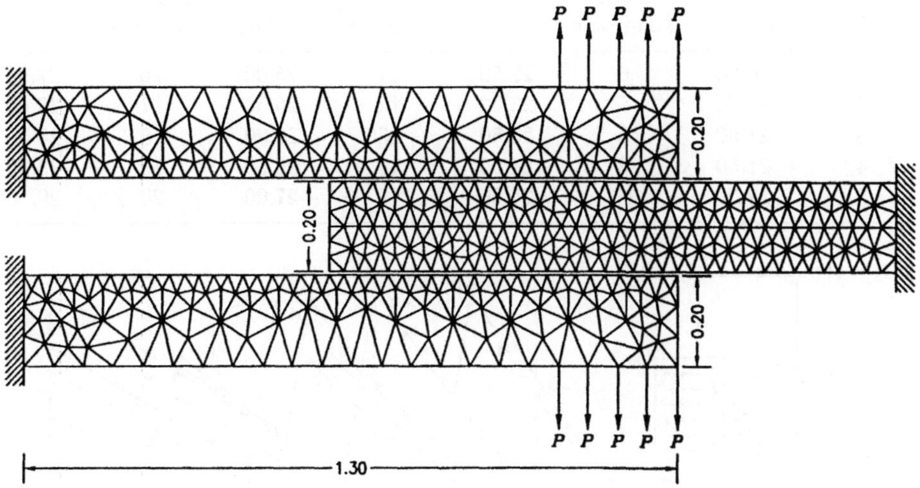

Figure 7.25. Laminated structure under cleavage load

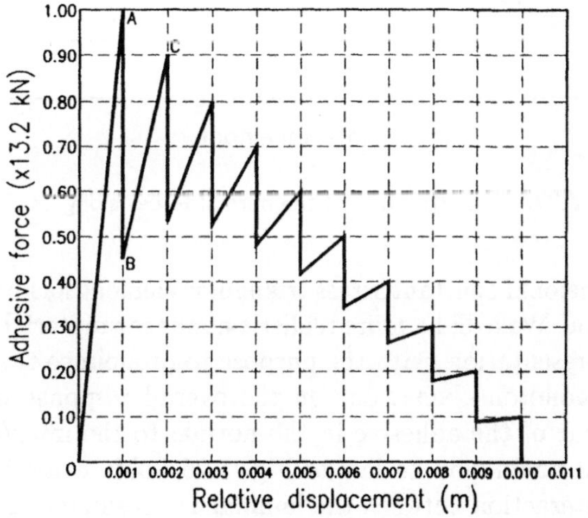

Figure 7.26. Nonmonotone interface response in the normal direction

Here Algorithm (6.4) is applied to the analysis of the two-dimensional simple structure of Fig. 7.25. The structure consists of three layers which are glued by an adhesive material. The material of the layers is assumed to be linear elastic

Table 7.3. The considered load cases

Load case	*P(kN)*	*Load case*	*P(kN)*	*Load case*	*P(kN)*	*Load case*	*P(kN)*
1	20.00	6	22.50	11	25.00	16	27.50
2	20.50	7	23.00	12	25.50	17	28.00
3	21.00	8	23.50	13	26.00	18	28.50
4	21.50	9	24.00	14	26.50	19	29.00
5	22.00	10	24.50	15	27.00	20	29.50

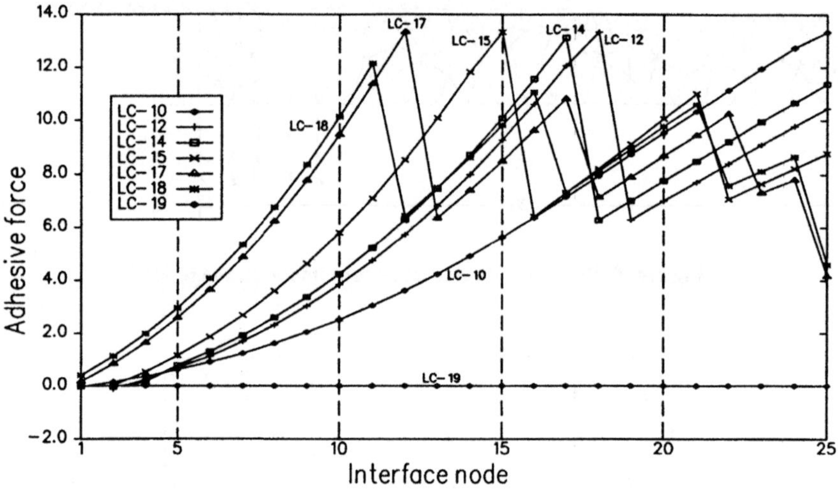

Figure 7.27. Distribution of the normal forces along the interface

and a two-dimensional constant stress triangular element has been employed for the discretization. We avoid a more realistic nonlinear material law and a three-dimensional representation, with the purpose to pronounce the influence of the interface's nonmonotone behaviour on the overall response of the structure. For the behaviour of the adhesive in the normal to the interface direction we assume that the nonmonotone law of Fig. 7.26 holds while in the tangential direction, the interaction between the laminae is considered as negligible. This fact is also dictated by the physical model (Davies and Benzeggagh, 1989). The structure is loaded with cleavage forces as shown in Fig. 7.25. This particular structure - loading configuration was selected because of the high sensitivity it exhibits at the initial loading stages, that is expected to magnify any numerical instabilities (cf. Williams et al., 1986 on the inherent instability of the "Double Cantilever Beam" tests). In order to follow the behaviour of the structure,

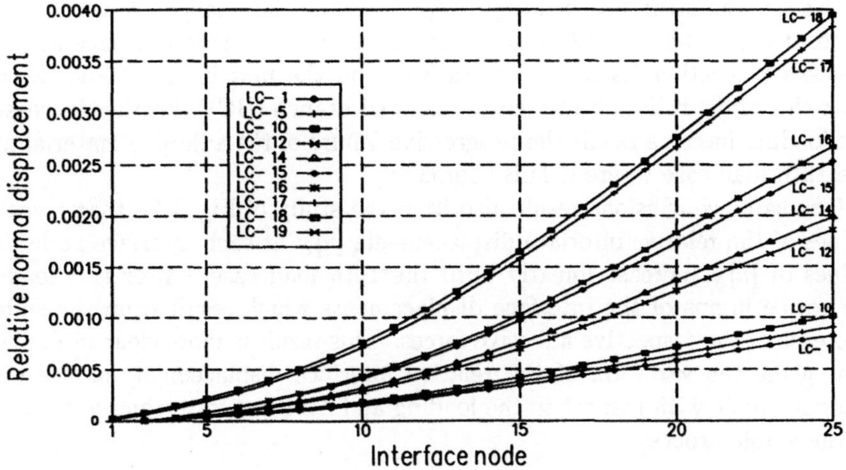

Figure 7.28. **Relative normal displacements along the interface**

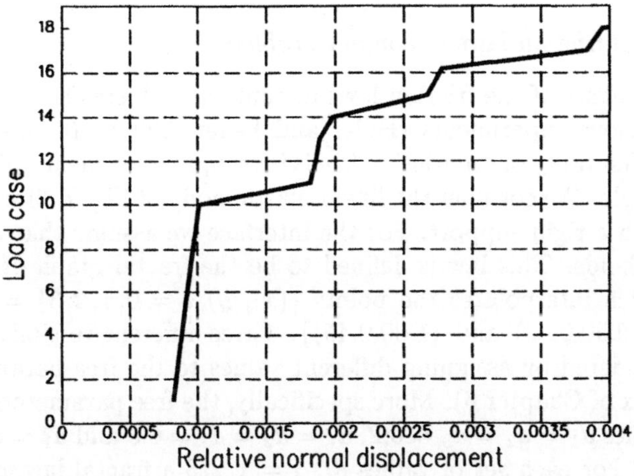

Figure 7.29. **Load–relative displacement diagram for the outer interface node**

twenty load cases are considered according to Table 7.3. Due to the symmetry with respect to the horizontal axis, only one half of the structure is analyzed.

Fig. 7.27 shows the distribution of the debonding forces along the interface. The traction peaks of the diagram are caused by the multiple vertical branches of the interface law. These abrupt changes have their nature to the changes of the adhesive behaviour, i.e. the passing from one to another branch of the nonmonotone law of Fig. 7.26, at each point of the interface. Until the 10th load case, the structure behaves linearly. After this load case, the adhesive forces decrease (cf. LC-12 in Fig. 7.27 where until the 18th interface node the adhesive's reaction-displacement state lies on the first branch (OA) whereas after this node, it lies on the second inclined branch (BC)). Further increase of the loading has as a result the progressive failure of the adhesive material until the 19th load case where it fails completely.

The same conclusions result also from the study of Fig. 7.28 that gives the values of the relative interface displacements $[u]_N$ at each interface node. The values of $[u]_N$ increase linearly until the 10th load case. After this loading, we notice jumps of the interface displacements which result from the sudden decrease of the respective adhesive forces. This result is more clear in Fig. 7.29 that gives the variation of the relative interface displacement for the outer interface node with respect to the loading and confirms the strong nonlinearity of the whole process.

7.4 STRUCTURES WITH FRACTAL INTERFACES AND/OR FRACTAL FRICTION LAWS

7.4.1 *Fractal friction laws in contact problems*

The fractal nature of the friction laws in contact problems has been confirmed lately by various experiments (Feder and Feder, 1991). In this Section, the effect of the fractality of the friction law is investigated by means of the following simple example. We consider the linear elastic body of Fig. 7.30 which forms an interface with a rigid support. For the interface we assume that a fractal type friction law holds. This law is defined to be the fractal graph of a continuous function which interpolates the points $\{(x_i, y_i), i = 0,1,2,3\} = \{(0.00,0.75),$ $(0.50,1.00), (0.50,0.20)$ and $(1.00,0.45)\}$. Three different versions of the fractal law are considered by assigning different values to the free parameters d_i (see the Appendix of Chapter 6). More specifically, the free parameters are taken to have the values $d_1 = d_2 = d_3 = 0.2$, $d_1 = d_2 = d_3 = 0.3$ and $d_1 = d_2 = d_3 = 0.4$ respectively. For each set of values d_i, $i = 1,2,3$ a fractal law φ_j, $j = 1,2,3$ is derived according to the procedure presented in the Appendix of Chapter 6. In this example the first five approximations of the fractal law are considered. Figures 7.31a,b,c depict the fifth approximations of the laws φ_j, $j = 1,2,3$ respectively. The friction coefficient μ is 0.2 for all the cases.

Three cases are considered by taking into account each one of the friction laws φ_1, φ_2 and φ_3. For each case, a sequence of subproblems was solved by considering the five approximations of the corresponding fractal friction law. As it is clear from the curves of Fig. 7.31, the arising subproblems are nonconvex, nonsmooth and for the numerical solution the Algorithm (6.5) of Chapter 6

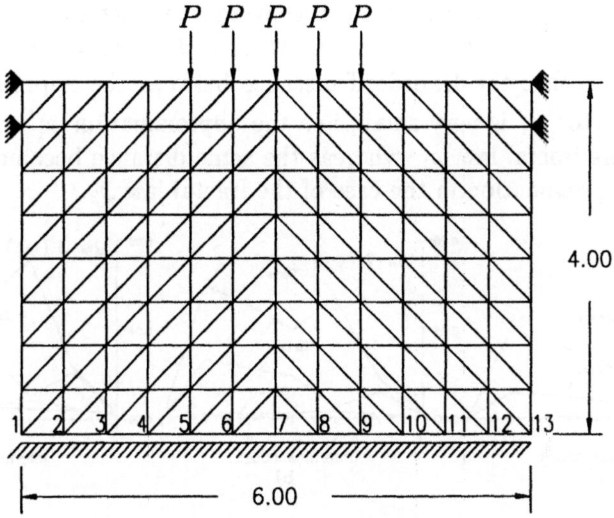

Figure 7.30. The analyzed structure

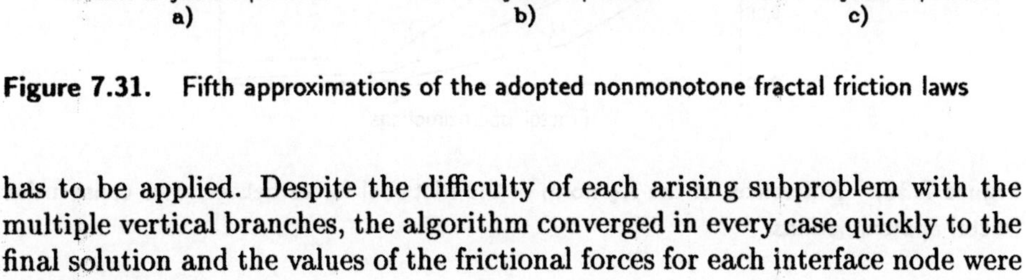

Figure 7.31. Fifth approximations of the adopted nonmonotone fractal friction laws

has to be applied. Despite the difficulty of each arising subproblem with the multiple vertical branches, the algorithm converged in every case quickly to the final solution and the values of the frictional forces for each interface node were determined.

Fig. 7.32 gives the differences of the frictional forces between two consecutive approximations of the fractal law. The results are given for each one of the three different fractal frictions laws. Due to the symmetry with respect to the vertical axis, the results are displayed only for the first 7 interface nodes. In these figures, curve i corresponds to the differences between the i and the $i-1$ fractal approximations. It is observed that these differences decrease as we proceed to higher approximations of the fractal laws. The convergence to the solution of the problem is achieved after the third approximation of the fractal law φ_1

where $d_1 = d_2 = d_3 = 0.2$, whereas this convergence is achieved after the fourth approximation of the fractal law φ_3 where $d_1 = d_2 = d_3 = 0.4$. This happens due to the fact that the Hausdorff distance between the approximation $\varphi_1^{(3)}$ and the attractor φ_1 is very small, i.e. the approximation $\varphi_1^{(3)}$ approximates sufficiently the fractal law φ_1, whereas the same distance becomes small after the fourth approximation in the case of the fractal law φ_3.

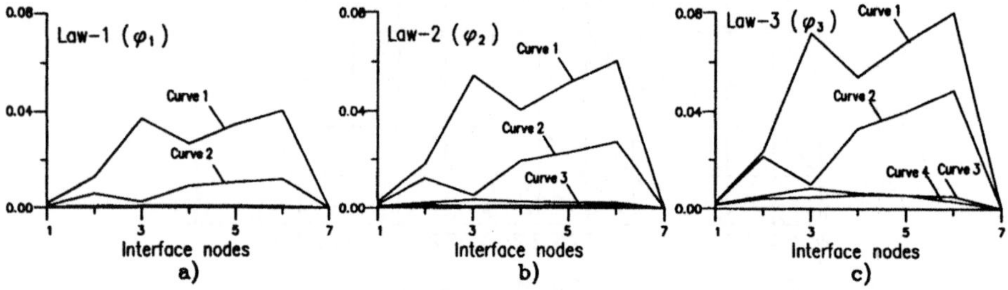

Figure 7.32. Differences of the friction forces between two consecutive approximations of the fractal law

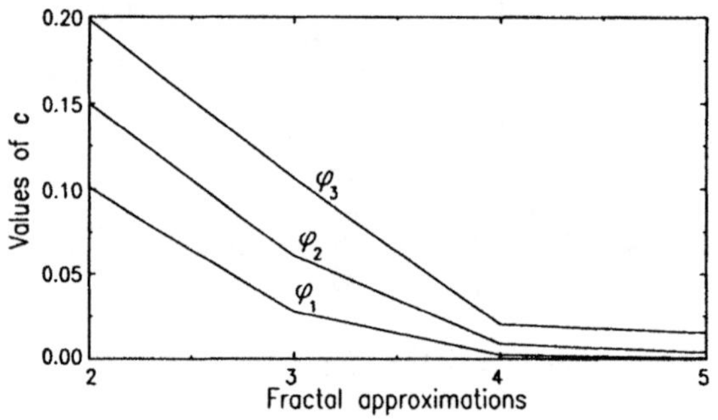

Figure 7.33. Differences of the L_2 norm of the frictional forces between two consecutive fractal approximations

The same results are derived from Fig. 7.33 where the L_2 norm of the differences of the frictional forces between two consecutive approximations of the three fractal laws are given. More specifically, the quantity c is displayed, where

$$c^{(n)} = \sqrt{\sum_l (S_{T_l}^{(n+1)} - S_{T_l}^{(n)})}, \quad l = 1, 13, \quad n = 1, \ldots, 4 \qquad (7.13)$$

and $S_{T_l}^{(n)}$ is the frictional force at node l (see Fig. 7.30) that corresponds to the $\varphi^{(n)}$ approximation of the respective fractal law. Numerical experiments have shown that the convergence to the final solution of the problem depends strongly on the fractal dimension of the fractal friction law. As this dimension becomes larger, a larger number of approximations is needed in order to have convergence.

7.4.2 Multifractured structures with fractal interfaces

It is well-known from the relevant literature (MacLeod and el Magd, 1980) that the crack interfaces in structures of brittle behaviour, e.g. in masonry structures, may have either a zig-zag shape, following the joints of the stones (or the blocks) or pass through the blocks. The first kind of cracks arises in lightly compressed structures with relatively weak mortars. On the contrary the second kind of cracks appears in structures constructed with mortars of better quality (strong mortars) and especially in highly compressed regions of the structure.

New theories concerning the simulation of brittle fracture (Takayasu, 1990) have been developed lately. According to the results of these theories, the fracture pattern in brittle materials may be considered to be a cluster of branches propagating in such a way that new branches in the $j+1$ step are successively created from a former branch at the j step (Xie, 1989, Xie, 1991). In other words, the fracture pattern can be modelled by an iterated function system (I.F.S.) (see the Appendix of Chapter 6). It is important to notice here that a softening interface behaviour has been observed in these problems, which has its nature to the partial cracking and crushing of the asperities of the interface.

The main purpose of this Section is the investigation of the influence of the fractality of a possible crack interface and of the interface stress - strain law on the stress and strain fields and therefore on the possible strength degradation of the structure.

We consider for example the structure of Fig. 7.34a subjected to loading in its plane. Linear elasticity and geometrical linearity are assumed. The modulus of Elasticity is $E = 2.1 \times 10^6 \text{kN/m}^2$ and the Poisson's ratio ν=0.33. The thickness of the structure is taken as equal to 0.10m. Two interfaces f_1 and f_2 of fractal geometry are formed in the structure as it is shown in Fig. 7.34a. f_1, f_2 are defined to be the graphs of the fractal interpolation functions $f_1, f_2 \in C^0$ interpolating the set of data $\{(0.0, 0.0), (3.5, 1.7), (5.0, 2.0)\}$, $\{(0.0, 3.5), (2.0, 2.25), (3.5, 1.7)\}$ respectively and they are:

$$Tf_1(x) = \begin{cases} 0.0286x + 0.80 f_1(1.4286x) & \text{if } x \in [0.0, 3.5] \\ -0.20x + 0.30 f_1(3.333x - 11.667) + 2.40 & \text{if } x \in [3.5, 5.0] \end{cases}$$

$$T' f_2(x) = \begin{cases} 0.0946x + 0.80 f_2(1.7513x) + 0.70 & \text{if } x \in [0.0, 2.0] \\ 0.352x + 0.60 f_2(2.331x - 4.662) - 0.55 & \text{if } x \in [2.0, 3.5]. \end{cases}$$

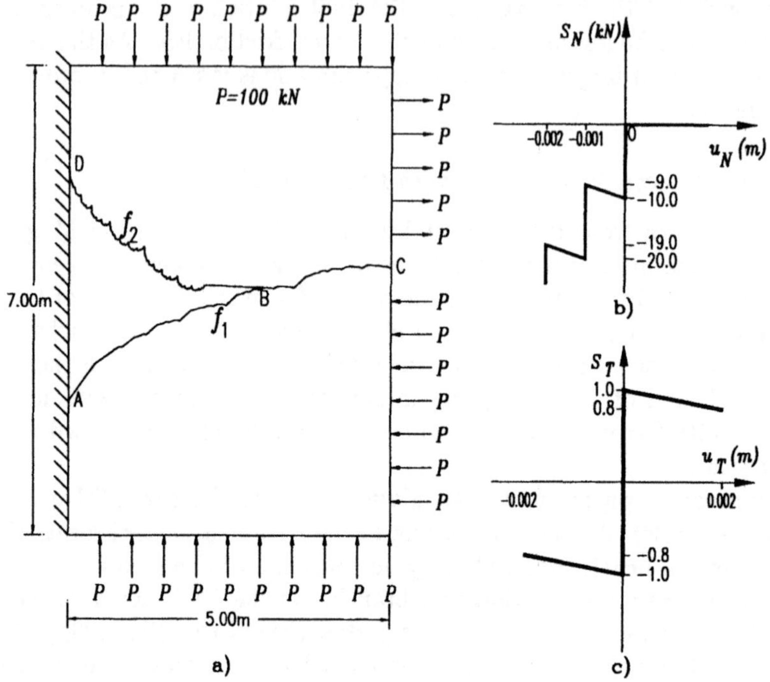

Figure 7.34. A multifractured masonry structure and the adopted interface laws

For the normal to the interface direction, the nonmonotone law of Fig. 7.34b is assumed to hold, whereas for the tangential to the interface direction, the normalized friction law of Fig. 7.34c holds. In order to obtain the frictional forces S_T we have to multiply the ordinate of the diagram by μS_N, where μ is the friction coefficient and in our case $\mu = 0.3$.

Based on the above fractal interpolation functions the different structures corresponding to the consecutive approximations of the fractal interfaces ($f_1^{(j)} = Tf_1^{(j-1)}$, $f_2^{(j)} = T'f_2^{(j-1)}$ $j = 1, 2, \ldots$) are calculated for the same kinematic conditions, the same loading and the interface laws of Fig. 7.34b,c.

In order to analyze each arising problem, Algorithm (6.6) presented in Chapter 6 is applied. The structure has been discretized by triangular constant stress elements. The finite element mesh has been refined near to the interfaces in order to take into account with accuracy the complicated geometry of the interfaces.

The whole algorithm proved to be stable although the complicated interfaces do not offer an ideal framework for a unilateral problem with nonmonotone contact and friction interface conditions. In Fig. 7.35 the major stresses for the

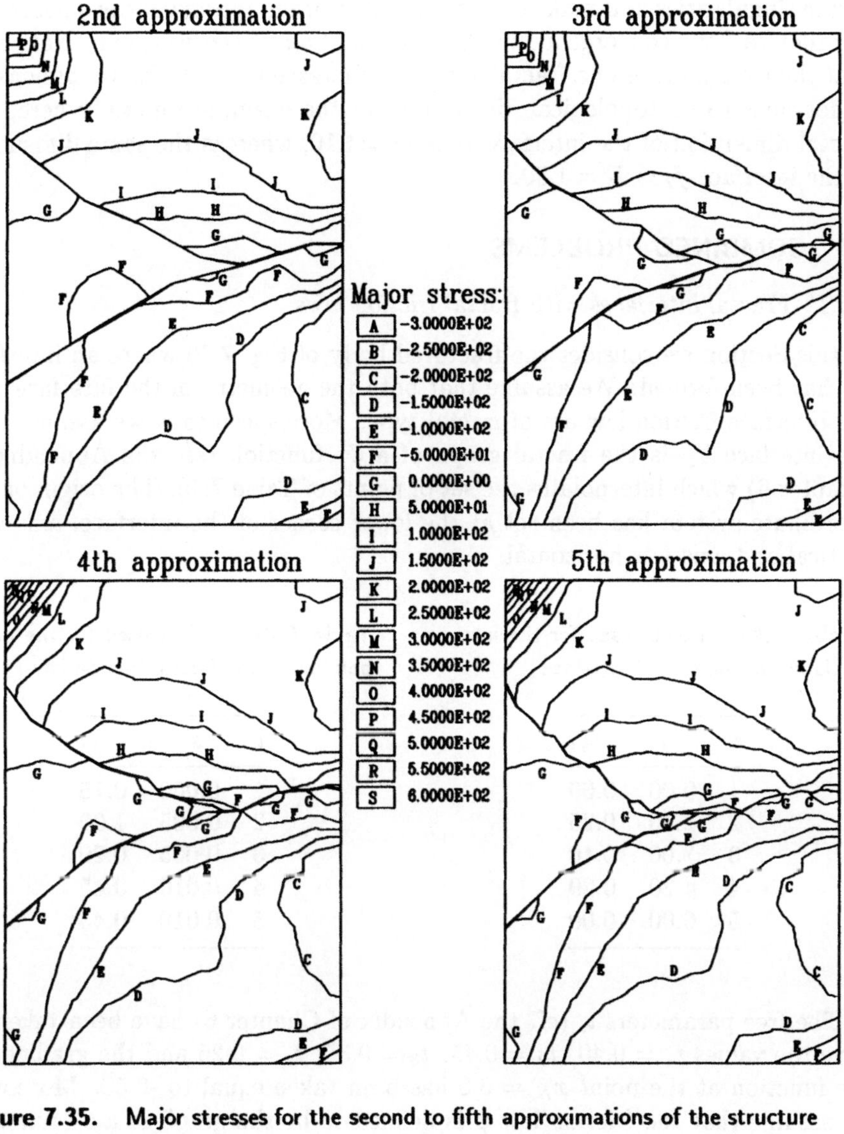

Figure 7.35. Major stresses for the second to fifth approximations of the structure

second to fifth approximations of the fractal interfaces f_1 and f_2 are given. It is observed that near to the interface f_1 the stress field becomes stable from the third approximation. On the contrary, it is obvious that near to the interface f_2 the same stress field becomes stable after the fourth approximation. This is due to the fact that this interface has many external and re-entrant corners, which create singularities and make the use of higher order approximations necessary in order to have convergence to the final solution of the problem. Note also that the irregularities of a fractal boundary increase when its fractal dimension is not near to its topological dimension. In the example we study here, the fractal dimension of the interface f_1 is $D = 1.10$, whereas the same dimension of the interface f_2 is $D = 1.50$.

7.5 COMBINED PROBLEMS

7.5.1 *Fractal interfaces with fractal friction laws*

In this Section we consider the fractured body of Fig. 7.36 where an interface Γ_S has been formed. We assume that both the geometry of the interface and the interface friction law are of fractal type. More specifically we assume that the interface Γ_S is the fractal graph of a C^1 function (see the Appendix of Chapter 6) which interpolates the set of points of Table 7.4a. The origin of the coordinate system has been set at the lower corner of the interface, axis x is vertical and axis y is horizontal.

Table 7.4a. Points used for the interpolation of the fractal interface

i	x_i	y_i
1	0.00	0.00
2	2.50	0.25
3	3.00	0.10
4	4.20	0.20
5	6.00	0.00

Table 7.4b. Points used for the interpolation of the fractal friction law

i	x_i	y_i
1	0.000	0.75
2	0.005	1.00
3	0.005	0.20
4	0.010	0.45
5	0.010	0.45

The free parameters t_i (cf. the Appendix of Chapter 6) have been taken to have the values $t_1 = 0.40$, $t_2 = 0.45$, $t_3 = 0.35$, $t_4 = 0.25$ and the gradient of the function at the point $x_N = 6.0$ has been taken equal to -0.50. Moreover, we assume that the friction law φ is defined to be the graph of a C^0 function which interpolates the points given in Table 7.4b. Taking as free parameters d_i, $i = 1, 2, 3$ the values $d_1 = d_2 = d_3 = 0.2$, the fractal law φ is derived, as it is shown in Fig. 7.37, where the first four approximation of the law are presented. The elasticity modulus is $E = 2.1 \times 10^7 \mathrm{kN/m}^2$, the Poisson's ratio $\nu = 0.3$ and the thickness of the structure is $t = 0.1$m. The friction coefficient μ is 0.2 for all the cases. Triangular constant stress elements have been used for the

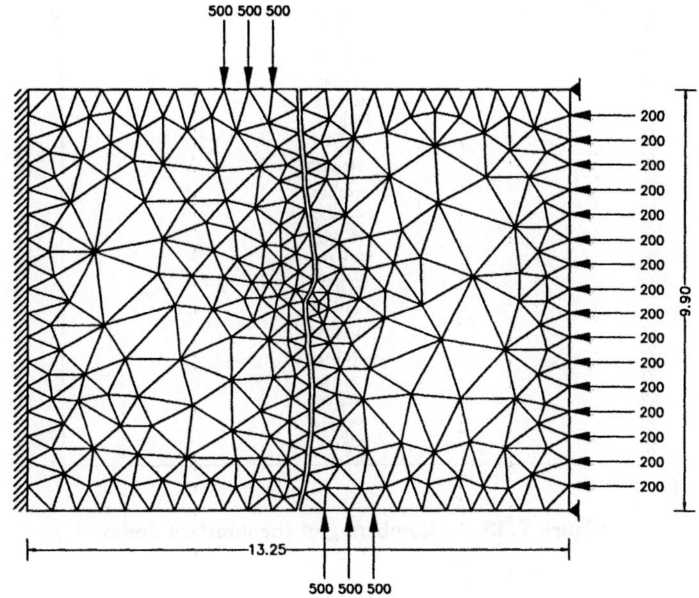

Figure 7.36. Finite element discretization of the analyzed structure

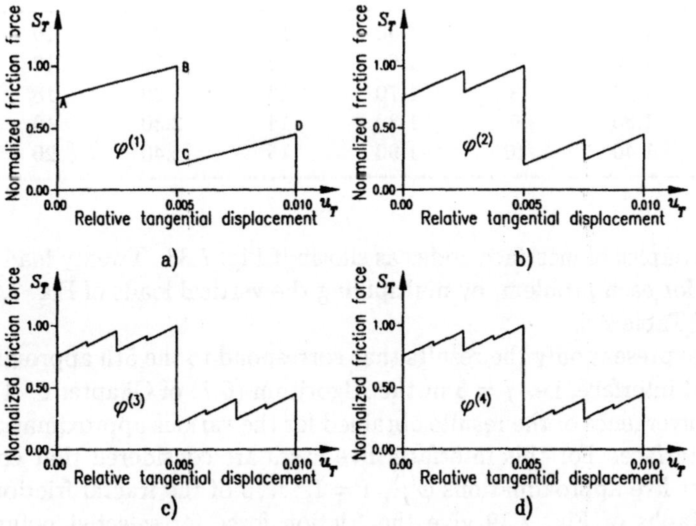

Figure 7.37. The approximations of the fractal friction law

discretization as it is shown in Fig. 7.36. The fractal interface is discretized

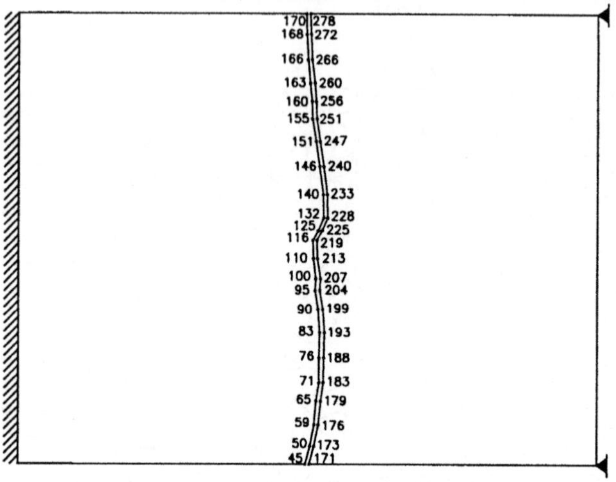

Figure 7.38. Numbering of the interface nodes

Table 7.5. The considered load cases

Load case	*Factor*	*Load case*	*Factor*	*Load case*	*Factor*	*Load case*	*Factor*
1	1.00	6	1.50	11	2.00	16	2.50
2	1.10	7	1.60	12	2.10	17	2.60
3	1.20	8	1.70	13	2.20	18	2.70
4	1.30	9	1.80	14	2.30	19	2.80
5	1.40	10	1.90	15	2.40	20	2.90

using 23 couples of interface nodes as shown if Fig. 7.38. Twenty load cases are analyzed for each problem, by multiplying the vertical loads of Fig. 7.36 by the factors of Table 7.5.

Here we present only the results that correspond to the 5th approximation of the fractal interface, i.e. $j = 5$ in the Algorithm (6.7) of Chapter 6. This is the case of convergence of the results obtained for the various approximations of the fractal interface. For this interface, five cases are considered that correspond to the first five approximations $\varphi^{(i)}$, $i = 1, \ldots, 5$ of the fractal friction law φ.

The graphs of Fig. 7.39 give the friction force for selected points on the interface with respect to the loading. We notice that the diagrams are almost identical for the 3rd, 4th and 5th approximations of the fractal friction law. This result is more clear in Fig. 7.40 which gives the differences between the values of the friction forces obtained for two consecutive approximations of the fractal

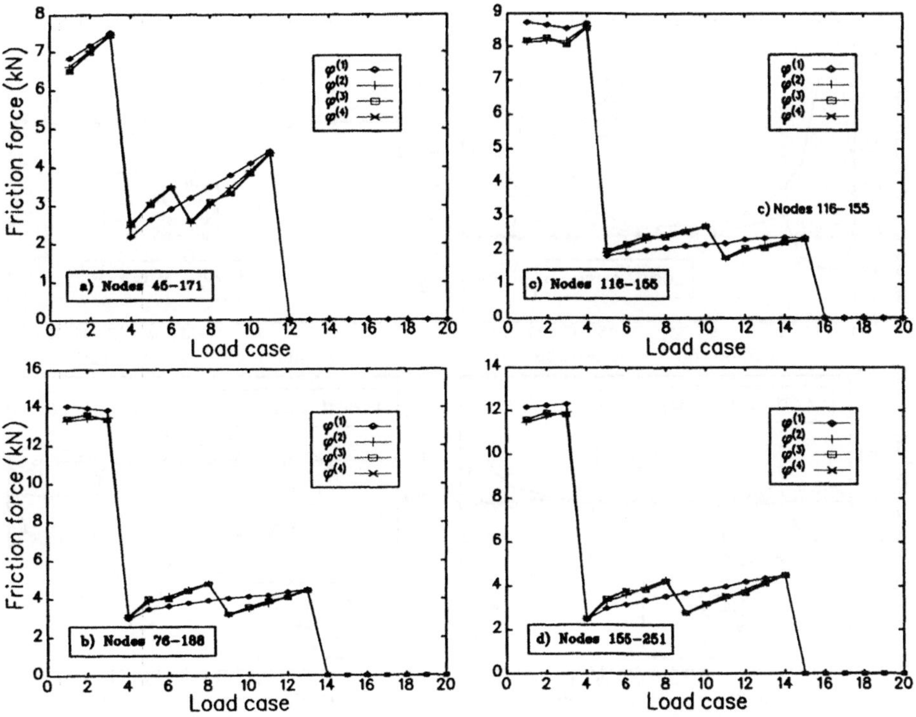

Figure 7.39. Values of the frictional forces for the various load cases

friction law, for selected load cases. We notice that these differences become smaller as we proceed to higher approximations of the fractal friction law. It is also clear that these differences become negligible for the 5th approximation, indicating that there is no reason to consider higher approximations for the fractal friction law.

In order to explain the abrupt jumps in the diagrams of Fig. 7.39, we have to consider also the diagram of Fig. 7.41, which gives the relative tangential displacements for selected interface nodes, with respect to the loading. The results correspond to the first approximation of the fractal friction law. In Fig. 7.39a we notice that we have a sudden decrease of the friction force value for load case 4. For this load case, from the first curve of Fig. 7.41, we get a relative tangential displacement greater than 0.005m, a fact that has as a result the realization of branch CD in the diagram of Fig. 7.37a and thus the sudden reduction of the friction force is verified. The same remarks hold also for the other interface nodes, for example, the "threshold" of the 0.005m is surpassed in the interface position that corresponds to the nodes 116-155 in the 5th load

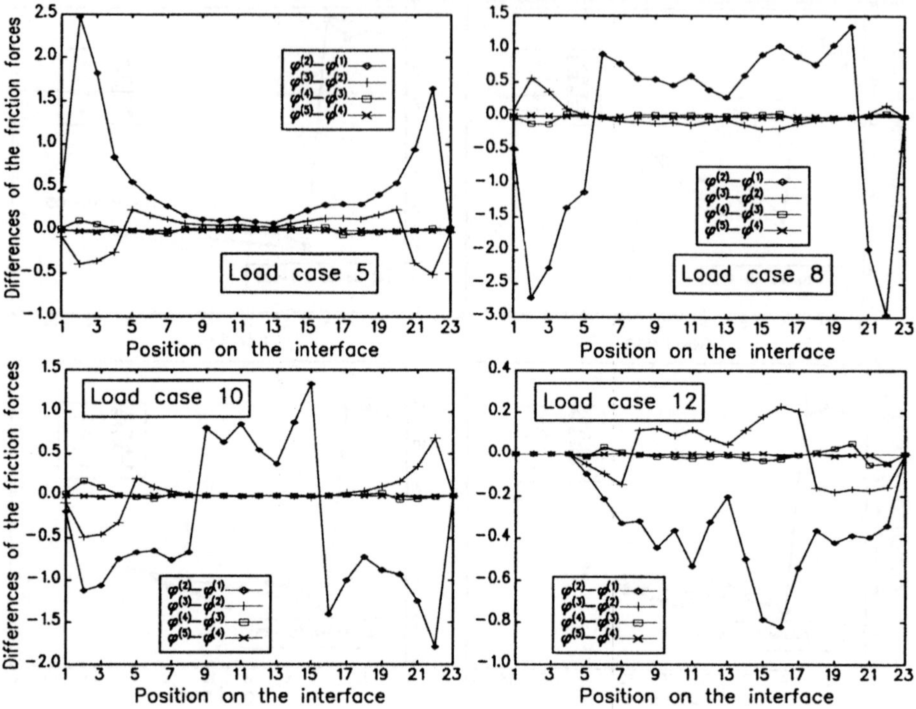

Figure 7.40. Differences of the frictional forces for the various interface positions

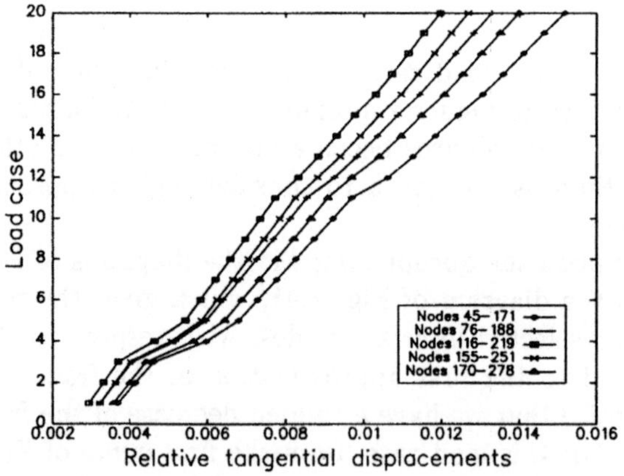

Figure 7.41. Relative tangential displacements for selected interface nodes

case (cf. Figs. 7.39, 7.41) and thus the value of the friction force for this load case is considerably lower.

Notice also, that the assumption of the fractal friction law of Fig. 7.37 leads to a "realistic" interface behaviour, with instant jumps in the load - relative displacements curve, which are caused by the partial cracking and crushing of the asperities of the interface.

7.5.2 Repair of a fractured body with adhesives

Experimental results indicate that the fracture in many structures is rough and irregular and depends strongly on their microstructures and loading situations. In many cases the problem is to repair structures involving such irregular cracks and interfaces. The aim of this Section is to contribute to the analysis of these problems by using the notion of fractal geometry in order to describe with great accuracy the irregular fracture phenomena. The general case of a mechanical behaviour described by nonmonotone possibly multivalued reaction - displacement laws both in the normal and tangential to the interface directions is also considered.

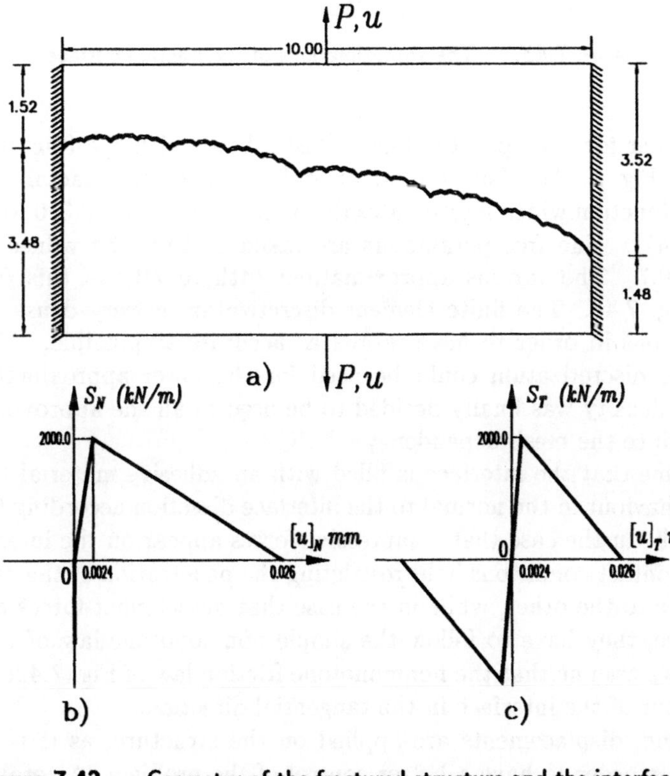

Figure 7.42. Geometry of the fractured structure and the interface laws

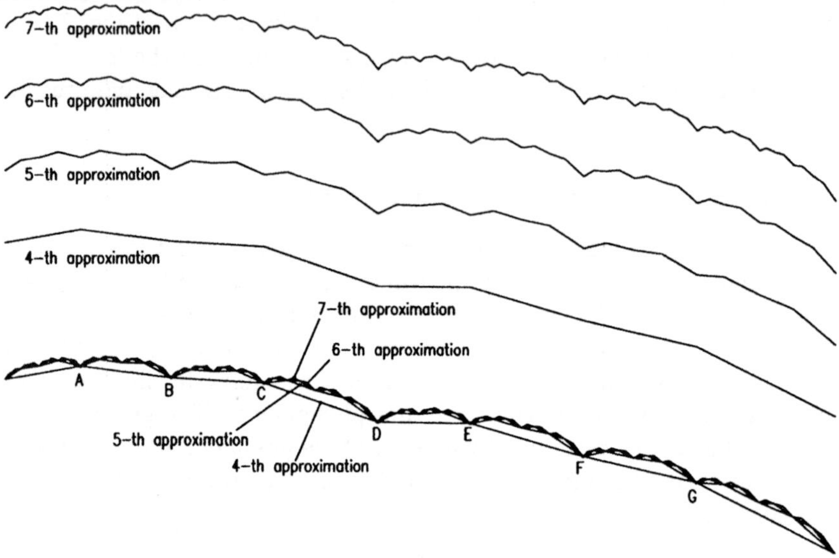

Figure 7.43. The approximations of the fractal interface

We consider for example, the linear elastic body with the fractal interface depicted in Fig. 7.42a. The interface is defined to be the fractal graph of a continuous function which interpolates the points $\{(0.0, 0.0), (0.3, 0.3), (0.8, 0.5)$ and $(1.0, 0.4)\}$. The free parameters are taken to have the values $d_1 = 0.6$ and $d_2 = 0.7$. The various approximations (4th to 7th) of this fractal are given in Fig. 7.43. The finite element discretization is very dense in all the approximations in order to have results as accurate as possible. Although a more coarse discretization could be used for the lower approximations, the same mesh density was finally decided to be used in all the approximations in order to reduce the mesh dependency.

We assume that the interface is filled with an adhesive material that guarantees a behaviour in the normal to the interface direction according to the law of Fig. 7.42b. In the case that compressive forces appear on the interface then unilateral contact conditions hold restricting the penetration of the one part of the interface to the other, while in the case that detachment forces appear on the interface, they have to follow the simple nonmonotone law of Fig. 7.42b. Moreover, we assume that the nonmonotone friction law of Fig. 7.42c describes the behaviour of the interface in the tangential direction.

As loading, displacements are applied on the structure, as it is shown in Fig. 7.42a, in order to have a better control of the problem. A total displacement of 0.0175 m is applied in 35 increments.

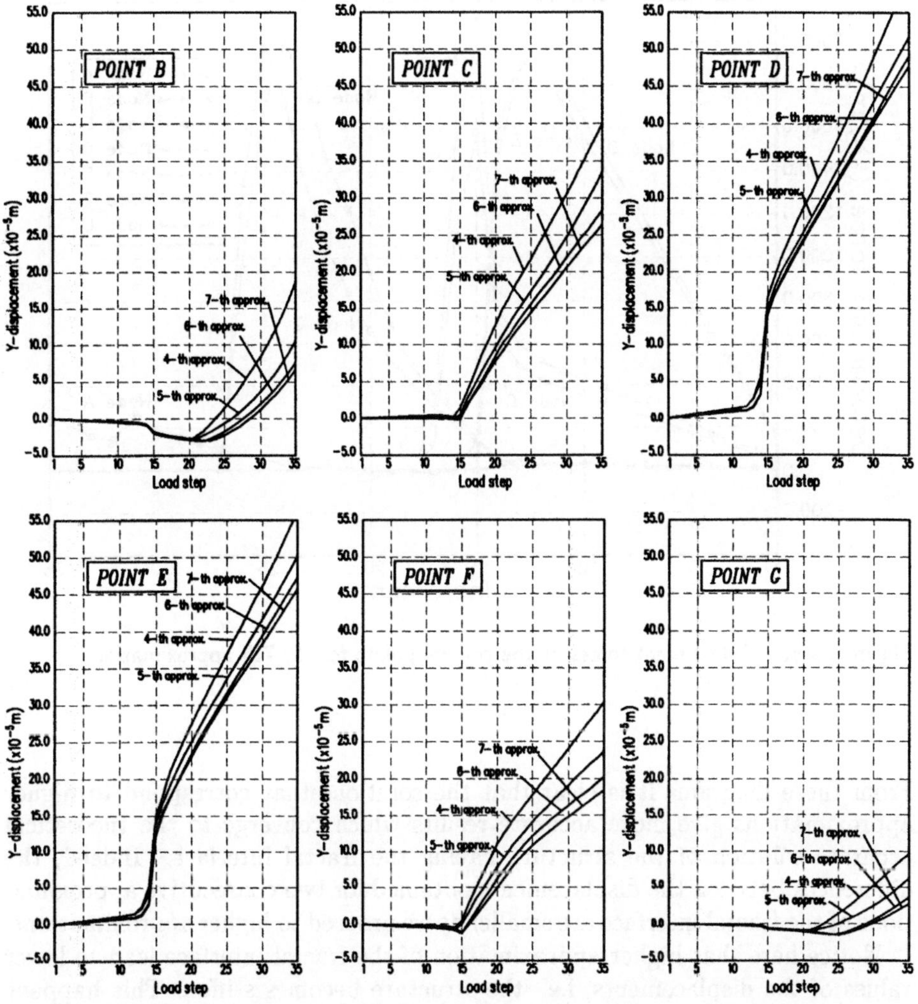

Figure 7.44. Y-displacements of the control points for the various approximations

In order to check the convergence of the results obtained for each approximation of the fractal interface, we define certain control points at which various quantities are checked. These points have to be the same for all the approximations, therefore, we select the common points of the various approximations of the fractal interface, i.e. the seven points (A to G) depicted in Fig. 7.43.

Fig. 7.44 gives the Y-displacements of each point (B to G) for the four structures that correspond to the various approximations of the fractal interface.

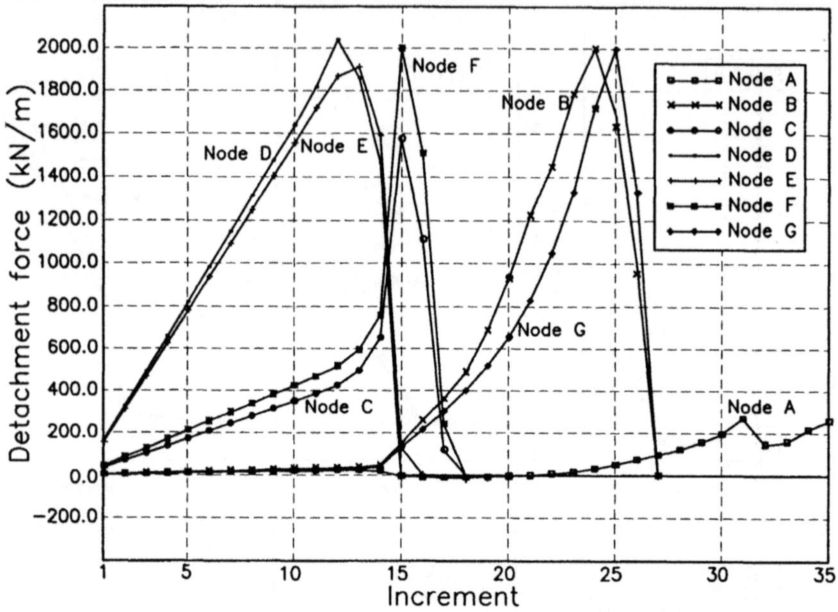

Figure 7.45. Detachment forces at the control points for the 7th approximation

From these diagrams it is clear that the solutions that correspond to higher approximations give more accurate results which converge to the theoretical accurate solution of the structure having the fractal interface. Indeed, the differences between the displacements obtained for two consecutive approximations of the fractal interface are smaller as we proceed to higher approximations.

Notice here that higher approximation of the fractal interface lead to lower values of the displacements, i.e. the structure becomes stiffer. This happens because the total length of the fractal interface increases as we proceed to higher approximations. As a result the total work of the interface increases giving lower values of the displacements.

It is also interesting to see the change of the reaction forces S_D at the selected characteristic points of the interface. Fig. 7.45 gives the normal detachment forces at nodes B to G for the 7th approximation. At each node the behaviour is almost the same, i.e. the reaction force reaches the maximum and then the strength of the adhesive gradually reduces. This behaviour results from the softening law of Fig. 7.42b. This gradual reduction of the capability of the adhesive to transmit stresses from the one part of the interface to the other, has as a result the gradual opening of the interface.

Until the 12th increment the behaviour of the structure is linear. The separation of the two parts of the interface starts at increment 13. This is verified from the curve of Fig. 7.45 that corresponds to node D. Indeed, the reaction force at this node decreases after the 12th increment. After the first separation, the reaction forces at the rest nodes of the interface increase and the separation is extended to node E at the 14th increment(cf. Fig. 7.45). Further increase of the load has as a result the almost complete separation of the two parts of the interface. The results obtained for the rest approximations of the fractal interface are similar and for the sake of brevity are not presented here.

7.6 THIN-WALLED STEEL BEAMS WITH SOFTENING BEHAVIOUR

Thin-walled steel members are commonly used in structural steelwork. The most commonly used thin-walled sections are those manufactured by cold forming. This fact makes necessary the existence of special regulations for their use in structural applications. A problem early recognized is that of the exploitation of the inelastic bending strength reserve that seems to exist, as well as the associated problem of bending moment redistribution. The main objection to the existence of such a strength reserve is based on the fact that thin-walled cross-sections are not compact enough for significant plastic strain to develop. The high $\frac{b}{t}$ ratios (b:width, t:thickness) of the cross-section and the consequent existence of local buckling corroborate in favour of this objection. However, experimental and theoretical research (Reck et al., 1976, Yener and Pekoz, 1980), has shown that some inelastic strength reserve in bending does indeed exist. This reserve mainly depends on the $\frac{b}{t}$ ratio of the compressed part of the cross-section and the way of longitudinal restraint of the member's walls. The response function of a thin-walled member usually ranges between one exhibiting a plastic-like plateau to one steeply softening right after having reached the maximum resistance moment M_u that is the worst case to be expected.

The value of M_u can be obtained either experimentally (Rasmussen and Hancock, 1993), or analytically by estimating the ultimate plastic moment of some "effective" cross-section. In both cases, the influence of (elastic) local buckling is taken into account. The important point is that the existence of such strength reserve has been accepted by several structural codes (AISI, 1986, ECCS, 1987, BS5950, 1987, Eurocode 3 – Part 1.3, 1993). Some researchers regard the redistribution of moments due to this inelastic strength reserve as possible (Unger, 1973, Wang and Yeh, 1975) this being also the point of view of the AISI Regulation. The BS 5950, part 5 also admits the use of the inelastic strength reserve under some additional requirements i.e. small $\frac{b}{t}$ ratii plus stiffened compression elements. For instance, these requirement states that for a steel with a yield strength $f_y = 280$ N/mm^2 there must be $\frac{b}{t} < 25$.

Eurocode 3, part 1.3 allows the consideration of the redistribution only when the inelastic strength reserve is based on $M-\varphi$ graphs that are experimentally derived baring strictly the use of the analytical method for the estimation of the strength reserve that the same regulations provide.

The main reservations for the application of redistribution is due to the unfamiliar form of the $M-\varphi$ diagram and the strong dependence of the $M-\varphi$ on the shearing forces that develop near the support points. The form of this nonlinear $M-\varphi$ diagram is due to the combination of material nonlinearity with severe local buckling, the softening branch being its main trait (Grundy, 1990).

The possibility therefore exists for a thin-walled member to operate in the softening regime, the classical plasticity analysis being unable to predict its behaviour. In view of this allowance, that the Eurocode only recently started to provide, the development of structural analysis methods dealing with softening moment resistance evolution diagrams is of importance. The Eurocode proposes a simple trial and error method to deal with two-equal-span continuous beams but the application of this method in arbitrary multi-span continuous beams is not possible.

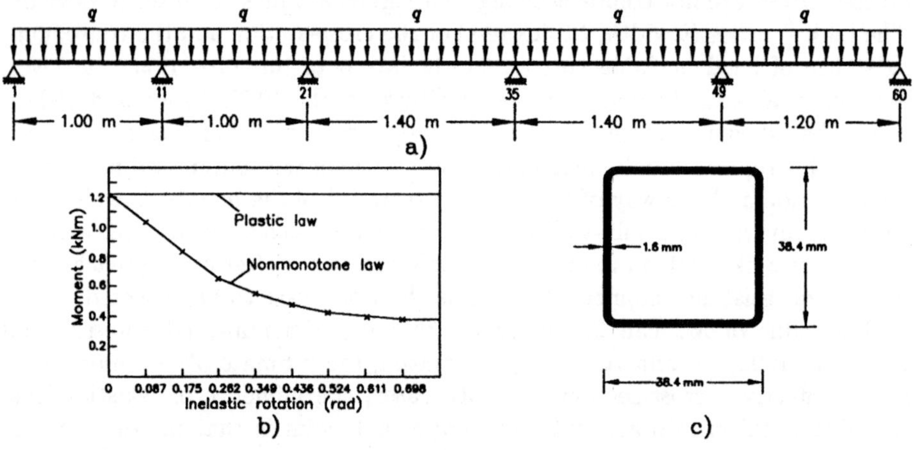

Figure 7.46. The analyzed structure and the considered moment - rotation curves

The history of research in this direction dates from 1973 when Unger (Unger, 1973) presented an approximation method based on elementary tools of beam theory for continuous beams. The method does not give reliable results for all types of beam geometry, actually it holds only for the cases numerically treated. Here, the method developed in Chapter 6 and more specifically, Algorithm (6.9) is applied in order to solve the above problem. As an example we consider the continuous beam depicted in Fig. 7.46a. The cross-section and the assumed moment - inelastic rotation relation (taken after the experimental results of Kecman, 1983) are shown in Fig. 7.46b,c. The finite element discretization density is 0.1m.

In order to easily produce some referencing data for the structure involving the softening effects, an exactly similar elastic-plastic structure with the same ultimate moment was also calculated. It is therefore provided a "reference solution" against which the new results can be judged. The vertical displacements of the beam are shown in Fig. 7.47a (nonmonotone elastic-softening) and Fig. 7.47b (elastic-plastic). The respective distributions of neutral axis rotations are presented in Fig. 7.48a,b. The jump in rotation that is due to the formation of inelastic hinges at support point 35 and 49, can be easily seen.

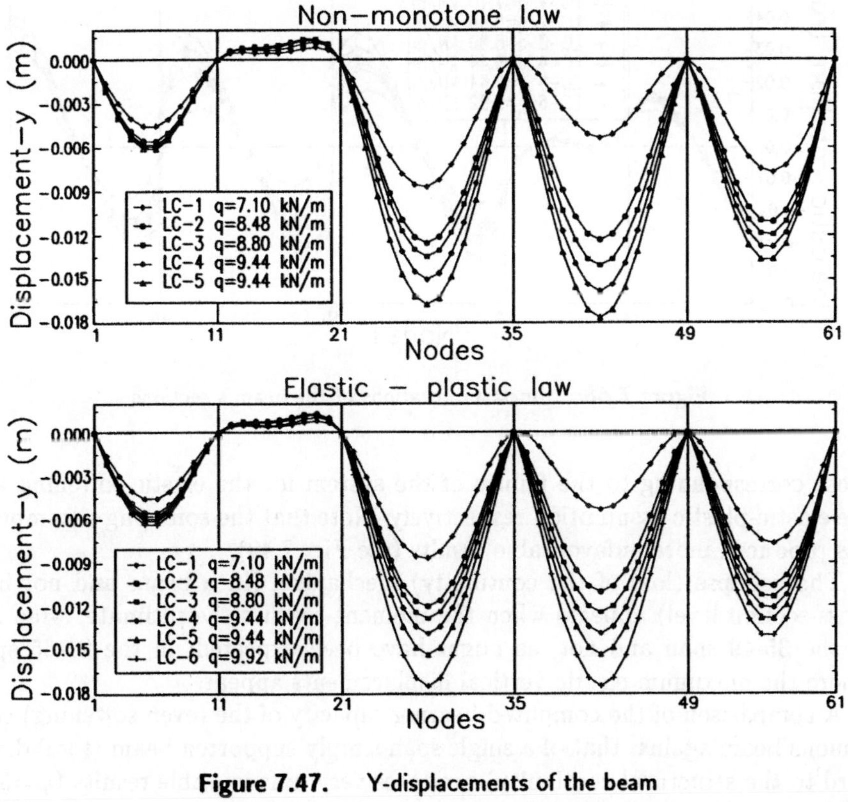

Figure 7.47. Y-displacements of the beam

Load case 1 corresponds to the maximum load before the appearance of nonlinearity ($q = 7.10$kN/m). Note the continuity of the distribution of rotations for this load case. Load cases 5 (q =9.44 kN/m) and 6 (q =9.92 kN/m) are

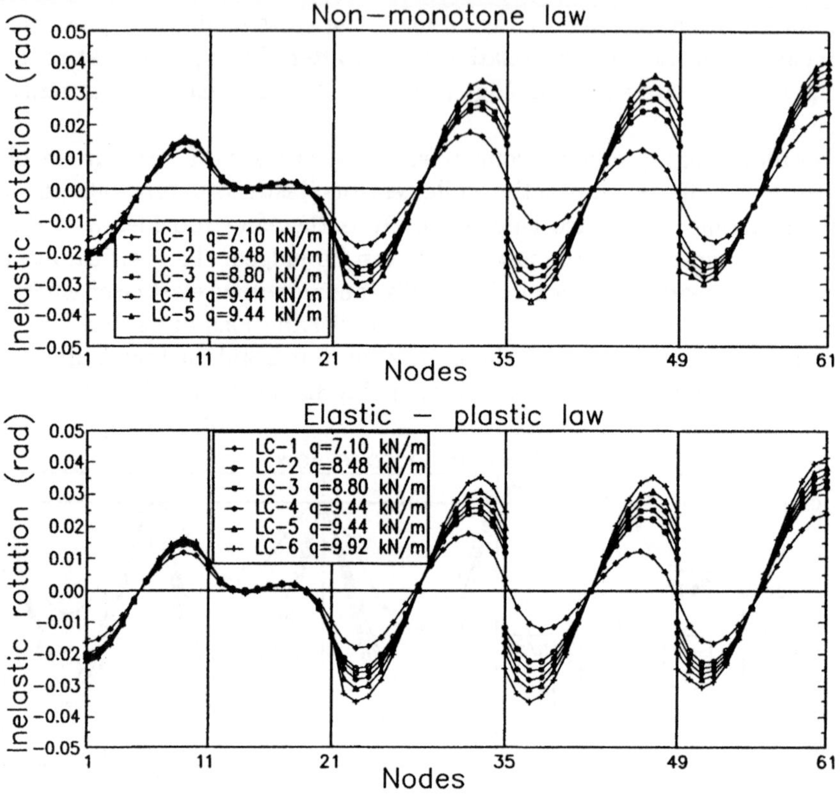

Figure 7.48. Inelastic rotations of the beam's sections

those corresponding to the failure of the system for the elastic-softening and the elastic-plastic assumption respectively. Note that the softening assumption has produced more unfavourable results (see Fig. 7.49).

The collapse (loss of the continuity) mechanism (in a beam and not in a cross-section level) appears when the moment reaches the ultimate value M_u at the 35-49 span and not, as might have been expected, in the 21-35 span where the maximum elastic vertical displacements appear.

A comparison of the computed bearing capacity of the (even softening) continuous beam against that of a single span simply supported beam (total disregard to the structural continuity) produces very unfavourable results (q =5.02 kN/m against q =9.44 kN/m of available ultimate bearing capacity). This is a gross margin of bearing capacity that must not go unexploited due to hesitation in the structural modelling assumptions. However, the most important point is that the ultimate load predicted by the softening schemes, is lower than

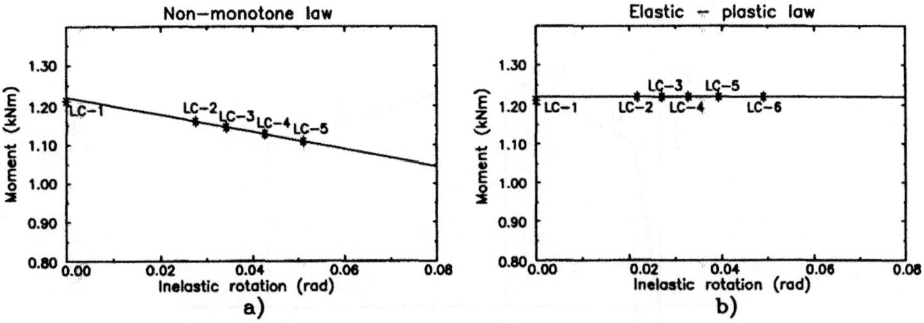

Figure 7.49. Moment - rotation diagrams for the node - 35

the "classical" estimate that elastoplasticity produces. Given, in addition, the simplicity of the proposed methods of computation, the authors feel that the novelty of the present approach must not withhold its technical exploitation prospects.

The algorithm converged in a quite fast way, i.e. four to five iterations were enough for the moments in the softening regions to stabilize. The numerical convergence was slower as the load increased, causing the points where softening appears to multiply in number. Softening appears in a manner similar to that of the plastic hinges i.e. at the maximum bending moment points.

7.7 STRUCTURES WITH SEMIRIGID CONNECTIONS

In the classical elastic-plastic approach of limit analysis methods, a bilinear form of the moment-rotation relationship is usually employed. Such a simplified moment-rotation relationship implies that when any section reaches its moment carrying capacity, it will carry the same maximum moment under increasing rotation. However, steel beams exhibit softening, right after having reached the maximum resistance moment (Fig. 7.50). This form of the $M - \varphi$ curve is due to a combination of material nonlinearity with severe local buckling of the plate elements which constitute the member cross-section and, if torsional restraints are not provided, with the occurrence of lateral-torsional buckling.

The existence of a softening branch in the $M - \varphi$ curve of steel beams has been confirmed experimentally (Mitani and Mahino, 1980) and a simplified form of $M - \varphi$ curve was proposed in Mazzolani and Piluso, 1996, Kato and Akiyama, 1981, Akiyama, 1985, allowing a distribution between strain-hardening and strain-softening parts.

Therefore, in this case the structure may operate in the softening regime, while the classical plastic analysis is unable to predict its behaviour. On the

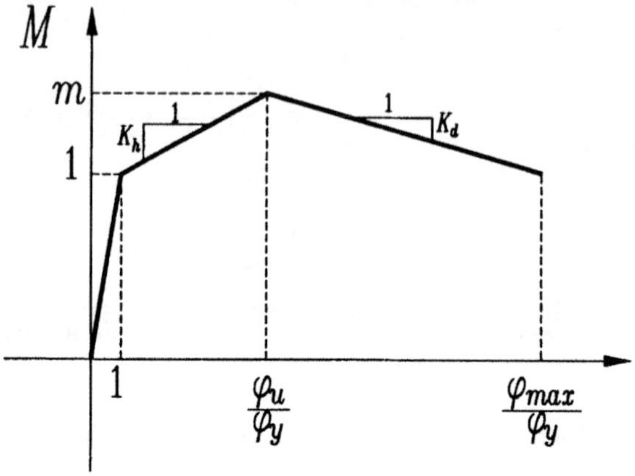

Figure 7.50. General moment-rotation relationship

other hand, the need of an accurate prediction of the collapse load does exist, particularly in the design of earthquake resistant structures, and the attainment of this prediction is related to the use of a realistic $M - \varphi$ curve, taking into account the existing softening branch.

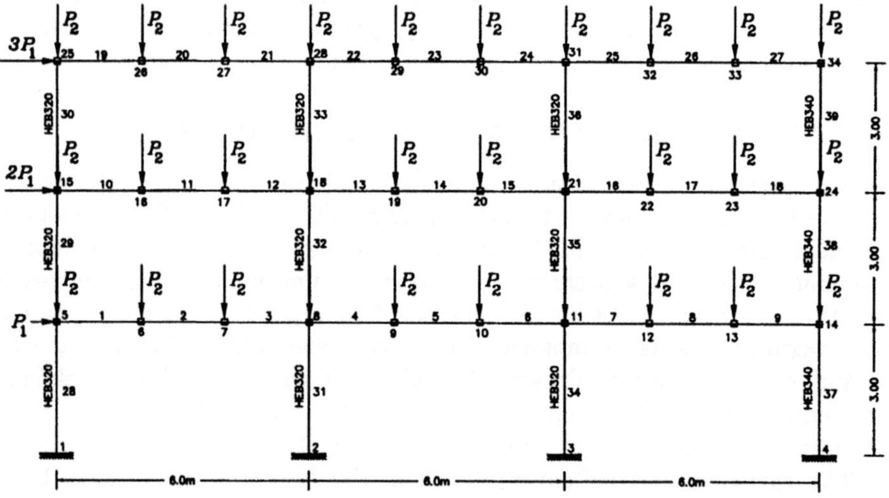

Figure 7.51. Steel frame with members exhibiting softening behaviour

In this Section, Algorithm (6.9) is applied in order to solve the softening problem. As an example we will consider the multistorey plane frame of Fig. 7.51. The structure consists of HEB320 and HEB340 columns and IPE220 beams. For the rotational capacity of the beams of the structure the general moment-rotation curve of Fig. 7.50 is assumed to hold while for the columns the classical elastoplastic diagram is used. In the diagram of Fig. 7.50, the parameters m, φ_y, φ_u, K_h, K_d are calculated according to Kato and Akiyama, 1981, Akiyama, 1985.

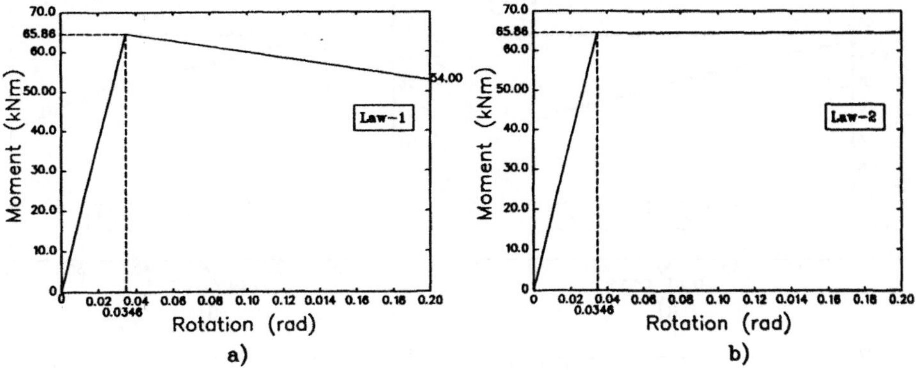

Figure 7.52. The adopted moment-rotation laws

In the particular case treated here, due to the lack of torsional restraint of the beams, flexural-torsional buckling has to be taken into account. According to Kato and Akiyama, 1981, Akiyama, 1985 this effect, together with the local buckling effect has as a result the reduction of the parameter m. In the case that the calculated parameter m is less than 1, the strain-hardening branch of the diagram of Fig. 7.52a is eliminated. Also, the coupling of the buckling modes leads to a higher slope of the softening branch (see Kato and Akiyama, 1981, Akiyama, 1985 for the calculation of the various parameters). The specific geometrical and elastic data used here lead to the moment-rotation diagram of Fig. 7.52a (law f_1). For comparison reasons, the simple law of Fig. 7.52b is also considered (law f_2) which is similar to the classical plasticity law. The steel grade is Fe360 with a yield stress of 235 N/mm^2.

Table 7.6. The considered load cases

Load case	*Factor*	*Load case*	*Factor*	*Load case*	*Factor*	*Load case*	*Factor*
5	1.15	20	1.60	35	2.05	50	2.50
10	1.30	25	1.75	40	2.20	55	2.65
15	1.45	30	1.90	45	2.35	60	2.80

The loading of the structure is shown in Fig. 7.51, where P_1 =30.0 kN and P_2 =20.0 kN. Sixty load cases are considered, by multiplying the horizontal loads with the factors of Table 7.6, while the vertical loads remain constant.

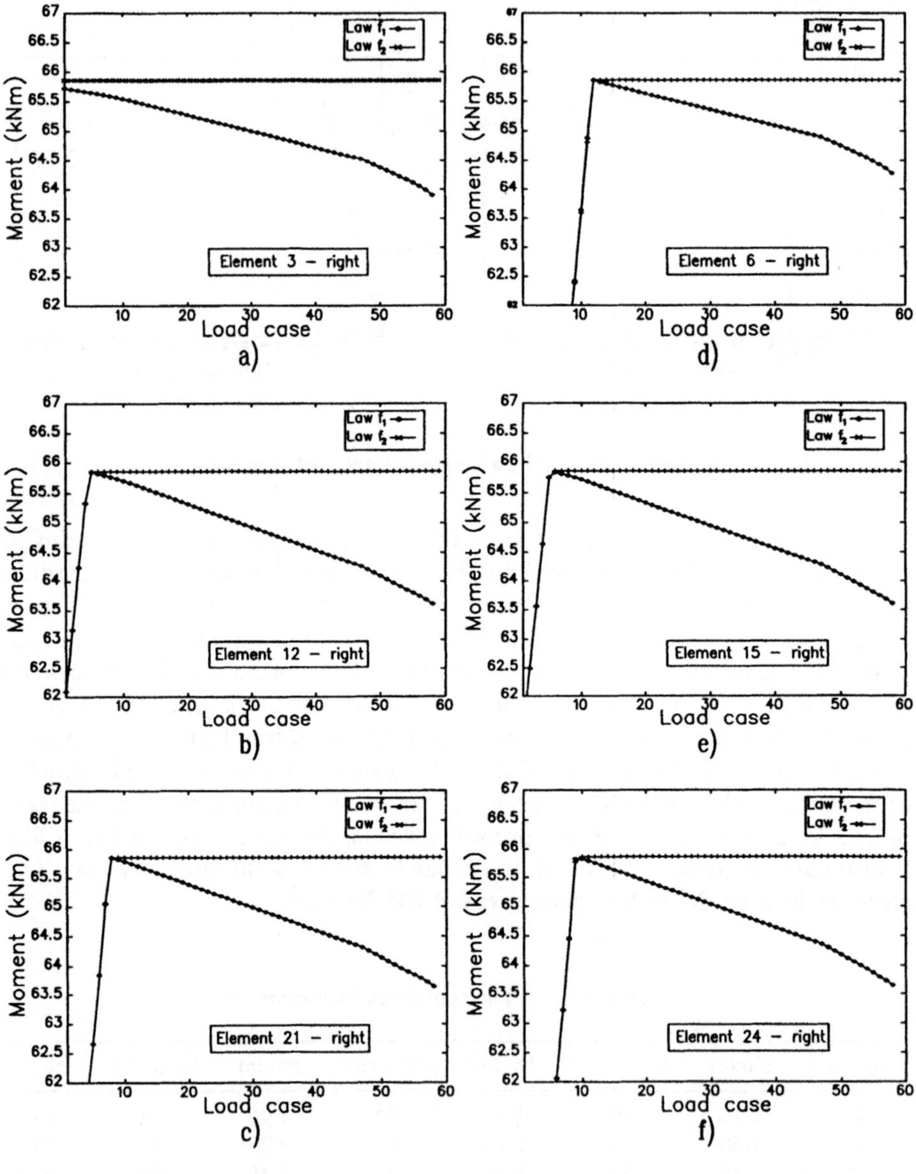

Figure 7.53. Load–end–moment curves for various elements of the structure

The structure is analyzed using the Algorithm (6.9). Figs 7.53a-f, give the load case–end-moment curves corresponding to the right end of elements 3,12,21,6,15 and 24 respectively, for the two laws f_1 and f_2.

As the horizontal loading increases, more and more beams enter into the softening region of the moment-rotation curve. This phenomenon has as a result the gradual reduction of the beams' moments, the redistribution of the stresses of the whole frame, and the increase of the moments in the columns. Further increase of the horizontal loading has as a result the plastification of the lower sections of the columns. Finally, as the plastification procedure develops under increasing loading, the whole frame becomes kinematically unstable and collapses (load case 58). Analogous is the situation if we assume that the moment-rotation law of Fig. 7.52b holds. In this case, the elastoplastic diagram has as a result the constant value of the moment after reaching the plastification moment M_p. Thus, the redistribution of the stresses is slower due to the increased rotational capacity of this law. The calculations verify that the collapse load for law f_2 is 3% higher than the collapse load for law f_1.

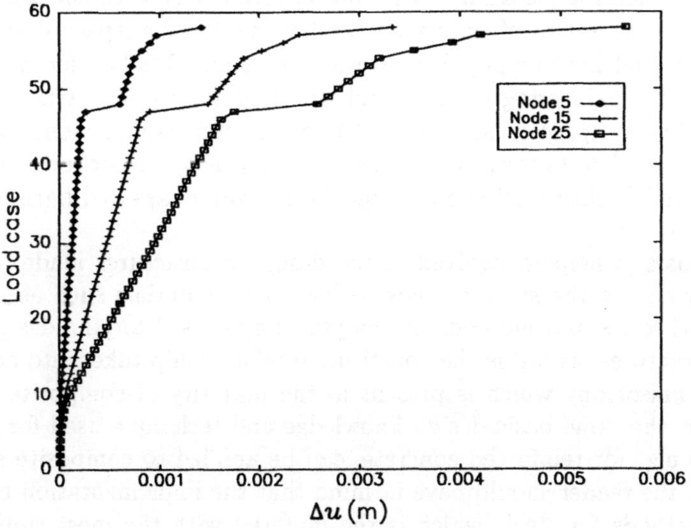

Figure 7.54. **Differences of the displacements resulting from the assumption of laws f_1 and f_2, with respect to the load**

The previous remarks are also verified by the diagram of Fig. 7.54 that gives the differences Δu of the displacements that correspond to the assumption of law f_1 from the ones that correspond to law f_2, with respect to the various load cases. The results are depicted for the characteristic nodes 5, 15, 25. It is noticed that the differences of the displacements increase together with the loading. The jumps of the diagrams that correspond to load cases 47–48 are explained as a result of the fact that the lower sections of the HEB320 columns are plastified at load case 47 with the assumption of law f_1, whereas they are plastified at load case 48 with the assumption of law f_2.

A significant, for the design rules, conclusion that arises from this example, is that the consideration of the softening branch of the moment-rotation curve leads to a lower collapse load compared to the one obtained with the assumption of the classical elastoplastic law. The fact that in the considered example, the first one is just 3% lower than the second one, leads to the conclusion that a lot of numerical experimentation is still needed in order to derive certain results concerning the effect of the softening branch to the ultimate resistance of steel structures under seismic loading.

7.8 INVESTIGATION OF THE BEHAVIOUR OF HYBRID LAMINATES MADE OF UNIDIRECTIONAL COMPOSITES

Composite materials is a rapidly maturing technology with emerging applications in a wide range of industries far beyond the aerospace domain where composites first became popular. Nowadays, composites become popular even in the Civil Engineering domain and are used for the construction of light bridges, domes, space trusses, etc. Moreover, composite elements are used in structures which are susceptible to electro-chemical actions or corrosion, such as underground facilities, off-shore oil platforms, waterways and harbours (Kim, 1995).

The basic principles involved in the design of structures made of composite materials are the same as those of isotropic materials such as steel. The classical theories and methods of analysis can be used for the design of composite structures, as far as the constitutive relationship takes into account the material anisotropy which is present in the majority of composite materials. Moreover, the same basic design knowledge and technique used for other materials as e.g. for reinforced concrete, can be applied to composite structures. However, the reader should have in mind that the implementation of accurate design methods for steel (which is the material with the most simple constitutive relationship used today in Civil Engineering structures) has required a century of research and experience. In this framework, it is very important to study these new materials not only at the materials level but it is also very important to know their behaviour in the structural level (cf. in this respect the combined effects in the study of beam-to-column connections in steel structures which lead to strong nonlinearities) and to know the phenomenological (macroscopic) response under certain types of loading.

A very important category of composite elements is that of unidirectional composites. In this category the fibres are aligned in only one direction thus achieving the maximum fibre alignment and the maximum fibre content. As in principle the strength of a composite structural element increases in proportion to increasing fibre content, this type of composites provides high strength to the direction of the fibres but very low transverse strength. Moreover, it is very common to combine unidirectional composite materials in a certain layered arrangement, in order to construct structural elements with improved overall behaviour, taking advantage of the properties of each of the constituents. These elements are called hybrid laminates.

In this Section we study the overall tensile and bending behaviour of such hybrid laminated elements which have as constituents unidirectional composites. Each unidirectional composite is assumed to have a nonmonotone, possibly multivalued constitutive law. Thus it is possible to study the behaviour of such elements using the methods and algorithms developed in Chapter 6. This treatment sheds light into the complex behaviour of the composite structural elements and, on the other hand, the arising overall laws can be used as phenomenological laws for structural analysis of large scale structures incorporating the studied composite elements.

More specifically, we consider here hybrid laminates where each layer is made of a material with a different constitutive (stress–strain) relation. Monotone and nonmonotone laws which may include complete vertical branches and which cover the cracking or crushing effects of the material in each layer are assumed. Due to the above mentioned local damage phenomena a highly nonlinear overall mechanical behaviour arises. In order to use the overall mechanical law for the phenomenological modelling and for the computation of large composite structures, a detailed investigation is first done on the structural element scale. This effort is undertaken in this Section for a few models of hybrid laminate elements.

7.8.1 Laminated composites under axial loading

Here we consider one-dimensional, hybrid layered composites where each layer is made of a unidirectional composite or steel. The following laminates are examined:

a) Hybrid laminate in tension made of a layer of Aramid Fiber Composite (AFC) backed up with steel plates.

b) Hybrid laminate in tension made of a layer of AFC backed up with Carbon Fiber Composite (CFC) plates.

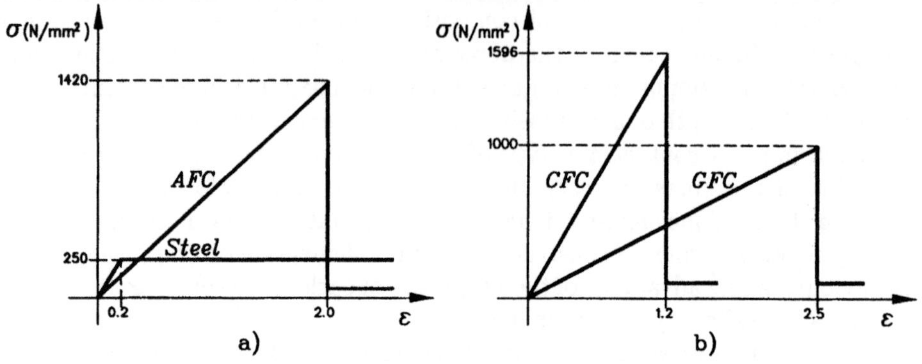

Figure 7.55. Stress-strain laws for the AFC, steel, CFC and GFC

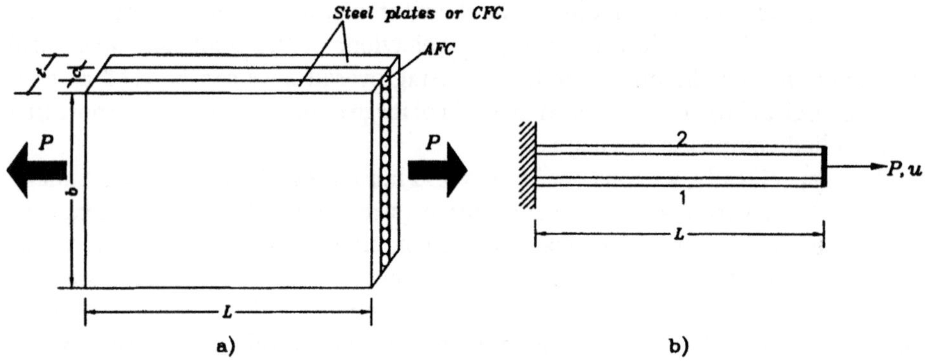

Figure 7.56. Laminated composite under axial loading and the corresponding analysis model

The elasticity modulus for the AFC is equal to $E_{A,L} = 71000\ \mathrm{N/mm^2}$, and the breaking point lies at $\varepsilon_{B,AFC} = 2\%$. For the steel plates we have $E_{st,L} = 125000\ \mathrm{N/mm^2}$ and yield limit $\sigma_{st,y} = 250\ \mathrm{N/mm^2}$. For the CFC we consider $E_{C,L} = 133000\ \mathrm{N/mm^2}$ and breaking point at $\varepsilon_{B,CFC} = 1.2\%$. The stress-strain laws for each material are depicted in Fig. 7.55.

The arrangement and dimensions of the specimen are depicted in Fig. 7.56a. In order to analyze the above problem, the model of Fig. 7.56b was used. The model consists of two elements, one representing the AFC core and the other representing the external CFC or steel plates. For each problem, several different cases are considered by changing the thickness of each layer. The characteristic parameter is the ratio c/t where c is the thickness of the AFC

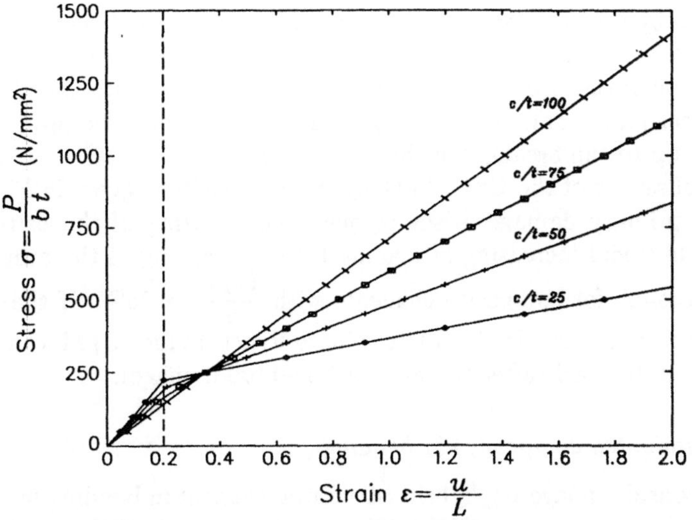

Figure 7.57. Stress-strain curves for the steel-AFC composite

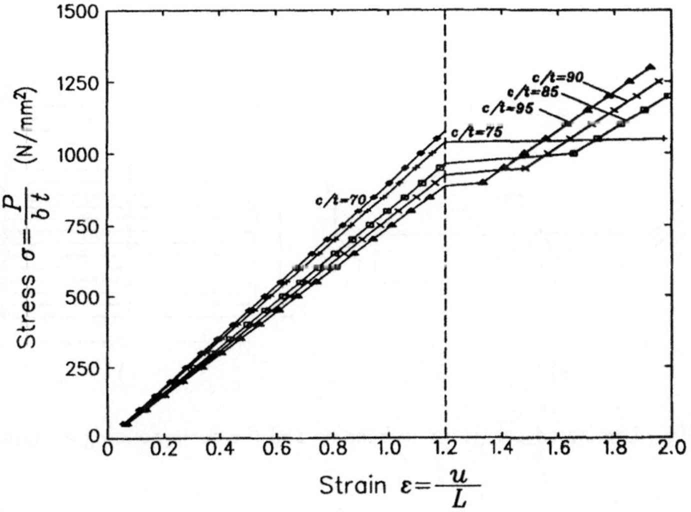

Figure 7.58. Stress-strain curves for the CFC-AFC composite

core and t is the total thickness of the three layers. For each case, the cross-sectional area of each element is accordingly assigned.

The problem was analyzed by applying an appropriate version of Algorithm (6.9) of Chapter 6 because here the longitudinal stress and the respective strain are related by the nonmonotone diagram. The results are depicted in Figs 7.57 and 7.58. Fig. 7.57 gives the one-dimensional overall $\sigma - \varepsilon$ response of the steel–AFC–combination. We notice that the damage initiates with the plastification of the steel plates and then, by constant loading, the structural element is deformed up to the breaking of the AFC–core.

The behaviour of the CFC–AFC–hybrid composite is given in Fig. 7.58. In this case primary damage arises through the breaking of the brittle carbon fibres. Additional increasing of loading till the breaking of the aramide fibres is only possible for composite elements with $\frac{t_{AFC}}{t} > 75\%$. If the structural element is used in a statically indeterminate structure then a reduced remaining strength can be used (after the first carbon–fibre damage).

7.8.2 Laminated composites in bending

Here we consider a layered, hybrid, structural element in bending made of CFC and glass fibre composite (GFC). The core is made of GFC and the external layers of CFC. The material constants for the GFC are $E_{G,L} = 40000$ N/mm^2 and $\varepsilon_{B,C}$ =2.5% (see Fig. 7.55b).

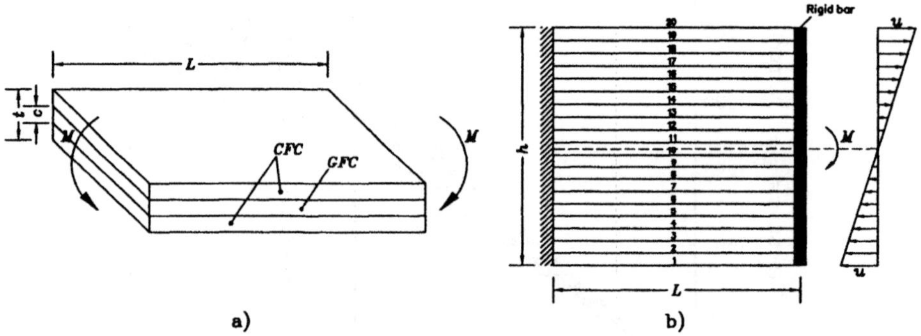

Figure 7.59. Laminated composite in bending and the corresponding analysis model

The arrangement and dimensions of the specimen are given in Fig. 7.59a. In order to analyze the problem, the model of Fig. 7.59b was used. The model consists of 20 parallel beam elements which are connected with a rigid bar. This model was selected in order to enforce the Bernoulli beam conditions (deformed cross-section remains plane). Each element has constant cross-sectional area equal to $ht/20$. In order to analyze different cases of c/t (core thickness to total thickness) we assign different material properties to each element. For example, in order to consider the case CFC-GFC-CFG, $c/t = 0.70$, we assign the material

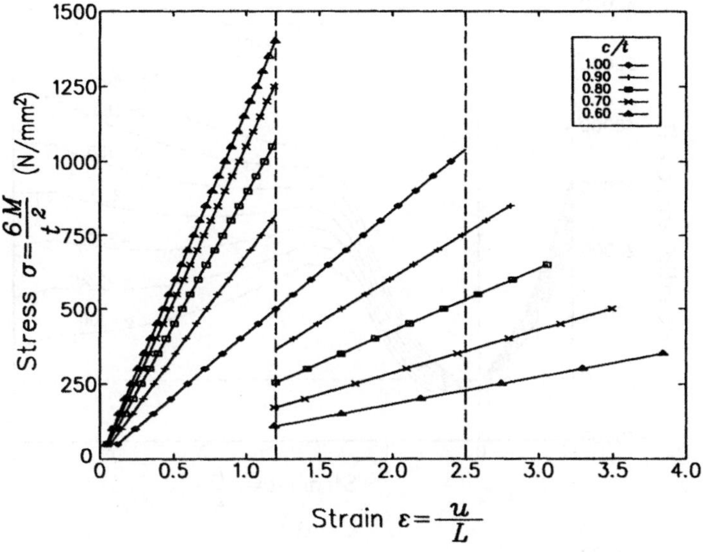

Figure 7.60. Stress-strain curves for the GFC-CFC composite

properties of the CFC to the elements 1,2,3,18,19,20 and the material properties of the GFC to the rest elements.

Each one of the above problems was analyzed using Algorithm (6.9) of Chapter 6. Due to the particular characteristics of the problem (multiple minima for the same load level), different starting points had to be defined in order to obtain the points beyond the breaking of the CFC ($\varepsilon = 1.2\%$) for all the cases of c/t except of the one that corresponds to $c/t = 1.00$.

The equivalent stress ($6M/t^2$) – extreme strain diagrams for the CFC-GFC-CFC combination are given in Fig. 7.60. The brittle carbon fibre composite, placed at the external layers, is the first constituent to break and to delaminate. A further increasing of the loading is only possible for relatively large core thickness (proportion of the glass laminate; i.e., $h/t > 0.9$).

It is very interesting for this problem to examine the potential energy functions. For this reason we considered the cases $c/t = 0.60$ and $c/t = 0.80$. Fig. 7.61a gives the potential energy curves for the structure with $c/t = 0.60$ and for several values of the external loading. For low values of the loading (e.g. $\sigma = 100$), the solution is unique as it is also verified from Fig. 7.60. For higher values of the loading ($\sigma = 150 - 350$), the substationarity problem has two solutions, one in the interval $0 \leq \varepsilon \leq 1.2$ and one for $\varepsilon > 1.2$. For even higher values of the loading, the problem has only one solution that corresponds to the structure with both materials assumed as linear elastic. The same results

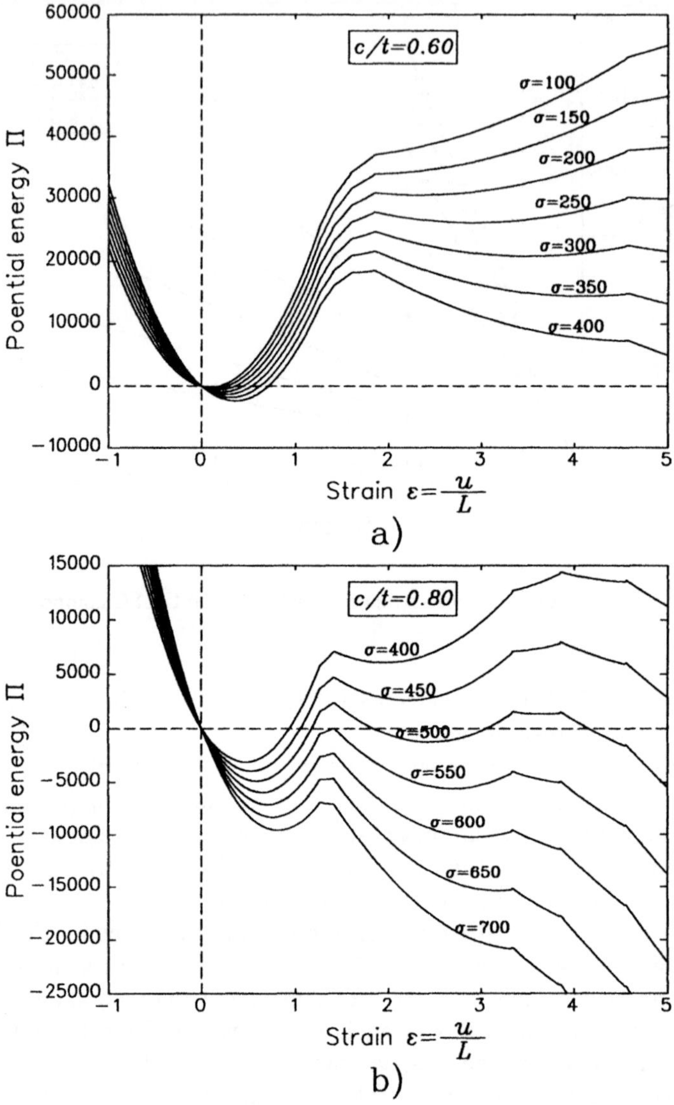

Figure 7.61. Potential energy curves for two GFC-CFC composites: a) $c/t = 0.60$, b) $c/t = 0.80$

hold for the case $c/t = 0.80$ (Fig. 7.61b). In this case, due to the greater proportion of the GFC (which has a higher ultimate strain than the brittle CFC), the undertaken loading is significantly higher.

7.9 OPTIMAL DESIGN EXAMPLES

In this Section optimal (stiff or flexible) structures will be designed by the Algorithm (5.1) presented in Chapter 5. For the stiff design we consider the power law (5.15) while for the more flexible design (primitive approach to damage) the linear law (5.19) is adopted. An application with cellular materials with variable (even negative Poisson's ratio) can be found in Theocaris and Stavroulakis, 1997.

For all the examples a two-dimensional plane stress plate is considered with dimensions equal to 11.00 in the horizontal direction and to 8.00 in the vertical direction. The plate is discretized by 857 nodes and 1590 constant strain triangular finite elements. The material elasticity constants are: modulus $E = 1.0 \times 10^6$ and Poisson's ratio $\nu = 0.30$, with all variables being given in compatible units.

Example 1

The right vertical boundary is fixed and a point load at the right-bottom edge is considered with vertical component equal to 100.0 and a small horizontal component equal to 20.0. The stiffest structure is considered by using the adaptive law (5.15) and a resource limit $V_{max} = 0.90\%$. The final plot of variable $\rho(x)$ is given in Fig. 7.62.

Example 2

The right vertical boundary is fixed and four tensile nodal loads are considered at the left boundary with each horizontal component equal to 100.0 and at the vertical positions (from the bottom) 6.60, 6.88, 7.20 and 7.72. The stiffest designs with law (5.15) and limits equal to $V_{max} = 0.95V_{total}$ and $V_{max} = 0.90V_{total}$ are given in Fig. 7.63 and Fig. 7.64 respectively.

Example 3

The same problem with Example 2 is considered, this time with five horizontal (tensile) loads posed at the centre of the right-hand-side vertical boundary. The stiffest designs with law (5.15) and limits equal to $V_{max} = 0.95V_{total}$ and $V_{max} = 0.90V_{total}$ are given in Fig. 7.65, Fig. 7.66 respectively.

Example 4

For the same plate structure we consider a bending loading case with vertical nodal loads equal to 20.0 at the right-hand-side boundary (a total of 26 boundary nodes and loads). The "damage-like" prediction of the flexible structure design with the linear adaptive material law (5.19) and resource limits equal to $V_{min} = 0.95V_{total}$, $0.90V_{total}$, $0.85V_{total}$ and $0.80V_{total}$ are given in Figures 7.67 to 7.70 respectively.

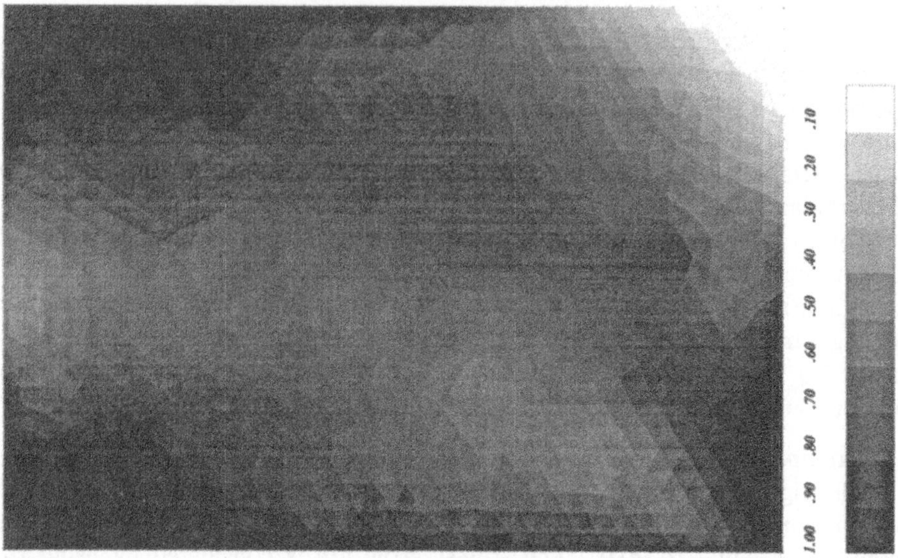

Figure 7.62. Stiffest design of a cantilever

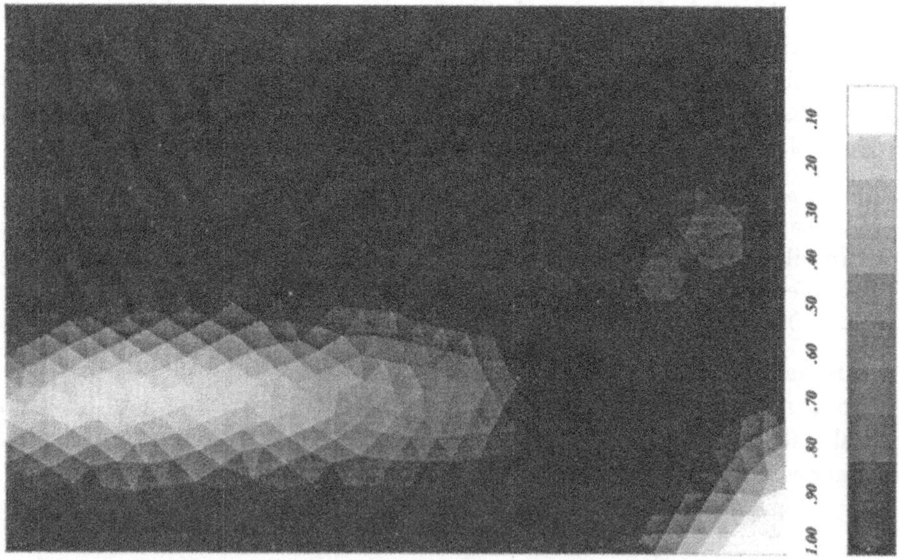

Figure 7.63. Stiffest design in tension (eccentrical loading, $V_{max} = 0.95V_{total}$)

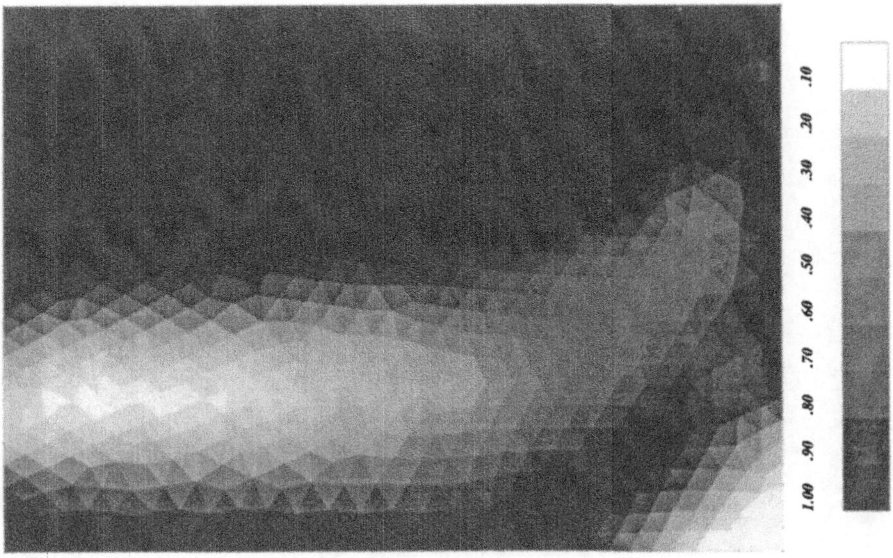

Figure 7.64. Stiffest design in tension (eccentrical loading, $V_{max} = 0.90V_{total}$)

Figure 7.65. Stiffest design in tension (central loading, $V_{max} = 0.95V_{total}$)

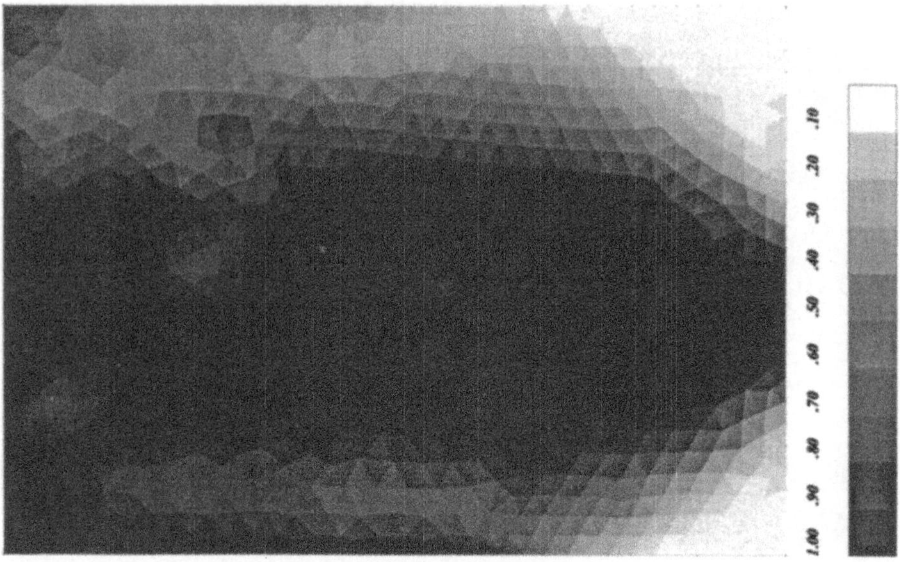

Figure 7.66. Stiffest design in tension (central loading, $V_{max} = 0.90V_{total}$)

Figure 7.67. Flexible design in bending, ($V_{min} = 0.95V_{total}$)

Figure 7.68. Flexible design in bending, $(V_{min} = 0.90V_{total})$

Figure 7.69. Flexible design in bending, $(V_{min} = 0.85V_{total})$

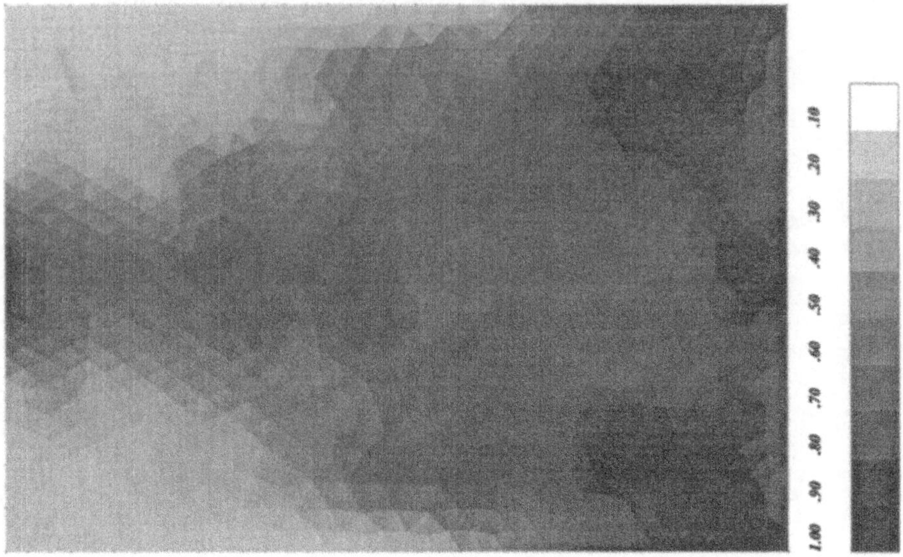

Figure 7.70. Flexible design in bending, $(V_{min} = 0.80 V_{total})$

References

AISI (1986). *Specification for the design of cold-formed steel structural members.* American Iron and Steel Institute, Washington, DC.

Akiyama, H. (1985). *Earthquake resistant limit state design for buildings.* University of Tokyo Press, Tokyo.

BS5950 (1987). *Code of practice for the design of cold-formed sections.* British Standards Institution, Part 5.

Crisfield, M. A. (1986). Snap-through and snap-back response in concrete structures and the dangers of under-integration. *Computer Method in Applied Mechanics and Engineering*, 22:751–767.

Crisfield, M. A. (1991). *Non-linear finite element analysis of solids and structures.* J. Wiley, Chichester.

Crisfield, M. A. and Wills, J. (1988). Solution strategies and softening materials. *Computer Method in Applied Mechanics and Engineering*, 66:267–289.

Davies, P. and Benzeggagh, M. L. (1989). Interlaminal mode-I fracture testing. In Friedrich, K., editor, *Application of fracture mechanics to composite materials.* Elsevier.

ECCS (1987). *European recommendations for the design of light gauge steel members.* European Convention for Constructional Steelworks, Brussels.

Feder, H. J. S. and Feder, J. (1991). Self-organized criticality in a stick-slip process. *Phys. Rev. Lett.*, 66:2669–2672.

Green, A. K. and Bowyer, W. H. (1981). The testing analysis of novel top-hat stiffener fabrication methods for use in GRP ships. In Marshall, I. H., editor, *Proc. 1st International Conf. on Composite Structures*, Barking-Essex. Applied Science.

Grundy, P. (1990). Effect of pre and post buckling behaviour on load capacity of continuous beams. *Thin-walled Structures*, 9(1-4):407–415.

Kato, B. and Akiyama, H. (1981). Ductility of members and frames subject to buckling. *ASCE Convention.*

Kecman, D. (1983). Bending collapse of rectangular and square tubes. *Int. J. of Mechanical Sciences*, 13(9-10):623–636.

Kim, D. H. (1995). *Composite structures for Civil and Architectural Engineering.* E & FN SPON, London.

MacLeod, J. A. and el Magd, S. A. A. (1980). The behaviour of brick walls under conditions of settlement. *The Structural Engineer*, 58A(9):279.

Mazzolani, F. and Piluso, V. (1996). *Theory and design of seismic resistant steel frames.* E & FN Spon, London.

Mitani, I. and Mahino, M. (1980). Post local buckling behaviour and plastic rotation capacity of steel beam-columns. In *7th World Conference on Earthquake Engineering*, Istanbul.

Moyson, E. and Gemert, D. V. (1985). Experimentelle Prüfung der Laminat theorie für faserverstärkte Verbunderwerkstoffe. In Ondraček, G., editor, *Proc. Verbunderwerkstoffe - Phasenverbindung und mechanische Eigenschaffen I*, pages 99–115, Karlsruhe.

Eurocode 3 – Part 1.3 (1993). *Cold-formed thin gauge members and sheeting.* European Committee for Standardisation.

Rasmussen, K. J. R. and Hancock, G. J. (1993). Design of cold-formed stainless steel tubular members. II: Beams. *J. Str. Div. ASCE*, 119 (ST8):2368–2386.

Reck, H., Pekoz, T., and Winter, G. (1976). Inelastic strength of cold-formed steel beams. *J. Str. Div. ASCE*, 101(ST11):2193–2204.

Roman, I., Harlet, H., and Marom, G. (1981). Stress intensity factor measurements in composite sandwich structures. In Marshal, I. H., editor, *Proc. 1st Conf. on Composite Structures*, pages 633–645, London. Applied Science Publishers.

Schwartz, M. M. (1984). *Composite materials handbook.* McGraw-Hill, New York.

Takayasu, H. (1990). *Fractals in physical sciences.* Manchester Univ. Press, Manchester.

Theocaris, P. S. and Stavroulakis, G. E. (1997). Multilevel iterative optimal design of composite structures, including materials with negative Poisson's ratio. *Structural Optimization*, (to appear).

Unger, B. (1973). Ein Beitrag zur Ermittlung der Tranglast von querbelasteten Durchlauftragern mit dünnwändigen Quershnitt, insbensondere von durchlaufenden Trapezblechen für Dack und Geschossdecken. *Der Stahlbau*, 42:20–24.

Wang, S. T. and Yeh, S. S. (1975). Post local-buckling behaviour of continuous beams. *J. Str. Div. ASCE*, 100(ST6):1169–1188.

Williams, J. G. and Rhodes, M. D. (1982). Effect of resin on impact damage tolerance of graphite/epoxy laminates. In Daniel, I., editor, *6th International Conf. on Composite Materials, Testing and Design.* ASTM STP, Philadelphia.

Williams, J. G., Stouffer, D. C., Ilic, S., and Jones, R. (1986). An analysis of delamination behavior. *Computers and Structures*, 5:203–216.

Xie, H. (1989). The fractal effect of irregularity of crack branching on the fracture toughness of brittle materials. *Int. Journal of Fracture*, 41:267–274.

Xie, H. (1991). Fractal nature on damage evolution of rock materials. In *2nd Int. Symposium of Mining Tech. and Science.* CUMT Press.

Yener, M. and Pekoz, T. (1980). Inelastic load-carrying capacity of cold-formed steel beams. In *Proceedings of the 5th International Speciality Conference on Cold-Formed Steel Structures.* University of Missouri-Rolla.

Index

Nonconvex Optimization and Its Applications

1. D.-Z. Du and J. Sun (eds.): *Advances in Optimization and Approximation.* 1994. ISBN 0-7923-2785-3
2. R. Horst and P.M. Pardalos (eds.): *Handbook of Global Optimization.* 1995 ISBN 0-7923-3120-6
3. R. Horst, P.M. Pardalos and N.V. Thoai: *Introduction to Global Optimization* 1995 ISBN 0-7923-3556-2; Pb 0-7923-3557-0
4. D.-Z. Du and P.M. Pardalos (eds.): *Minimax and Applications.* 1995 ISBN 0-7923-3615-1
5. P.M. Pardalos, Y. Siskos and C. Zopounidis (eds.): *Advances in Multicriteria Analysis.* 1995 ISBN 0-7923-3671-2
6. J.D. Pintér: *Global Optimization in Action.* Continuous and Lipschitz Optimization: Algorithms, Implementations and Applications. 1996 ISBN 0-7923-3757-3
7. C.A. Floudas and P.M. Pardalos (eds.): *State of the Art in Global Optimization.* Computational Methods and Applications. 1996 ISBN 0-7923-3838-3
8. J.L. Higle and S. Sen: *Stochastic Decomposition.* A Statistical Method for Large Scale Stochastic Linear Programming. 1996 ISBN 0-7923-3840-5
9. I.E. Grossmann (ed.): *Global Optimization in Engineering Design.* 1996 ISBN 0-7923-3881-2
10. V.F. Dem'yanov, G.E. Stavroulakis, L.N. Polyakova and P.D. Panagiotopoulos: *Quasidifferentiability and Nonsmooth Modelling in Mechanics, Engineering and Economics.* 1996 ISBN 0-7923-4093-0
11. B. Mirkin: *Mathematical Classification and Clustering.* 1996 ISBN 0-7923-4159-7
12. B. Roy: *Multicriteria Methodology for Decision Aiding.* 1996 ISBN 0-7923-4166-X
13. R.B. Kearfott: *Rigorous Global Search: Continuous Problems.* 1996 ISBN 0-7923-4238-0
14. P. Kouvelis and G. Yu: *Robust Discrete Optimization and Its Applications.* 1997 ISBN 0-7923-4291-7
15. H. Konno, P.T. Thach and H. Tuy: *Optimization on Low Rank Nonconvex Structures.* 1997 ISBN 0-7923-4308-5
16. M. Hajdu: *Network Scheduling Techniques for Construction Project Management.* 1997 ISBN 0-7923-4309-3
17. J. Mockus, W. Eddy, A. Mockus, L. Mockus and G. Reklaitis: *Bayesian Heuristic Approach to Discrete and Global Optimization.* Algorithms, Visualization, Software, and Applications. 1997 ISBN 0-7923-4327-1
18. I.M. Bomze, T. Csendes, R. Horst and P.M. Pardalos (eds.): *Developments in Global Optimization.* 1997 ISBN 0-7923-4351-4
19. T. Rapcsák: Smooth Nonlinear Optimization in R^n. 1997 ISBN 0-7923-4680-7

Nonconvex Optimization and Its Applications

20. A. Migdalas, P.M. Pardalos and P. Värbrand (eds.): *Multilevel Optimization: Algorithms and Applications*. 1998 ISBN 0-7923-4693-9
21. E.S. Mistakidis and G.E. Stavroulakis: *Nonconvex Optimization in Mechanics*. Algorithms, Heuristics and Engineering Applications by the F.E.M. 1998 ISBN 0-7923-4812-5

KLUWER ACADEMIC PUBLISHERS – DORDRECHT / BOSTON / LONDON

GPSR Compliance

The European Union's (EU) General Product Safety Regulation (GPSR) is a set of rules that requires consumer products to be safe and our obligations to ensure this.

If you have any concerns about our products, you can contact us on ProductSafety@springernature.com

In case Publisher is established outside the EU, the EU authorized representative is:

Springer Nature Customer Service Center GmbH
Europaplatz 3
69115 Heidelberg, Germany

Batch number: 10755284

Printed by Printforce, the Netherlands